Glutamine: Metabolism, Enzymology, and Regulation

MEXICO
1929-1979

Proceedings of the International Symposium on Glutamine: Metabolism, Enzymology, and Regulation sponsored by the UNIVERSIDAD NACIONAL AUTÓNOMA DE MÉXICO held in La Mansión Galindo, Querétaro, México, November 25–28, 1979.

Glutamine: Metabolism, Enzymology, and Regulation

Edited by

Jaime Mora

Rafael Palacios

Departamento de Biología Molecular
Instituto de Investigaciones Biomédicas
Universidad Nacional Autónoma de México
México, D. F. México

1980

ACADEMIC PRESS

A Subsidiary of Harcourt Brace Jovanovich, Publishers

New York London Toronto Sydney San Francisco

ACADEMIC PRESS, INC.
111 Fifth Avenue, New York, New York 10003

United Kingdom Edition published by
ACADEMIC PRESS, INC. (LONDON) LTD.
24/28 Oval Road, London NW1 7DX

Library of Congress Cataloging in Publication Data

International Symposium on Glutamine: Metabolism, Enzymology, and Regulation, Querétaro, Mexico, 1979.
Glutamine—metabolism, enzymology, and regulation.

Proceedings of the symposium sponsored by the Universidad Nacional Autónoma de México held in La Mansion Galindo, Querétaro, México, Nov. 25–28, 1979.

1. Glutamine metabolism—Congresses. I. Mora, Jaime. II. Mexico (City). Universidad Nacional. III. Title. [DNLM: 1. Glutamine—Congresses. 2. Glutamine—Metabolism—Congresses. QU 60 I607g 1979]
QP562.G55157 1979 612'.398 80-24655
ISBN 0-12-506040-8

PRINTED IN THE UNITED STATES OF AMERICA

80 81 82 83 9 8 7 6 5 4 3 2 1

CONTENTS

CONTRIBUTORS

Numbers in parentheses indicate the pages on which authors' contributions begin.

Fernando Bastarrachea (107, 123), Departmento de Biología Molecular, Instituto de Investigaciones Biomédicas, Universidad Nacional Autónoma de México, Apartado postal 70228, México 20, D. F. México

David M. Bedwell (79), Department of Biological Sciences, Purdue University, W. Lafayette, Indiana 47907

Lourdes Blanco (167), Departamento de Biología Molecular, Instituto de Investigaciones Biomédicas, Universidad Nacional Autónoma de México, Apartado postal 70228, México 20, D. F. México

Fredric Bloom (69), Department of Biology, Massachusetts Institute of Technology, Cambridge, Massachusetts 02139

Francisco Bolivar (107, 123), Departamento de Biología Molecular, Instituto de Investigaciones Biomédicas, Universidad Nacional Autónoma de México, Apartado postal 70228, México 20, D. F. México

Jean E. Brenchley (79), Department of Biological Science, Purdue University, W. Lafayette, Indiana 47907

Susana Brom (107), Departamento de Biología Molecular, Instituto de Investigaciones Biomédicas, Universidad Nacional Autónoma de México, Apartado postal 70228, México 20, D. F. México

Doris Burton (95), Department of Bacteriology, University of California, Davis, California 95616

Edmundo Calva (167), Departamento de Biología Molecular, Instituto de Investigaciones Biomédicas, Universidad Nacional Autónoma de México, Apartado postal 70228, México 20, D. F. México

Miguel Campomanes (167), Departamento de Biología Molecular, Instituto de Investigaciones Biomédicas, Universidad Nacional Autónoma de México, Apartado postal 70228, México 20, D. F. México

Ofelia Chávez (167), Departamento de Biología Molecular, Instituto de Investigaciones Biomédicas, Universidad Nacional Autónoma de México, Apartado postal 70228, México 20, D. F. México

P. Boon Chock (41), Laboratory of Biochemistry, National Heart, Lung, and Blood Institute, National Institutes of Health, Bethesda, Maryland 20205

Philip P. Cohen (315), Department of Physiological Chemistry, University of Wisconsin, Madison, Wisconsin 53706

Julio Collado (167), Departamento de Biología Molecular, Instituto de Investigaciones Biomédicas, Universidad Nacional Autónoma de México, Apartado postal 70228, México 20, D. F. México

Alejandra Covarrubias (107, 123), Departamento de Biología Molecular, Instituto de Investigaciones Biomédicas, Universidad Nacional Autónoma de México, Apartado postal 70228, México 20, D. F. México

Robert A. Darrow (139), Charles F. Kettering Research Laboratory, 150 East South College Street, Yellow Springs, Ohio 45387

Guillermo Dávila (185), Departamento de Biología Molecular, Instituto de Investigaciones Biomédicas, Universidad Nacional Autónoma de México, Apartado postal 70228, México 20, D. F. México

José Ma. de la Peña (167), Departamento de Biología Molecular, Instituto de Investigaciones Biomédicas, Universidad Nacional Autónoma de México, Apartado postal 70228, México 20, D. F. México

Susan M. Dendinger (79), Department of Biological Science, Purdue University, W. Lafayette, Indiana 47907

Guadalupe Espin (185), Departamento de Biología Molecular, Instituto de Investigaciones Biomédicas, Universidad Nacional Autónoma de México, Apartado postal 70228, México 20, D. F. México

Alan J. Garber (259), Internal Medicine Department, Division of Endocrinology and Metabolism, Baylor College of Medicine, Houston, Texas 77030

Emilio Garcia (95), Department of Bacteriology, University of California, Davis, California 95616

Alicia González (185), Departamento de Biología Molecular, Instituto de Investigaciones Biomédicas, Universidad Nacional Autónoma de México, Apartado postal 70228, México 20, D. F. México

Josefina Guzmán (185), Departamento de Biología Molecular, Instituto de Investigaciones Biomédicas, Universidad Nacional Autónoma de México, Apartado postal 70228, México 20, D. F. México

Georgina Hernández (185), Departamento de Biología Molecular, Instituto de Investigaciones Biomédicas, Universidad Nacional Autónoma de México, Apartado postal 70228, México 20, D. F. México

Gisela Hummelt (185), Departamento de Biología Molecular, Instituto de Investigaciones Biomédicas, Universidad Nacional Autónoma de México, Apartado postal 70228, México 20, D. F. México

Hans Krebs (319), Metabolic Research Laboratory, University of Oxford, Radcliffe Infirmary, Woodstock Road, Oxford, United Kingdom

John M. Kuchta[1] (79), Department of Biological Science, Purdue University, W. Lafayette, Indiana 47907

Sidney Kustu (95), Department of Bacteriology, University of California, Davis, California 95616

Miguel Lara (185), Departamento de Biologia Molecular, Instituto de Investigaciones Biomédicas, Universidad Nacional Autónoma de México, Apartado postal 70228, México 20, D. F. México

Peter J. Lea (213), Biochemistry Department, Rothamsted Experimental Station, Harpenden, Herts, United Kingdom

Paul Linser (299), Laboratory for Developmental Biology, Cummings Life Science Center, University of Chicago, Chicago, Illinois 60637

Susana López (167), Departamento de Biología Molecular, Instituto de Investigaciones Biomédicas, Universidad Nacional Autónoma de México, Apartado postal 70228, México 20, D. F. México

Edmundo Lozoya (123), Departamento de Biología Molecular, Instituto de Investigaciones Biomédicas, Universidad Nacional Autónoma de México, Apartado postal 70228, México 20, D. F. México

Linda McCarter (95), Department of Bacteriology, University of California, Davis, California 95616

Nancy McFarland (95), Department of Bacteriology, University of California, Davis, California 95616

Boris Magasanik (61), Department of Biology, Massachusetts Institute of Technology, Cambridge, Massachusetts 02139

Esperanza Martínez (185), Departamento de Biología Molecular, Instituto de Investigaciones Biomédicas, Universidad Nacional Autónoma de México, Apartado postal 70228, México 20, D. F. México

Paula Mayerson (299), Laboratory for Developmental Biology, Cummings Life Science Center, University of Chicago, Chicago, Illinois 60637

Alton Meister (1), Department of Biochemistry, Cornell University Medical College, New York, New York 10021

Benjamin J. Miflin (213), Biochemistry Department, Rothamsted Experimental Station, Harpenden, Herts, United Kingdom

Jaime Mora (185), Departamento de Biología Molecular, Instituto de Investigaciones Biomédicas, Universidad Nacional Autónoma de México, Apartado postal 70228, México 20, D. F. México

Yolanda Mora (185), Departamento de Biología Molecular, Instituto de Investigaciones Biomédicas, Universidad Nacional Autónoma de México, Apartado postal 70228, México 20, D. F. México

[1]Present address: Department of Biology, Duquesne University, Pittsburg, Pennsylvania 15219.

Aron A. Moscona (299), Laboratory for Development Biology, Cummings Life Science Center, University of Chicago, Chicago, Illinois 60637

Malka Moscona (299), Laboratory for Development Biology, Cummings Life Science Center, University of Chiçago, Chicago, Illinois 60637

Umberto Mura[2] (41), Laboratory of Biochemistry, National Heart, Lung, and Blood Institute, National Institutes of Health, Bethesda, Maryland 20205

Aurora Osorio (107), Departamento de Biología Molecular, Instituto de Investigaciones Biomédicas, Universidad Nacional Autónoma de México, Apartado postal 70228, México 20, D. F. México

Greg Pahel (69), Department of Biology, Massachusetts Institute of Technology, Cambridge, Massachusetts 02139

Rafael Palacios (167), Departamento de Biología Molecular, Instituto de Investigaciones Biomédicas, Universidad Nacional Autónoma de México, Apartado postal 70228, México 20, D. F. México

Carmen Quinto (167), Departamento de Biología Molecular, Instituto de Investigaciones Biomédicas, Universidad Nacional Autónoma de México, Apartado postal 70228, México 20, D. F. México

Sue Goo Rhee (41), Laboratory of Biochemistry, National Heart, Lung, and Blood Institute, National Institutes of Health, Bethesda, Maryland 20205

David Romero (185), Departamento de Biología Molecular, Instituto de Investigaciones Biomédicas, Universidad Nacional Autónoma de México, Apartado postal 70228, México 20, D. F. México

David M. Rothstein (61), Department of Biology, Massachusetts Institute of Technology, Cambridge, Massachusetts 02139

Federico Sánchez (167), Departamento de Biología Molecular, Instituto de Investigaciones Biomédicas, Universidad Nacional Autónoma de México, Apartado postal 70228, México 20, D. F. México

Ray Sánchez-Pescador (123), Departamento de Biología Molecular, Instituto de Investigaciones Biomédicas, Universidad Nacional Autónoma de México, Apartado postal 70228, México 20, D. F. México

Earl R. Stadtman (41), Laboratory of Biochemistry, National Heart, Lung, and Blood Institute, National Institutes of Health, Bethesda, Maryland 20205

Ricardo Tapia (285), Departamento de Neurociencias, Centro de Investigaciones en Fisiologia Celular, Universidad Nacional Autónoma de México, México 20, D. F. México

[2]Present address: Laboratorio di Biochimica, Facoltá di Scienze Matematiche, Fishiche e Naturali, Pisa, Italy.

Bonnie Tyler (69), Merck, Sharp and Dohme Research Laboratories, Rahway, New Jersey

Roger M. Wallsgrove (213), Biochemistry Department, Rothamsted Experimental Station, Harpenden, Herts, United Kingdom

Herbert G. Windmueller (235), Laboratory of Nutrition and Endocrinology, National Institute of Arthritis, Metabolism, and Digestive Diseases, National Institutes of Health, Bethesda, Maryland 20205

PREFACE

A book was published seven years ago, edited by Stanley Prusiner and Earl R. Stadtman, that for the first time integrated the available information about the enzymes that catalyze the formation and breakdown of glutamine. The benefits of this book were self-evident since the amino acid glutamine lies at the center of cellular nitrogen metabolism.

Since that time new developments in the regulation of nitrogen metabolism in prokaryotes, eukaryotic microorganisms, plants, and higher animals have appeared, as a result of the effort of several laboratories, and new roles for glutamine in cellular metabolism have been established. The aim of this book is to present significant contributions on the metabolism, enzymology, and regulation of glutamine, with a scope ranging from microorganisms to higher animals.

Contributions to this volume have been organized starting with an overview of glutamine metabolism centered mainly in enzymatic studies on glutamine synthetase and covering also the enzymes that utilize glutamine; this is followed by a model of cascade control of glutamine synthetase in *Escherichia coli*. Seven chapters are devoted to a highly controversial topic: the mechanisms that control gene expression of glutamine synthetase and its relation to nitrogen metabolism in prokaryotes, including *Klebsiella, Escherichia, Salmonella,* and *Rhyzobium*. The relationship between the structure and function of glutamine synthetase in eukaryotic microorganisms is presented in two chapters. For the first time a contribution on nitrogen and glutamine metabolism in plants is included. Four chapters cover the latest information about the role of glutamine in the animal body, including the production of glutamine in muscle, its utilization as an energy donor in intestine, its role as a precursor of neurotransmitters, and as part of a mechanism to eliminate fenilacetic acid. Also, a chapter is centered in the study of glutamine synthetase gene expression as a marker of cell differentiation in retina.

This book ends with a special lecture by Professor Sir Hans Krebs, which, starting with his discovery of the synthesis of glutamine in animal tissues, gives an integral view of glutamine metabolism and functions in the animal body.

This symposium was one of the scientific events to commemorate the fiftieth anniversary of the granting of the autonomy of the University of

Mexico, and was held in Hacienda Galindo, Querétaro, México, on November 25–28, 1979.

We want to express our gratitude to Dr. Philip P. Cohen and Dr. Alton Meister for their advice and help in the organization of this symposium. We want to acknowledge Dr. Edmundo Calva, Elena Palacios, and Yolanda Mora for editorial assistance.

The camera ready copy set of this volume was made by the Dirección General de Publicaciones of the University of México.

INTRODUCTORY REMARKS

The symposium "Glutamine: Metabolism, Enzymology, and Regulation" is one of the scientific events organized by the University of Mexico to commemorate the fiftieth anniversary of its autonomy.

The University of Mexico, the oldest in Continental America, was founded 428 years ago according to the pattern of the University of Salamanca. It developed high academic standards and served as a starting point for launching the Universities of Havana, Guatemala, and Manila. After difficult periods throughout the nineteenth century, our University was redefined in 1910 to become the National University of Mexico. Autonomy was granted in 1929 after a struggle between students and the civil government. In 1945 a new law was enacted to establish the present academic and administrative organization.

The size of our Institution, 280,000 students, 26,000 academic personnel, and 18,000 administrative workers, astonishes everybody. However, our Institution is more than a conventional type of university, and has evolved into a rather complex university system. It includes senior high school, which accounts for about one-half of the students, located in 14 different sites in the city; we have six campuses in the metropolitan area for college and postgraduate levels; and we also have agencies in 15 states of the Republic.

The establishment of our research structure started in 1929 when four existing institutions were incorporated into the University. These were: the National Astronomical Observatory, the National Library, the National Institute of Biology and the Directorate of Geological Studies, which in time became the Institute of Astronomy, the Institute of Bibliographic Research, the Institute of Biology, and the Institute of Geology, respectively. Since that time, other research groups have emerged; at present we are proud of our 21 institutes, 8 centers and 7 divisions of research. Institutes, created by the University Council, enjoy a large degree of independence from the administration. The research institutes are Astronomy, Biology, Physics, Geophysics, Geography, Geology, Engineering, Anthropology, Bibliography, Biomedical Sciences, Economics, Fine Arts, Philology, Philosophy, History, Law, Applied Mathematics and Systems, Sociology, Mathematics, Materials Research, and Chemistry. Research centers, created by the Rector, are either in the process of evolving toward the status of an institute or else they are devoted to support research activities. Three can be found in the last category: the Center of Computation Services, the Center of Scientific and Humanistic Information, and the Center of Instruments. The other research centers are Sciences of the Atmosphere, Marine Sciences and Limnology, Nuclear Studies, Cell Physiology, and Studies of the University.

The research divisions are made up of research departments located in the schools. We have established these in the Schools of Sciences, Political Sciences, Administration, Engineering, Medicine, Psychology, and Chemistry.

The University of Mexico has also been very active in establishing research organizations throughout different cities of the country. We have to mention separately those research centers which belong entirely to our institution but require a special arrangement to carry out specific research programs. They are the Astronomical Observatories in Tonanzintla, Puebla and in San Pedro Mártir, Baja California; the Geological Research Stations in Hermosillo, Sonora, and Tapachula, Chiapas; the Marine Research Stations in Mazatlán, Sinaloa, and Ciudad del Carmen, Campeche; and the Tropical Ecology Reserves in Chamela, Jalisco, and Los Tuxtlas, Veracruz.

We have likewise joined efforts with the National Council of Sciences and Technology and with local state governments to create other research centers like the Center of Scientific Research and Teaching of Ensenada, Baja California; the Ecological Research Center of San Cristobal, Las Casas, Chiapas; the Applied Chemistry Research Center of Saltillo, Coahuila; and the Experimental Biology Research Center of La Paz, Baja California. Last week we started another research center in Mérida, Yucatán cooperating with the National Council of Science and Technology, the State Government, the Ministry of Agriculture, and the Center of Advanced Studies of the National Politechnic Institute.

Next year we intend to launch a Mathematics Research Center in Guanajuato, Guanajuato; a Mathematics Research Regional Center in Aguascalientes, Aguascalientes; a Nitrogen-Fixation Center in Cuernavaca, Morelos; an Optics Research Center in León, Guanajuato; a Marine Research Center in Can Cun, Quintana Roo; and a Physics Center in Ensenada, Baja California.

More than 1000 research workers (not including technical and administrative help) work in the above-mentioned Institutions. Thus our research effort represents from one-quarter to one-third of the total national research structure. In the present year we will spend 100 million dollars for research purposes. This represents about one-sixth of the total national budget for research. On the other hand, we publish annually almost one-half of the research papers produced in Mexico. It should be mentioned that approximately one-third of our research expenses are supported by funds derived from contracts with government agencies, which means that our work is closely linked to problems of national interest.

Research is organized into two different systems: one for humanities and one for sciences; each is effected under the supervision of a coordinator appointed by the Rector; the Directors of the Research Institutions integrate, respectively, the Council of Humanities and the Council of Sciences and each council is presided over by the corresponding coordinator. We have established a policy whereby priority is given to programs focused on interdisciplinary research, designed to solve national problems or aimed to develop research centers outside Mexico City.

To refer specifically to the symposium, I should now focus on the Biomedical Research Institute, which started in 1941 with the aim of providing facilities for Spanish research fellows who emigrated to Mexico immediately after the Spanish Civil War. The institute is now involved in basic and applied research within the framework of six departments: Physiology, Neurosciences, Molecular Biology, Biomathematics, Developmental Biology, and Biotechnology.

I would now like to describe how the interest in glutamine arose in the Mexican research groups responsible for the organization of this symposium.

In 1957, biochemical research started in Mexico as a consequence of the return of about a dozen scientists trained abroad in laboratories noted for their fundamental contributions to sciences.

One of these pioneer groups, our own, at the Biochemistry Department of the National Institute for Nutrition, got involved in the study of nitrogen metabolism in the animal body. My personal interest in the subject originated during my stay in the laboratory of Prof. Philip P. Cohen at the University of Wisconsin, where the urea cycle was one of the main problems investigated for many years. Although I was not personally involved in such studies, since I was working on uric acid oxidation, I naturally shared the excitement of the people around me.

Once we were established in Mexico, we inquired into the distribution of ammonia loads in the animal body and found a key role of muscle in the fixation of this compound. The capacity of the urea cycle to fix ammonia was also determined. The phylogenetic operation of the urea cycle was studied, and this contributed to our knowledge of the regulation of the enzymes that participate in the cycle.

Arginase was studied in some detail and the existence of two different kinds of the enzyme was demonstrated: ureotelic and uricotelic arginase. The two types of arginase were correlated with different patterns of nitrogen excretion. The studies led also to the discovery of a new enzyme, guanido ureo hydrolase, different from arginase, able to hydrolyze arginine, as well as other guanido compounds, producing urea.

Another problem we approached was related to the physical integration of the urea cycle enzymes. A functional integration of the biosynthesis and the hydrolysis of arginine was found; furthermore, the existence of different molecular forms of ureotelic arginase was shown. In the course of these studies we moved to the Institute of Biomedical Research where we created the first Department of Molecular Biology in the University and in Mexico.

A new lead of our research group on nitrogen metabolism started when the existence of the urea cycle in eukaryotic microorganisms was challenged, in spite of the fact that all intermediate enzyme activities are present. Indeed, the studies of Mora *et al.* demonstrated that the urea cycle does not operate in *Neurospora crassa*. As a consequence of this discovery a new problem arose: the relationship and coordination between the synthesis and the catabolism of arginine in this organism. It was found that, in fact, a correlation exists to the extent that when arginine biosynthesis is operating, its catabolism is turned off, and vice versa.

In the course of these studies it was found that arginine, glutamine, and other amino acids accumulate when *Neurospora* growth is restricted. These findings led to the study of cellular metabolism in microorganisms and raised the question as to the meaning of metabolic activities in the absence of growth. Since under these conditions there is an accumulation of glutamine, which can be considered to be both a substrate for arginine biosynthesis and a product of arginine catabolism, research was directed toward the study of the regulation of the synthesis of glutamine. It was found that glutamine participates in the regulation of arginine catabolism by preventing the induction of arginase. This finding suggested a general role of glutamine in the regulation of nitrogen catabolism in *Neurospora*.

A third period in the development of the research group in the Department of Molecular Biology at the University of Mexico occurred with the return of Dr. Rafael Palacios who, together with Dr. Mora, initiated a collaborative study on the genetic factors involved in the regulation of the enzymes affecting the assimilation and metabolism of nitrogen compounds in *Neurospora crassa*. This effort has now expanded to include collaborative studies with other research groups.

Because of the many special roles which glutamine is now known to have in the metabolism of prokaryotic and eukaryotic organisms, Dr. Mora and Dr. Palacios initiated a proposal to hold a symposium to review and discuss current developments in this important, basic area of research. In developing plans for this symposium, they enlisted the cooperation of Professor Alton Meister, a pioneer investigator in glutamine metabolism, and of Professor Philip P. Cohen, who together with Dr. Mora and Dr. Palacios served as the organizing committee.

On behalf of the University of Mexico, I wish to thank the organizing committee for their efforts in arranging the program of this symposium and to thank the participants from Mexico and abroad who by their presentations and discussion will provide a critical summary of current knowledge of glutamine metabolism, and its enzymology and regulation. I would conclude by stressing once more that throughout more than 20 years now, we have enjoyed, through a close relationship, the friendship and support of Prof. Cohen to whom we are fully indebted, not only those of us who belong to the research effort that motivated the idea of the symposium, but most Mexican biochemists as well.

Thank you.

Guillermo Soberón
Rector, National University of Mexico
November 25, 1979.

1. Catalytic Mechanism of Glutamine Synthetase; Overview of Glutamine Metabolism

ALTON MEISTER

Department of Biochemistry
Cornell University Medical College
1300 York Avenue
New York, New York 10021 U.S.A.

SUMMARY

Glutamine synthetase, an enzyme that is widely distributed in microorganisms, animal tissues, and higher plants, catalyzes the reversible formation of glutamine from glutamate and ammonia at the expense of cleavage of ATP to ADP and P_i. Studies with highly purified enzyme preparations have shown that glutamine synthetases from different sources vary in amino acid composition and also in subunit structure. Most have subunits of 44,000-50,000; the enzymes from brain, liver, peas, and certain microorganisms are octameric and show cubelike morphology whereas the *E. coli* enzyme is dodecameric. The *E. coli* and pea enzymes contain 2 and 4 half-cystines, respectively, compared to 12 half-cystines in the mammalian enzymes. The *E. coli* enzyme occurs in adenylylated forms, which have not been found in the enzymes isolated from certain other microorganisms, peas or mammalian tissues. Nevertheless, the mechanism of the reaction catalyzed by glutamine synthetases isolated from various cells seems to be basically similar.

Studies on the several partial reactions catalyzed by the enzyme (e.g., γ-glutamyl transfer, arsenolysis, formation of 5-oxoproline, formation of ATP from ADP and acyl phosphates, formation of acyl phosphates), on its interaction with glutamate analogs, and on certain kinetic and binding phenomena show that γ-glutamyl phos-

ISBN: 0-12-506040-8

phate is an enzyme-bound intermediate. Stereochemical mapping studies of the active site show that L-glutamate attaches to the active site in an extended conformation in which the carboxyl groups are as far apart as possible and in which the α-hydrogen atom of L-glutamate is directed away from the enzyme. Computer-aided calculations have identified the binding sites for the γ-carboxyl atom, the α-amino nitrogen atom, the γ-carboxyl oxygen atom which is phosphorylated, the γ-carboxyl oxygen atom which is not phosphorylated, and the nitrogen atom of the tetrahedral intermediate formed in the attack of ammonia on enzyme-bound γ-glutamyl phosphate. Only one of the four isomers of methionine sulfoximine (the L, S-isomer) is phosphorylated on the sulfoximine nitrogen to form methionine sulfoximine phosphate, which (together with ADP) binds tightly to the enzyme producing irreversible inhibition. When L-methionine-S-sulfoximine binds to the enzyme, its sulfoximine nitrogen atom occupies the same site as that occupied by the oxygen of glutamate that is phosphorylated. These studies show that L-methionine-S-sulfoximine is an inhibitory analog of the intermediate or transition state formed in the normal catalytic reaction. L-Methionine-S-sulfoximine also inhibits γ-glutamylcysteine synthetase by a similar mechanism. Selective sulfoximine inhibitors of glutamine synthetase and γ-glutamylcysteine synthetase were prepared; α-ethylmethionine sulfoximine specifically inhibits glutamine synthetase, and buthionine sulfoximine specifically inhibits γ-glutamylcysteine synthetase. L-Methionine-S-sulfoximine and α-ethylmethionine sulfoximine produce convulsions in animals, whereas buthionine sulfoximine does not, indicating that the convulsant activity of methionine sulfoximine is due to its effect on the synthesis of glutamine rather than to its effect on glutathione synthesis.

The synthesis of glutamine provides many types of cells with a mechanism for removal of ammonia and for its storage. Although not essential in the diet of mammals, glutamine plays a central role in amino acid, protein, and nitrogen metabolism; it is required as a building block of proteins, and also provides a nitrogen source for a number of biosynthetic pathways. Glutamine is the most prevalent amino acid in mammalian blood plasma; it is usually required in the medium for the growth of cells in tissue culture. Glutamine synthesis appears to be essential for brain function. Glutamine and its hydrolysis by renal glutaminase play a key role in acid-base balance.

The α-amino group of glutamine is utilized for the formation of amino acids: thus, transamination of glutamine with α-keto acids by specific transaminases leads to the formation of the corresponding amino acids and α-ketoglutaramate, which is hydrolyzed to α-ketoglutarate and ammonia by a specific ω-amidase. These reactions form part of a homeostatic metabolic mechanism in mammalian tissues for the preservation of amino acid balance in which glutamine and other dietary non-essential amino acids function to maintain the tissue levels of amino acids and to prevent loss of essential carbon chains.

The utilization of the amide nitrogen atom of glutamine, catalyzed by the glutamine amidotransferases, leads to the synthesis of such important biological compounds as the pyridine nucleotide coenzymes, purines, pyrimidines, glucosamine-6-phosphate, and asparagine. Glutamine amide nitrogen is also used in bacteria and plants for the biosynthesis of histidine, p-aminobenzoate, tryptophan, and glutamate. Glutamine amide nitrogen is used in some cells for synthesis of carbamyl phosphate, which serves as a precursor of both the pyrimidine ring and of arginine; the enzyme that catalyzes this reaction (glutamine-dependent carbamyl phosphate synthetase) and a number of the other glutamine amidotransferases have certain properties in common. These include a binding site for glutamine that contains an SH-group, inherent glutaminase activity, some ability to use ammonia in place of glutamine, and susceptibility to irreversible inhibition by certain reagents that bind at the glutamine site such as 6-diazo-5-oxonorleucine and 2-amino-4-oxo-5-chloropentanoate.

Thus, the glutamine molecule is used intact for protein synthesis and for certain special processes (e.g., phenylacetyl-glutamine formation). The α-amino and the amide nitrogen moieties of glutamine are used in a multitude of significant processes that include ammonia formation, purine and pyrimidine biosynthesis, many aspects of amino acid biosynthesis and degradation, and the biosynthesis of amino sugars and certain essential cofactors. The carbon chain of glutamine is available to many cells as a source of energy. Glutamine and the enzyme that catalyzes the synthesis of this ubiquitous amino acid amide are therefore of major importance in cellular metabolism and function.

INTRODUCTION

The existenece of glutamine seems to have been first considered by Hlasiwetz and Habermann, who suggested in 1873 that the ammonia found in hydrolysates of protein arose by hydrolysis of protein linked glutamine and asparagine (1). Ten years later Schulze and Bosshard first isolated glutamine from a natural source (beet juice) (2). It was not until 1932 that glutamine was isolated from a protein hydrolysate; this was accomplished in Chibnall's laboratory in studies in which glutamine was obtained from an enzymatic hydrolysate of gliadin (3). Indeed by 1932 all of the other common amino acids of proteins except for threonine had been isolated from protein hydrolysates (4). During the century or so that has elapsed since the early speculations of Hlasiwetz and Habermann, we have learned that glutamine is widely distributed in virtually all species in protein-bound and free forms (5). Glutamine is present in many proteins, and is the most abundant free amino acid constituent of mammalian blood plasma. Glutamine performs a crucial function in the metabolism of ammonia and in the biosynthesis of the building blocks of nucleic acids, certain amino acids, various cofactors, and still other compounds of biological significance. Glutamine is not a dietary essential amino acid for animals nor is it required for growth by plants and most microorganisms. It is evident therefore that this amino acid amide must be synthesized by a wide variety of living cells. Indeed, the synthesis of glutamine is a reaction of major importance in the cellular metabolism of animals, microorganisms and plants. I will consider here some of the properties of glutamine synthetase, especially the mechanism of the reaction catalyzed by this important enzyme, and its inhibition by transition state inhibitors such as methionine sulfoximine and α-ethylmethionine sulfoximine. This paper also offers an overview of the metabolism of glutamine and brief consideration of some of the enzymes that are involved in the utilization of glutamine.

SYNTHESIS OF GLUTAMINE

The synthesis of glutamine was achieved in 1933 by Bergmann and his collaborators (6) who made use of Bergmann's then new carbobensoxy method. Carbobenzoxy glutamate was treated with acetic anhydride in glacial acetic acid to obtain carbobenzoxy glutamic acid anhydride. The anhydride was opened in the presence of benzyl alcohol to obtain the α-benzyl ester; this was converted to carbobenzoxy-γ-glutamyl chloride α-benzyl ester. The acid chloride was treated with ammonia to give the corresponding derivative of glutamine. After removal of the blocking groups by catalytic hydrogenation, Bergmann and his colleagues obtained the first synthetic glutamine.

The first studies on the biosynthesis of glutamine were undertaken by Krebs in 1935, who found that the disappearance of ammonia in guinea pig slices was increased when L-glutamate was added, and that the disappearance of ammonia was associated with the formation of a compound that gave ammonia on acid hydrolysis (7, 8). This compound was shown to be glutamine by isolation as the crystalline hydrochloride. Krebs also examined the synthesis of glutamine in slices of brain, retina, and liver and concluded that the synthesis of glutamine was an endothermic reaction. Later studies by Elliott (9-12), Speck (13-15), and Leuthardt and Bujard (16, 17) revealed that the synthesis of glutamine in cell free systems required the hydrolysis of ATP. The glutamine synthesis reaction was shown to be freely reversible (18) and from the equilibrium constant experimentally determined, the value for the standard free energy of hydrolysis of ATP was obtained (18-20). Glutamine synthetase was first isolated in essentially homogeneous form from sheep brain (21). Studies in several laboratories in the late 1940's and early 1950's established the reaction catalyzed by glutamine synthetase:

$$\text{L-Glutamate} + \text{ATP} + NH_3 \rightleftharpoons \text{L-glutamine} + \text{ADP} + P_i \quad (1)$$

A divalent cation (Mg^{++}, Mn^{++}, Co^{++}) is required.

When glutamine synthetase is incubated with 0.01 M concentrations of L-glutamate, ammonium ions, and ATP (in the presence of Mg^{++} at pH 7.0 and 37°) the reaction reaches equilibrium when about 90% of the glutamate added is converted to glutamine (18). When hydroxylamine is substituted for ammonia in this system greater than 99% of the glutamate is converted to γ-glutamyl hydroxamate.

These results are consistent with the apparent free energy change associated with the hydrolysis of glutamine (about 3500 cal/mole) and with that of γ-glutamyl hydroxamate (about 700 cal/mole) (22). Although glutamine synthetase activity is generally measured at its optimal pH range (7.0-7.4) in the presence of L-glutamate, ATP, Mg^{++}, and either ammonium ions or hydroxylamine, the pH optimum of the reaction may vary from about pH 4.8 to 8.5 depending on the nature and concentration of the divalent cation present. In the pH range 7.0-7.4 magnesium ions are more effective than are manganese or cobalt ions, but at their respective pH optima, virtually equivalent activities are observed with Mg^{++}, Mn^{++}, and Co^{++} (23, 24). A number of studies have been carried out on the specificity of glutamine synthetase toward various nucleotides; only ATP and deoxy-ATP are appreciably active although some activity has been reported with adenosine tetraphosphate, ITP, GTP, and UTP (25). In addition to ammonia and hydroxylamine, which are about equally active when tested with L-glutamate, certain other nucleophilic agents such as hydrazine, monomethylhydrazine, methylamine, ethylamine, and glycine ethyl ester are active in place of ammonia and therefore yield the corresponding γ-glutamyl products (26, 27). In addition to studies on the specificity of the enzyme with respect to nucleoside triphosphates, metal ions, and analogs of ammonia, extensive studies have been carried out with glutamate analogs; these are considered below.

Some properties of the glutamine synthetases isolated from various sources are collected in Table I. Glutamine synthetase from sheep brain is composed of 8 apparently identical subunits which exhibit a cube-like appearance on electron microscopy. In the model for the subunit structure of this enzyme proposed by Haschemeyer (28, 29) the enzyme is considered to possess D_4 symmetry. The octamer is formed by isologous association of 2 heterologously bonded tetramers. A number of studies have been carried out on the dissociation of the octameric form of the enzyme to the tetramer and monomer (30, 31). The glutamine synthetases isolated from rat liver (32) and chinese hamster liver (33) appear to have similar subunit structures. Glutamine synthetase from peas and certain other glutamine synthetases also have 8 subunits whereas a number of the bacterial glutamine synthetases such as that from *Escherichia coli* and *Bacillus subtilis* have 12 subunits (26). Electron microscopic studies of the glutamine synthetase from *E. coli* have shown that the subunits of this dodecameric enzyme are arranged in the form of 2 stacked hexagons (34). The enzyme from *E. coli* and certain related organisms, but not that from *B. subtilis* and certain other microorganisms, may exist in an adenylylated form which is much more sensitive to feedback inhibition (34). It seems notable that all of the glutamine synthetases have subunits of about the same size (44,000-50,000). The amino acid compositions of the mammalian glutamine synthetases that have been isolated are remarkably similar (26, 35). Some similarities in amino acid composition between the enzymes of mammalian tissues and those isolated from plants and bacteria have been noted, but there are some striking differences. For example, the glutamine synthetases from pea and *E. coli* contain, respectively, 2 and 4 half-cystine residues per subunit as compared to 12 for the enzymes isolated from mammalian sources.

MECHANISM OF ACTION OF GLUTAMINE SYNTHETASE

Enzyme-bound γ-Glutamyl Phosphate

For a number of years the mechanism of the reaction catalyzed by glutamine synthetase was considered enigmatic. Later some controversy developed as to whether there are individual steps in the synthesis of glutamine or whether the enzyme catalyzes a concerted reaction. There is now very strong evidence that the reaction is not concerted and that it takes place by formation of enzyme-bound intermediates including γ-glutamyl phosphate. The controversy about the mechanism was very useful in that it stimulated a number of different experimental approaches which ultimately

TABLE I *

Comparison of glutamine synthetases from various sources

Properties	Rat Liver	Ovine Brain	Pea	*E. coli*	*B. subtilis*
No. of subunits (subunit MW)	8 (44,000)	8(49,000)	8 (45,000)	12 (50,000)	12 (50,000)
Approx. molecular weight	352,000	392,000	360,000	592,000	600,000
$S_{20,w}$ (S)	15.0	15.0	14.0	20.3	19.3
Specificity: D-glutamate	Yes	Yes	Yes	Yes	Yes
α-Methyl-L-glutamate	Yes	Yes	Yes	Yes	
cis-Cycloglutamate	Yes	Yes	Yes	Yes	
threo-γ-Methyl-L-glutamate	Yes	Yes	Yes	Yes	
Inhibition: L-methionine-S-sulfoximine	Yes	Yes	Yes	Yes	Yes
Glycine, alanine	Yes	Slight	Yes	Yes	
Carbamyl phosphate	Yes	Yes	Slight	Yes	
Tryptophan, histidine	No	No	Slight	Yes	Yes
AMP	No	No	Yes	Yes	Yes
Glucosamine 6-phosphate	No	No	Slight	Yes	
Glutamine	No	No		No	Yes
Activation: α-ketoglutarate	Yes	Slight	Slight	No	
Adenylylated form	No	No	No	Yes	No

* Adapted from Ref (26).

led not only to resolution of the issue but also to much detailed information about the reaction mechanism. Those favoring the idea of a concerted mechanism (36-38) were influenced by observations made in several laboratories that the enzyme does not catalyze exchange reactions between (a) inorganic phosphate and ATP, (b) ADP and ATP, (c) glutamate and glutamine, or (d) ammonia and glutamine in the absence of the other components of the synthesis system. The inability of the enzyme to catalyze such exchange reactions is consistent with a mechanism in which there are enzyme-bound intermediates (26). For example, the fact that the enzyme does not catalyze ADP-ATP exchange in the presence of glutamine is explicable in terms of an enzyme-glutamyl-phosphate-ADP complex from which ADP does not readily dissociate (39). Steady-state and equilibrium exchange kinetic studies of the glutamine synthetases from brain, *E. coli*, and peas have been carried out (38, 40, 41); these now seem to be consistent with a number of sequential mechanisms and also with the conclusion (40) that the reactions catalyzed by the glutamine synthetases from brain, pea, and *E. coli* proceed by similar, if not identical, pathways. Thus, although most of the experiments that have been carried out on the mechanism of this reaction have been performed with the sheep brain enzyme, it seems safe to conclude that the mechanism of catalysis is virtually identical for the glutamine synthetases isolated from different sources.

Very useful information about the reaction mechanism has come from studies on partial reactions catalyzed by the enzyme (reactions (*2*)-(*9*)) (26). In the γ-glutamyl transfer reaction (reaction (*2*) the γ-glutamyl moiety of glutamine

$$\text{L-Glutamine} + NH_2OH \xrightarrow[M^{2+},\ P_i\ (As_i)]{\text{ADP (ATP)}} \gamma\text{-glutamylhydroxamate} + NH_3 \qquad (2)$$

is transferred to hydroxylamine to form γ-glutamyl hydroxamate. The reaction requires the presence of either ATP or ADP, divalent metal ions, and either phosphate or arsenate. There is now little doubt that the γ-glutamyl transfer reaction is a catalytic activity of the synthetase itself. However, a number of studies have shown that the ratio of synthetase to transferase activity exhibited by the enzyme depends upon the condition employed for the activity determinations. Severe complications are possible when unpurified tissue preparations are studied because these may contain other enzyme activities such as glutaminases and glutamine amidotransferases that can also catalyze formation of γ-glutamyl hydroxamate from glutamate and hydroxylamine. It is worth noting that glutaminases can catalyze the formation of γ-glutamyl hydroxamate from glutamate and hydroxylamine (22). Thus, the use of the γ-glutamyl transfer reaction for measurement of tissue glutamine synthetase activity should be approached with caution. Other pitfalls also exist, especially if the formation of hydroxamate is measured only by a spectrophotometric method and it is not shown that γ-glutamyl hydroxamate is the actual product being measured (42).

Glutamine synthetase also catalyzes an arsenolysis of glutamine (reaction (3)) which, like the γ-glutamyl transfer reaction, requires catalytic quantities of nucleotide.

$$\text{L-Glutamine} + H_2O \xrightarrow[\text{As}_1]{\text{ADP, M}^{2+}} \text{L-glutamate} + NH_3 \qquad (3)$$

Although it was clear from early studies that the synthesis of glutamine involves a coupling between the exergonic cleavage of ATP to ADP and inorganic phosphate and the endergonic synthesis of glutamine from glutamate and ammonia, free intermediates such as amidophosphate and γ-glutamyl phosphate were not found, and it therefore became clear that the intermediates, if any, must be bound to the enzyme. Although glutamine formation was observed when the enzyme was incubated with N-acetyl-γ-glutamyl phosphate, amino acid-N-acyclase and ammonia, it was not possible to distinguish, under the conditions employed, between enzymatic and nonenzymatic synthesis (43). These studies did, however, clearly demonstrate the high reactivity of γ-glutamyl phosphate with ammonia and this reaction thus serves as a chemical model for the enzymatic reaction. Studies with ^{18}O-labeled glutamate showed that the synthesis of glutamine is accompanied by transfer of oxygen from glutamate to inorganic phosphate (44, 45); this finding, while in accord with an intermediate γ-glutamyl phosphate, can also be explained by other mechanisms involving intermediate oxygen transfer to and from the enzyme.

An early clue concerning the mechanism of the reaction was obtained in studies on glutamine synthetase from peas in which it was found that both optical isomers of glutamate are substrates. Whereas the rates of synthesis of D- and L-γ-glutamyl hydroxamate were found to be similar, the synthesis of D-glutamine occurred at a significantly lower rate than that of L-glutamine (46). These observations led to the suggestions that the enzyme catalyzes an initial activation of glutamate, a reaction of low optical specificity, followed by a more specific reaction of the activated glutamate with ammonia which becomes rate limiting in the case of D-glutamate. The known tendency of γ-glutamyl compounds to cyclize led to studies in which the enzyme was incubated with ATP and glutamate in the absence of ammonia (39). It was postulated that if an enzyme-bound γ-carboxyl-activated glutamate derivative were formed, such a derivative might exhibit a greater tendency to cyclize than either glutamate or glutamine. It was found that glutamine synthetase catalyzes a partial reaction in the absence of ammonia in which glutamate cyclizes with stoichiometric formation of 5-oxoproline, ADP and inorganic phosphate (reaction (4)). Both isomers of glutamate are cyclized and the rates of these reactions are similar.

$$\text{L- (or D-) Glutamate} + \text{ATP} \xrightarrow{M^{2+}} \text{5-oxo-L- (or D-) proline} + \text{ADP} + P_i \qquad (4)$$

Additional evidence for the activation of glutamate was obtained in the course of isotope trapping experiments in which the enzyme was incubated with ATP, Mg^{++}, and [^{14}C]glutamate; a mixture containing hydroxylamine and a large excess of unlabeled glutamate was then added, and after deproteinization the formation of labeled γ-glutamyl hydroxamate was determined (39). These studies demonstrated preferential conversion of the labeled glutamate to γ-glutamyl hydroxamate, showing that equilibration between the labeled and unlabeled glutamates did not occur. The isotope trapping ("pulse-chase") approach to the study of enzyme mechanisms has also been successfully applied to other enzymes (47-53) . Other studies in which high speed centrifugation and ultrafiltration were employed showed that glutamate becomes attached to the enzyme only when both ATP and Mg^{++} are present. It was also found that the binding of glutamate to the enzyme under these conditions is associated with cleavage of ATP to ADP, and that the phosphorus obtained from the terminal phosphate group of ATP remains attached to the enzyme complex.

Since both D-glutamate and L-glutamate are substrates of the enzyme, it is evident that the amino group of the substrate need not be attached to a specific position on the glutarate carbon chain. Indeed, β-glutamate was found to be a good substrate; the enzyme amidates a specific carboxyl group of this symmetrical substrate to form only D-β-glutamine (54, 55). Since the β-aminoglutaryl phosphate intermediate that would be formed in the enzymatic synthesis of β-glutamine is much more stable than γ-glutamyl phosphate it could be prepared by chemical synthesis and tested. It was found that the enzyme can utilize β-glutamyl phosphate for the synthesis of ATP from ADP (reaction (*5*)). The enzyme was subsequently found to be reactive with other acyl phosphates such as acetyl phosphate (reaction (*6*)) and carbamyl phosphate (reaction (7)) (56). The evidence indicates that β-glutamyl phosphate, acetyl phosphate, and carbamyl phosphate interact at the active site of the enzyme. Additional support for the formation of an intermediate γ-glutamyl

$$\beta\text{-Glutamyl phosphate} + \text{ADP} \xrightarrow{M^{2+}} \text{ATP} + \beta\text{-glutamate} \qquad (5)$$

$$\text{Acetyl phosphate} + \text{ADP} \xrightarrow{M^{2+}} \text{ATP} + \text{acetate} \qquad (6)$$

$$\text{Carbamyl phosphate} + \text{ADP} \xrightarrow{M^{2+}} \text{ATP} + CO_2 + NH_3 \qquad (7)$$

phosphate came from studies with a cyclic analog of glutamate, *cis*-1-amino-1, 3-dicarboxycyclohexane (57). This compound cannot cyclize to form an analog of 5-oxoproline. When the enzyme was incubated with this compound in the presence of ATP and Mn^{++} in the absence of ammonia, an enzyme complex was formed which contained cycloglutamyl phosphate (reaction (*8*)).

$$\text{ENZYME} + \text{cycloglutamate} + \text{ATP} \xrightarrow{M^{2+}} \text{ENZYME[cycloglutamyl phosphate] [ADP]} \qquad (8)$$

In these studies it was shown that cycloglutamate binds to the enzyme only in the presence of ATP and metal ions (58).

The studies reviewed above indicate that the enzyme catalyzes a reaction in which glutamate and ATP bind to the enzyme in a process in which ATP is cleaved to ADP and in which glutamate is activated; the phosphate formed in ATP cleavage remains bound to the enzyme. In the absence of ammonia, there is spontaneous dissociation of ADP, inorganic phosphate, and 5-oxoproline from the enzyme. The enzyme can synthesize ATP from ADP and acyl phosphates such as β-aminoglutaryl phosphate, carbamyl phosphate, and acetyl phosphate. The enzyme can synthesize the acyl phosphate derivative of *cis*-1-amino-1,3-dicarboxycyclohexane, a reaction which is directly analogous to formation of enzyme-bound γ-glutamyl phosphate in the normal catalytic reaction. Additional strong evidence for the acyl phosphate mechanism has come from studies on the inhibition of glutamine synthetase by methionine sulfoximine; this compound is phosphorylated by glutamine synthetase in the presence of ATP (reaction (*9*)); (see below).

$$\text{ENZYME} + \text{L-methionine-S-sulfoximine} + \text{ATP} \xrightarrow{M^{2+}}$$

$$\text{ENZYME [L-methionine-S-sulfoximine phosphate] [ADP]} \quad (9)$$

Further evidence consistent with the intermediate formation of γ-glutamyl phosphate has come from studies by Todhunter and Purich (59) who reported the borohydride trapping of γ-glutamyl phosphate in the absence of ammonium ions by isolation of α-amino-δ-hydroxyvalerate. The recent development of an isotope scrambling method for the detection of complexes of the type: enzyme-ADP-X-P by Midelfort and Rose (60) has provided an elegant proof for the formation of enzyme bound γ-glutamyl phosphate by glutamine synthetase. These authors incubated glutamine synthetase with glutamate and ATP labeled with ^{18}O in the βγ-bridge oxygen. Reversible cleavage of the $P_{\beta}O\text{-}P_{\gamma}$ bond was detected by the appearance of isotopic oxygen in the β-nonbridge oxygens of ATP. The relative rates of the scrambling observed with glutamine synthetases from sheep brain and *E. coli* were approximately equal to the maximum velocities of the normal enzyme catalyzed reactions. These studies thus provide kinetic evidence consistent with the hypothesis that γ-glutamyl phosphate is an enzyme-bound intermediate and also support a mechanism in which phosphorylation of glutamate precedes reaction with ammonia to form glutamine.

Mapping of the Active Site

Our studies on glutamine synthetase showed that the enzyme catalyzes the synthesis of both L- and D-glutamine (46). On the other hand, we found that the enzyme acts only on the L-isomer of α-methylglutamate to form α-methyl-L-glutamine (61). Another observation that intrigued us was that the enzyme catalyzes the amidation of β-glutamate to form only D-β-glutamine (55). In our effort to explain this unusual optical specificity we were able to elucidate certain interesting features of the active site of the enzyme. In these studies, we were aided by the use of space filling and skeletal models and later by application of a computer technique. The detailed experimental findings have been published (26, 55, 57, 61, 62, 63, 64, 65).

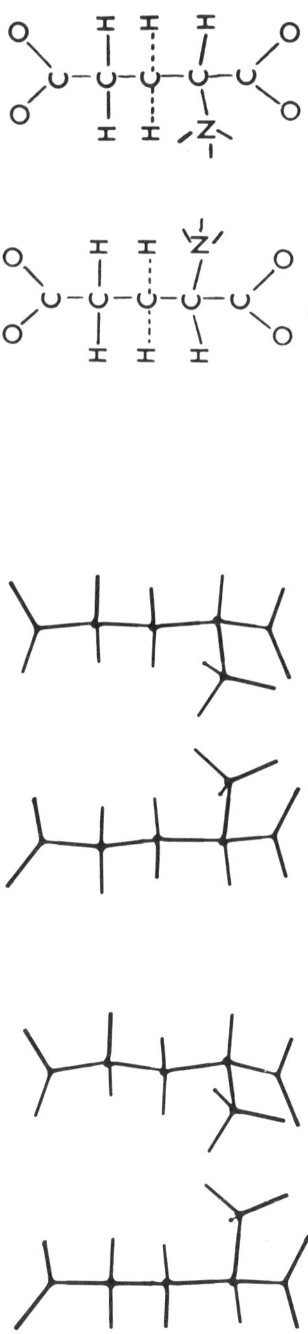

Fig. 1. Stereophotographs of Dreiding models of L-glutamic acid (left) and D-glutamic acid (right). (63)

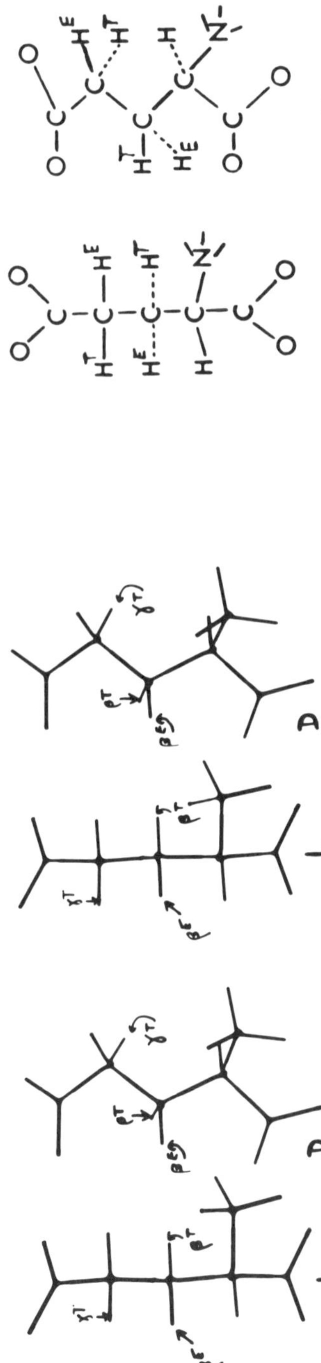

Fig. 2. Stereophotographs of Dreiding models of L-glutamic acid (left) and D-glutamic acid (right) rotated as described in the text. (63)

Since the enzyme does not act significantly on monocarboxylic amino acids or on glutaric acid derivatives that lack an amino group, we postulated that the enzyme has binding sites for both carboxyl groups of L-glutamate as well as for its amino group. Since aspartate is neither a substrate nor an effective inhibitor, we postulated that the distance between the carboxyl binding sites on the enzyme must be greater than the maximum possible intercarboxyl distance for aspartate (about 4 A). Thus, L-glutamate is in the fully extended, or almost fully extended, form when attached to the enzyme.

If one assumes that the enzyme sites which interact with the carboxyl and amino groups of L-glutamate also interact with the corresponding groups of the other substrates, it becomes evident that D-glutamate cannot be oriented on the enzyme in the same manner as L-glutamate (Fig. 1). However, the amino nitrogen atom of D-glutamate can be brought to the same relative position as that of L-glutamate by rotating the D-glutamate molecule 69° about an axis formed by a straight line intersecting the centers of carbon atoms 1, 3 and 5 (Fig. 2). In this way the α-hydrogen atom of D-glutamate can be moved to a position which is essentially opposite to that occupied by the α-hydrogen atom of L-glutamate. If L-glutamate is attached to the enzyme in such a manner that its α-hydrogen atom is directed away from the enzyme, then it would be expected that substitution of the α-hydrogen atom of L-glutamate by a methyl group would not interfere with its binding to the enzyme. However, replacement of the α-hydrogen atom of D-glutamate by a methyl group would provide interference with attachment of this substrate to the active site. As further consequence of the proposed orientations of L-glutamate and D-glutamate on the enzyme, it may be noted that both β-hydrogen atoms of L-glutamate and the erythro-β-hydrogen atom of D-glutamate are on the same side of the molecule as the α-hydrogen atom of D-glutamate. On the other hand, the threo-β-hydrogen atom of D-glutamate is in a position very close to that of the α-hydrogen atom of L-glutamate. These observations led us to predict that substitution by a methyl group of either of the β-hydrogen atoms of L-glutamate or of the erythro-β-hydrogen atom of D-glutamate would lead to loss or marked reduction in enzymatic activity. Similarly, we predicted that substitution of the threo-β-hydrogen atom of D-glutamate by a methyl group would not lead to loss of activity. These predictions were found to be correct; thus, of the four stereoisomers of β-methyl-glutamate, only threo-β-methyl-D-glutamate was found to be a substrate. Consideration of the γ-substituents of the enzyme-bound L- and D-glutamates reveals that the erythro-γ-hydrogen atoms of both L- and D-glutamate occupy about the same position in space and lie just between the γ-carboxyl and amino groups of these molecules. An erythro-γ-methyl group would be expected to provide considerable steric hindrance. The threo-γ-hydrogen atom of D-glutamate is close in position to the α-hydrogen atom of D-glutamate. We therefore predicted that threo-γ-methyl-D-glutamate, like α-methyl-D-glutamate, would not be a substrate. In contrast, the threo-γ-hydrogen atom of L-glutamate is close to the α-hydrogen atom of L-glutamate suggesting that threo-γ-methyl-L-glutamate would be a substrate whereas the other three γ-methyl glutamates would not be substrates. Experiments showed that of the four stereoisomers of γ-methylglutamate, only threo-γ-methyl-L-glutamate is a substrate.

There are ten possible monomethyl glutamates; of these, only three are substrates of glutamine synthetase. Examination of the three susceptible methyl-substituted substrates (α-methyl-L-glutamate, threo-β-methyl-D-glutamate, and threo-γ-methyl-L-glutamate) shows that the methyl groups of these substrates (when they are bound to the enzyme) are located on the same side of the molecule. This led us to design a substrate containing a cyclohexane ring consisting of carbon atoms 2, 3, and 4 of L-glutamate and a chain of three carbon atoms attached to carbon atoms 2 and 4 of glutamate. The stereoisomer of this cyclohexane compound that corresponds to L-glutamate can exist in a form possessing a relatively rigid 5-carbon chain that is identical to that of L-glutamate (Fig. 3). *Cis*-1-amino-1, 3-dicarboxycyclohexane was synthesized and found to be an excellent substrate (57) (Table II), thus supporting the mapping hypothesis discussed above.

The investigations briefly reviewed above indicate that L-glutamate attaches to the active site of the enzyme in an extended conformation in which the carboxyl groups are as far apart (or almost so) as possible, and in which the α-hydrogen atom of L-glutamate is directed away from the enzyme. In an attempt to define the individual points of attachment between the substrates and the active site of the enzyme, the three dimensional coordinates of the substrates and of the inhibitors of the enzyme were calculated, and an active site was mathematically designed using a computer (66). This approach led to some additional conclusions that were not immediately evident from consideration of the space filling and skeletal models. The computer program was designed to manipulate the coordinates of the molecules so that one could, for example, calculate the structure of the tetrahedral addition compound formed in the reaction of γ-glutamyl phosphate with ammonia. The details of these studies including various assumptions that were made, have been published (66, 67).

The computer studies showed that when the extended conformation of L-glutamate is reflected across a plane containing the α- and γ-carboxyl carbon atoms and the α-amino nitrogen atom, D-glutamate is obtained (Fig. 4). The α- and γ-carboxyl carbon atoms and the nitrogen atom of each molecule are in the reflecting plane; since these atoms are common to both molecules and are coincident they may bind to the same enzyme sites. It is possible to place one, but not both of the oxygen atoms of each γ-carboxyl group in the reflecting plane. Making the reasonable assumption that the phosphate groups of L-γ-glutamyl phosphate and D-γ-glutamyl phosphate occupy the same enzyme site, we may designate the oxygen atom in the reflecting plane as the oxygen atom that is phosphorylated. The position of the remaining oxygen atom of the γ-carboxyl group of L-glutamate is that of the oxygen atom which is not phosphorylated. In the stereograph shown in Fig. 5, the left-hand site labeled ENZ is defined as the "oxygen binding site" (oxygen that is not phosphorylated); the right-hand site labeled ENZ is the binding site

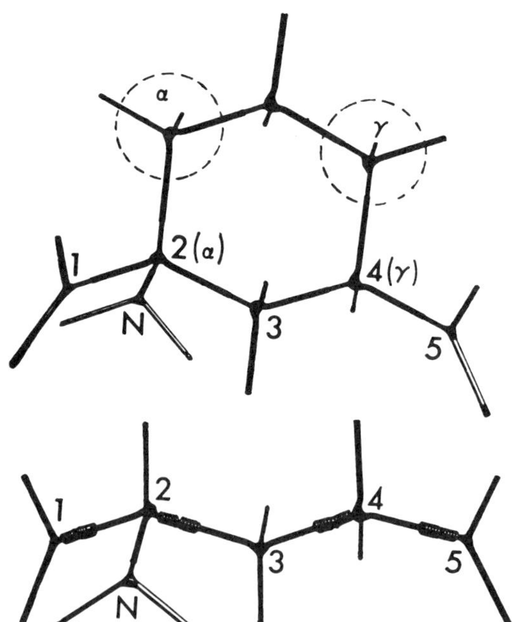

Fig. 3. Dreiding models of L-glutamate and *cis*-L-1-amino-1,3-dicarboxy-cyclohexane (*cis*-L-cycloglutamate). (57)

TABLE II *

Relative activities of various substrates of ovine brain glutamine synthetase with ammonia and hydroxylamine

Amino acid substrate	Relative Vmax NH_2OH	Relative Vmax NH_3
L-Glutamate	100	100
D-Glutamate	54	27
α-Methyl-L-glutamate	67	75
threo-β-Methyl-D-glutamate	46	2.2
threo-γ-Methyl-L-glutamate	63	27
threo-β-Hydroxy-D-glutamate [a]	48	22
threo-γ-Hydroxy-L-glutamate	89	100
threo-γ-Hydroxy-D-glutamate	1.6	<0.08
erythro-γ-Hydroxy-L-glutamate	64	81
erytro-γ-Hydroxy-D-glutamate	29	38
β-Glutamate	46	18
L-α-Aminoadipate	22	0.63
D-α-Aminoadipate	11	0.19
cis-L-1,3-dicarboxycyclohexane	102	20.0

[a] The values for the *erythro*-L, *erythro*-D, and *threo*-L isomers of β-hydroxyglutamate (with NH_2OH) are 3, 1.5, and 6, respectively.
* From Ref. (26).

of the oxygen atom of L-glutamate that is phosphorylated; the central site labeled ENZ is the position of the nitrogen atom in the tetrahedral intermediate ("ammonia binding site").

Fig. 5 gives the computer-drawn stereograph of the enzyme-bound tetrahedral intermediate formed in the reaction of ammonia with L-γ-glutamyl phosphate. The conformation of the tetrahedral intermediate was calculated on the assumption that the oxygen atoms and phosphate group of the intermediate are very close to their respective positions in the planar acyl phosphate; the conformation of the tetrahedral intermediate calculated was one in which the distances between its two oxygen atoms and the two oxygen reference points on the enzyme (derived from the binding sites of the carboxyl oxygens of glutamate) are at a minimum. In the stereograph, the positions of the nitrogen atom of the tetrahedral intermediate is designated as the ammonia binding site. The mapping studies thus identified 5 points on the enzyme derived from the positions of

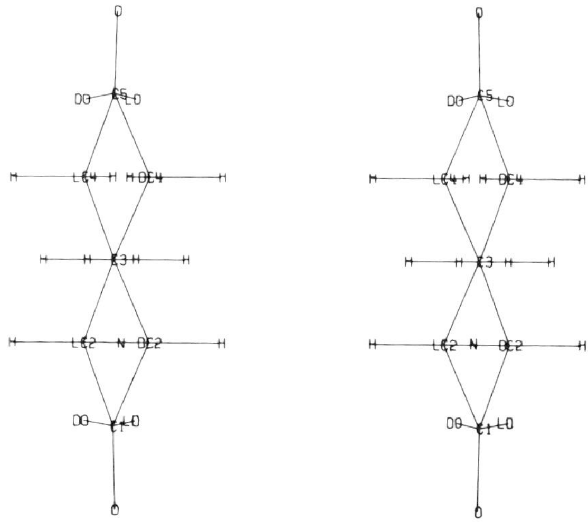

Fig. 4. Computer-drawn stereograph of D- and L-glutamate in an extended conformation. The C-1 and C-5 carbon atoms and the C-2 nitrogen atom of D- and L-glutamate have been made coincident. (66)

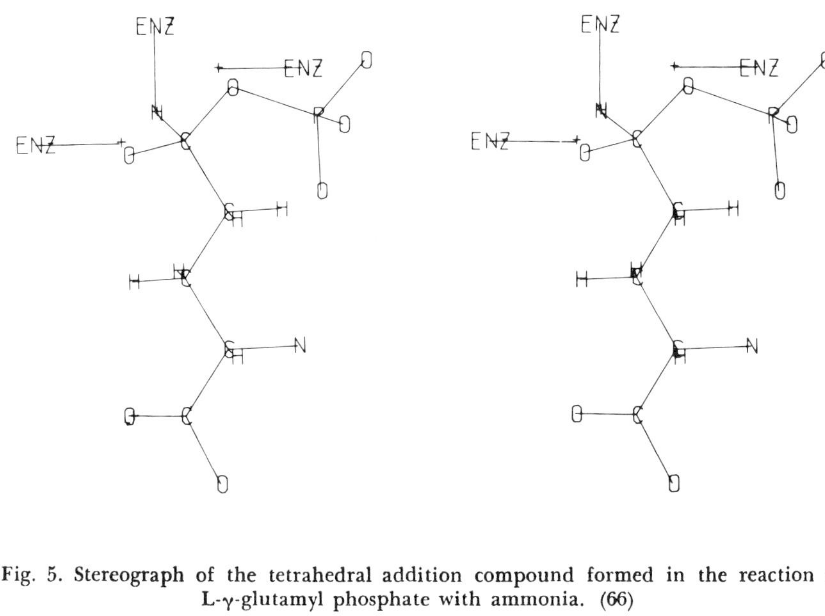

Fig. 5. Stereograph of the tetrahedral addition compound formed in the reaction of L-γ-glutamyl phosphate with ammonia. (66)

particular atoms of L-glutamate. These are the α-carboxyl carbon atom, the α-amino nitrogen atom, the γ-carboxyl oxygen atom which is phosphorylated, the γ-carboxyl oxygen atom which is not phosphorylated, and the nitrogen atom of the tetrahedral intermediate derived from L-glutamate. Other aspects of the reaction mechanism deduced from the computer-mapping work have been considered in detail elsewhere (66).

Inhibition by Methionine Sulfoximine, a Transition State Analog

Convincing evidence for the view that γ-glutamyl phosphate is an intermediate in the reaction catalyzed by glutamine synthetase came from investigations on the mechanism by which glutamine synthetase is irreversibly inhibited by methionine sulfoximine (68-72). These studies have elucidated the structure of the transition state or intermediate formed on the enzyme in the attack of ammonia on enzyme-bound γ-glutamyl phosphate, and have also led to further understanding of the convulsant action of methionine sulfoximine and to the development of a potent and specific inhibitor of γ-glutamylcysteine synthetase and therefore of glutathione synthesis.

When glutamine synthetase is incubated with methionine sulfoximine in the presence of ATP and divalent metal ions the enzyme becomes irreversibly inhibited. Such inhibition requires the presence of ATP and divalent metal ions. Ammonia does not decrease the rate at which the enzyme is inhibited under these conditions, but addition of L-glutamate decreases the rate at which inhibition is established. Inhibition is associated with cleavage of ATP to ADP, and the inhibited enzyme, after isolation by gel filtration, was found to contain close to 1 mole each of a phosphorylated derivative of methionine sulfoximine and ADP per subunit. Both methionine sulfoximine phosphate and ADP are tightly bound to the enzyme, but may be released from the enzyme by brief heating at 100° or by treatment with perchloric acid. The position of the phosphoryl moiety of methionine sulfoximine phosphate (Fig. 6) was established by chemical and infrared spectral studies; methionine sulfoximine phosphate was also prepared by chemical synthesis. Chemically synthesized methionine sulfoximine phosphate inhibits glutamine synthetase in the absence of added nucleotide and metal ions. However, such inhibition is more effective when magnesium ions and ADP are added. The phosphorylation of methionine sulfoximine by glutamine synthetase seems to reflect the phosphorylation of glutamate during the normal catalytic reaction.

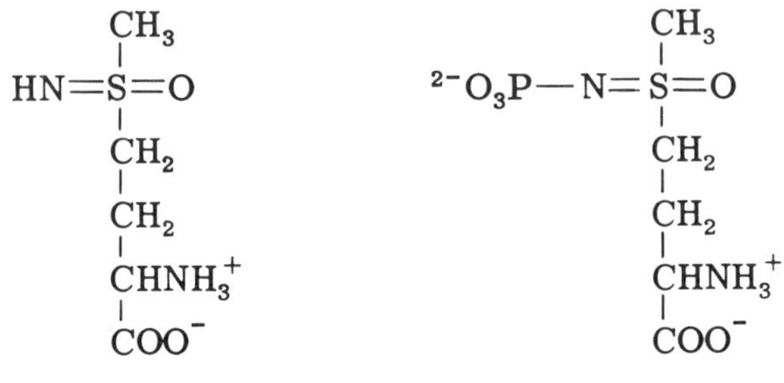

Fig. 6. Structures of methionine sulfoximine and methionine sulfoximine phosphate.

L-Methionine-SR- sulfoximine inhibits glutamine synthetase, but D-methionine-SR-sulfoximine does not. Studies on the two diastereoisomers of L-methionine sulfoximine, facilitated by a knowledge of their absolute configurations (73) and the capability of separating the isomers by chromatographic means (71), led to the finding that only L-methionine-S-sulfoximine inhibits glutamine synthetase irreversibly. It is of interest that only this same diastereoisomer of methionine sulfoximine also inhibits γ-glutamylcysteine synthetase and produces convulsions in animals (see below).

The mapping studies have offered considerable insight into the interaction of the enzyme with methionine sulfoximine. Thus, the computer studies indicate that the sulfoximine oxygen atom of L-methionine-S-sulfoximine binds to the enzyme-binding site for the glutamate carboxyl oxygen atom that is not phosphorylated, that the sulfoximine nitrogen atom binds to the binding site of the enzyme that normally binds the glutamate oxygen atom that is phosphorylated, and that the methyl group of the methionine sulfoximine binds to the ammonia binding site of the enzyme (Fig. 7) (66). Thus L-methionine-S-sulfoximine is bound so that its sulfoximine nitrogen atom is in a good position to be phosphorylated. It would appear that the sulfoximine oxygen atom of methionine sulfoximine is too acidic to react with ATP or, alternatively, that the sites on the enzyme that normally bind the oxygen atoms of glutamate cannot bind the sulfoximine oxygen and nitrogen atoms of L-methionine-R-sulfoximine. The active site mapping study thus supports the conclusion that L-methionine-S-sulfoximine is an inhibitory analog of the enzyme-bound intermediate or transition state formed in the normal catalytic reaction.

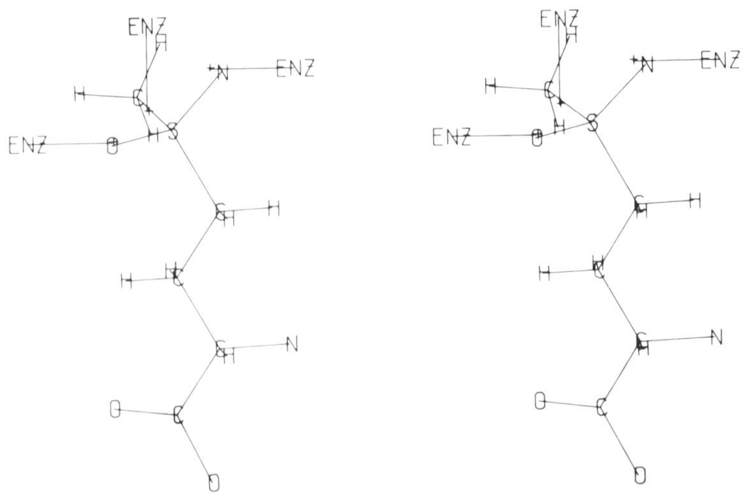

Fig. 7. Stereograph of L-methionine-S-sulfoximine (see the text). (66)

The studies on the mapping of the active site of glutamine synthetase have served usefully in the design of selective sulfoximine inhibitors of both glutamine synthetase and γ-glutamylcysteine synthetase. As stated above, methionine sulfoximine effectively inhibits γ-glutamylcysteine synthetase (74) as well as glutamine synthetase. It is notable that the same diastereoisomer of methionine sulfoximine (the L, S-isomer) inhibits both enzymes. Thus, after administration of methionine sulfoximine to animals, the tissue levels of both glutamine and glutathione decrease substantially. Studies on the mechanism of action of γ-glutamylcysteine synthetase, an enzyme that catalyzes a reaction (reaction *(10)*) closely analogous to that

$$\text{L-Glutamate} + \text{L-cysteine (or L-}\alpha\text{-aminobutyrate)} + \text{ATP} \overset{M^{2+}}{\rightleftharpoons} \text{L-}\gamma\text{-glutamyl-L-cysteine} + \text{ADP} + P_i \qquad (10)$$

of glutamine synthesis (reaction *(1)*), indicate that, like glutamine synthetase, enzymebound γ-glutamyl phosphate is an intermediate (75). Furthermore, inhibition of γ-glutamylcysteine synthetase by methionine sulfoximine is also associated with formation of enzyme-bound methionine sulfoximine phosphate (74), so that it may be concluded that methionine sulfoximine is a transition-state inhibitor of both glutamine synthetase and of γ-glutamylcysteine synthetase.

Since the mechanisms of the reactions catalyzed by these synthetases are closely analogous it appeared that successful design of reagents that would inhibit only one of these enzymes would have to be directed to differences in the substrate binding sites of the respective enzymes. Since α-methylglutamic acid is a good substrate of glutamine synthetase, we expected that α-methylmethionine sulfoximine would bind effectively to glutamine synthetase. Although α-methylglutamate is also a substrate of γ-glutamylcysteine synthetase, it is rather less effective as a substrate for this enzyme than for glutamine synthetase, suggesting that there is some steric restriction in this region of the substrate binding site of the γ-glutamylcysteine synthetase. α-Ethylmethionine sulfoximine was therefore prepared with the expectation that this compound would bind to glutamine synthetase, but that it would not bind to γ-glutamylcysteine synthetase effectively (76); such expectations were realized (see below).

We also reasoned that substitution of a bulkier group in place of the methyl group of methionine sulfoximine might interfere with binding to glutamine synthetase. Studies on the mapping of the active site of glutamine synthetase showed that the S-methyl group of methionine sulfoximine attaches to the enzyme site that normally binds ammonia. We thought that the comparable region of the active center of γ-glutamyl-cysteine synthetase would bind a group somewhat bulkier than a methyl group since acceptor amino acids for γ-glutamylcysteine synthetase (cysteine, α-aminobutyrate) are substantially larger than ammonia or the methyl group. A number of analogs of methionine sulfoximine were prepared in which the S-methyl moieties were varied (77, 78). It was found that prothionine sulfoximine (S-n-propylhomocysteine sulfoximine) and buthionine sulfoximine (S-n-butylhomocysteine sulfoximine) are excellent inhibitors of γ-glutamylsynthetase. Prothionine sulfoximine produced only slight inhibition of glutamine synthetase, whereas buthionine sulfoximine did not inhibit glutamine synthetase to a detectable extent. It is notable that prothionine sulfoximine is an extremely weak convulsant and that buthionine sulfoximine is not a convulsant. In contrast, α-ethylmethionine sulfoximine induces convulsions (76).

These enzymatic and physiological findings may be interpreted in the light of our knowledge of the active sites of the two synthetases (Fig. 8). Thus, the acceptor site of γ-glutamylcysteine synthetase (which can use α-aminobutyrate) can readily bind the n-butyl moiety of buthionine sulfoximine. On the other hand, the comparable site of glutamine synthetase, the ammonia binding site, can accept the S-methyl group of methionine sulfoximine but does not bind a molecule containing a much bulkier substitutent on the sulfur atom

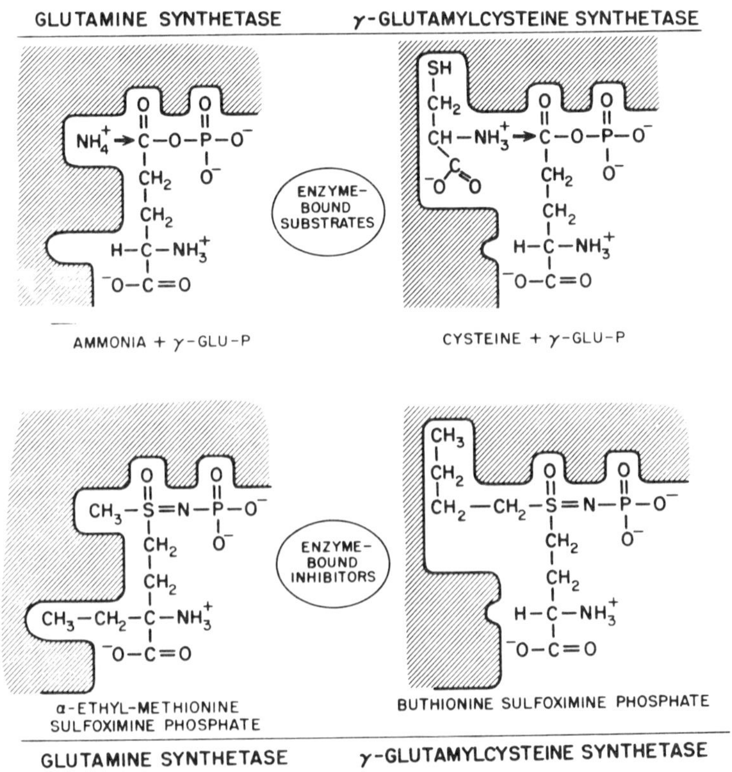

Fig. 8. Active sites of glutamine and γ-glutamylcysteine synthetases; see the text. Modified from (77).

such as buthionine sulfoximine. As stated above, the site on γ-glutamylcysteine synthetase adjacent to the α-hydrogen atom of the substrate appears to be much more sterically hindered than the comparable region of the active site of glutamine synthetase. Thus, although both synthetases can bind α-methylglutamate and α-methylmethionine sulfoximine, only glutamine synthetase can accommodate the α-ethyl moiety. Thus, only glutamine synthetase is inhibited by α-ethylmethionine sulfoximine. A knowledge of the topology of the active sites of these two enzymes has therefore made it possible to design selective sulfoximines inhibitors for each of the enzymes. This makes it possible to carry out studies in which glutathione synthesis is selectively blocked without appreciable disturbance of the synthesis of glutamine; the reciprocal task of inhibiting glutamine synthesis without inhibiting the synthesis of glutathione has also been achieved. The findings also indicate that the convulsant activity of methionine sulfoximine is due to its effect on glutamine synthetase rather than to an effect on the synthesis of glutathione (76, 79).

Summary of the Mechanism of Action of and the Partial Reactions Catalyzed by Glutamine Synthetase

The scheme given in Fig. 9 summarizes the steps considered to be involved in the synthesis of glutamine. It also explains the partial reactions catalyzed by glutamine synthetase. The evidence indicates that the enzyme has an active center equipped with binding sites for L-glutamate, ATP, and ammonia and that these sites are arranged in such a manner as to facilitate interactions between the substrates on the enzyme. There seems to be no evidence for interaction between substrates and enzyme that involve covalent bond formation. The scheme (which omits the divalent metal ions required for enzymatic activity) includes the activated form of glutamate, γ-glutamyl phosphate, and the tetrahedral intermediate or transition state formed in the reaction of ammonia with enzyme-bound γ-glutamyl phosphate. The scheme also takes into account the various experimental findings indicating that nucleotide is required for the binding and reaction of the substrates; thus, ADP (or ATP) is a functional portion of the active site. The scheme indicates that the enzyme does not interact with glutamate until ATP is bound, nor does the enzyme bind glutamine in the reverse direction until ADP is bound. Alternative possibilities such as the binding of glutamate prior to the binding of ATP or a random order of addition of glutamate and ATP to the enzyme are also conceivable. The rearrangement of complex III to yield IV (reaction (5), Fig. 9) may occur (whether reactions 1 and 3 occur as shown in Fig. 9 or in the reverse order or randomly) to yield an enzyme complex containing ADP and γ-glutamyl phosphate. As indicated in the legend to Fig. 9, the scheme accounts for the several partial reactions catalyzed by the enzyme such as cyclization of glutamate, phosphorylation of glutamate analogs, and utilization of acyl phosphates for ATP synthesis. It is notable that glutamate affords some protection against inhibition by methionine sulfoximine (reactions 1, 17, 18), but that ammonia does not; on the other hand, addition of both ammonia and glutamate provides complete protection against inhibition by methionine sulfoximine. This finding is consistent with the mapping studies which indicate that methionine sulfoximine attaches to both glutamate and the ammonia binding sites of the enzyme. The failure of ammonia to protect against inhibition by methionine sulfoximine in the absence of glutamate suggests that the binding of ammonia to the

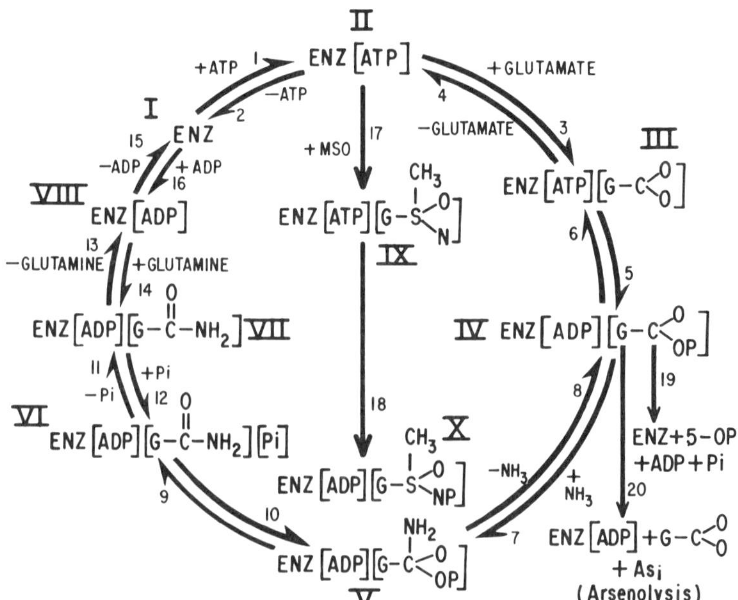

$G = {}^{-}OOCCH(NH_3^+)(CH_2)_2-$

Glutamine synthesis: 1, 3, 5, 7, 9, 11, 13, 15

Reversal of synthesis: 16, 14, 12, 10, 8, 6, 4, 2

Partial reactions:

γ-Glutamyl transfer: 16, 14, 12, 10, 8, 7 (+NH_2OH), 9, 11, 13

Arsenolysis of glutamine: 16, 14, 12 (+As_i), 10, 8, 20

Phosphorylation of L-methionine-S-sulfoximine (MSO): 1, 17, 18

Cyclization; formation of 5-oxoproline (5-OP): 1, 3, 5, 19

Phosphorylation of glutamate analog (cycloglutamate): 1, 3, 5

Acyl phosphate (β-glutamyl∼P, carbamyl∼P, acetyl∼P) + ADP ⟶ ATP: 6, 4, 2

Fig. 9. Scheme for the mechanism of action of glutamine synthetase. (26)

enzyme may require prior binding of glutamate. This is reflected in the scheme according to which ammonia reacts after formation of enzyme-bound γ-glutamyl phosphate. The data show clearly that γ-glutamyl phosphate can be formed in the absence of ammonia, although it is possible for ammonia to bind before formation of the acyl phosphate. The relatively slow rate of formation of 5-oxoproline observed when the enzyme is incubated with glutamate, ATP, and divalent metal ions may reflect the tight binding of the acyl phosphate intermediate to the enzyme; such binding undoubtedly confers stability on the acyl phosphate. Other aspects of the reaction mechanism are discussed elsewhere (26).

OVERVIEW OF GLUTAMINE METABOLISM

Glutamine is of central importance in the nitrogen metabolism of many species (5). In mammals, substantial concentrations of glutamine are found in many tissues and in the blood plasma. In man, glutamine accounts for about 20% of the total plasma amino acid content; the body pool of glutamine is of the same order as that of glucose. The synthesis of glutamine is a physiologically important pathway for ammonia utilization in the brain and in other tissues. Glutamine seems to be a storage and transport form of both glutamate and ammonia. In animals the tissue and body fluid levels of ammonia are maintained at relatively low levels, and elevated levels of ammonia are associated with toxicity. In general, glutamine is transported across cell membranes more effectively than is glutamate; for instance, glutamine readily crosses the blood-brain barrier, whereas glutamate does not (80, 81).

Plasma glutamine is probably derived from a number of tissues including liver and muscle. The amount of glutamine that arises from the skeletal muscle in mammals is substantial (82, 83). Amino acid metabolism in muscle provides precursors for the synthesis of alanine and glutamine. Alanine is extensively used by the liver for gluconeogenesis. Both glutamine and alanine serve as transport forms of nitrogen which arise in the muscle, and which are used ultimately in other tissues such as liver, kidney, and intestine. Although glutamine is probably synthesized in muscle, there is relatively little published information on muscle glutamine synthetase. In recent studies, Rowe (84) purified glutamine synthetase from rat and rabbit skeletal muscle and from rat heart. The total activities found in crude tissue preparations of skeletal muscle were of an order of magnitude consistent with the observed *in vivo* production of glutamine by muscle. Rowe found that highly purified preparations of rat skeletal muscle glutamine synthetase were markedly inhibited by ADP, in fact apparently much more so than found with sheep brain glutamine synthetase. For this reason, his assays of activity were carried out in the presence of an ATP-regenerating system such as creatine kinase.

There is evidence that the kidney (85), small intestine (86), and probably other organs use plasma glutamine as important source of energy. Presumably, the enzymes involved include glutaminase, glutamine transaminases, and glutamate transaminases. In the kidney, the hydrolysis of glutamine by glutaminase provides part of a physiologically important mechanism for the formation of ammonia; in acidosis, not only the amide nitrogen of glutamine but also its α-amino group are extensively used for ammonia formation (87). Phosphate-activated glutaminase (88) is most certainly a major catalyst in this process, but other enzymes including transaminases and glutamate dehydrogenase probably also play a role.

Brain glutamine synthetase, now known to be highly concentrated in the glial cells (89), constitutes the major system for utilization of ammonia in this organ. Studies in which toxic doses of [^{15}N]ammonia were given via the carotid artery to cats showed that the incorporation of ^{15}N into the α-amino group of brain glutamine was greater than that of glutamate suggesting that glutamate used for glutamine synthesis was in a separate compartment and thus not in equilibrium with another glutamate pool (90). In recent studies on rats in which $^{13}NH_3$ was administered via the carotid, the relative specific activities of the brain glutamate, the α-amino group of glutamine, and of the amide N of glutamine were found, respectively, to be in the ratio 1:5:400 (91). These and related findings and considerations suggest that the small glutamate pool used for glutamine synthesis lies within the glial cells. There is evidence that the glutamate which is released from glutamatergic neurons may be taken up by the glia. Glial glutamine is thought to be transported to the neurons where it may serve as a precursor of glutamate, and thus also of γ-aminobutyrate (92-94). Synaptosomes have substantial levels of glutaminase activities. Thus, it would appear that glial glutamine synthetase and glutaminase present in nerve endings play a significant role in the formation and disposition of glutamate in glutamatergic synapses.

The convulsant action of methionine sulfoximine and related sulfoximines has been pinpointed, as discussed above, to an effect closely related to inhibition of brain glutamine synthetase (76-79). A straight-forward interpretation of this phenomenon is that decreased synthesis of glutamine leads to increases in glutamate levels at specific neuronal sites leading to increased stimulation. Inhibition of glutamine synthetase by methionine sulfoximine might then be expected to disturb the normally balanced excitatory effects of glutamate (79). Although it is conceivable that metabolites of methionine sulfoximine induce convulsions by another mechanism, the finding that α-ethylmethionine sulfoximine (which would not be expected to undergo extensive metabolism) is a convulsant tends to render this possibility less likely (76).

Fig. 10 gives an outline of the metabolism of glutamine and glutamate in mammalian tissues; the chart emphasizes the several rather significant interrelationships between glutamate and gluta-

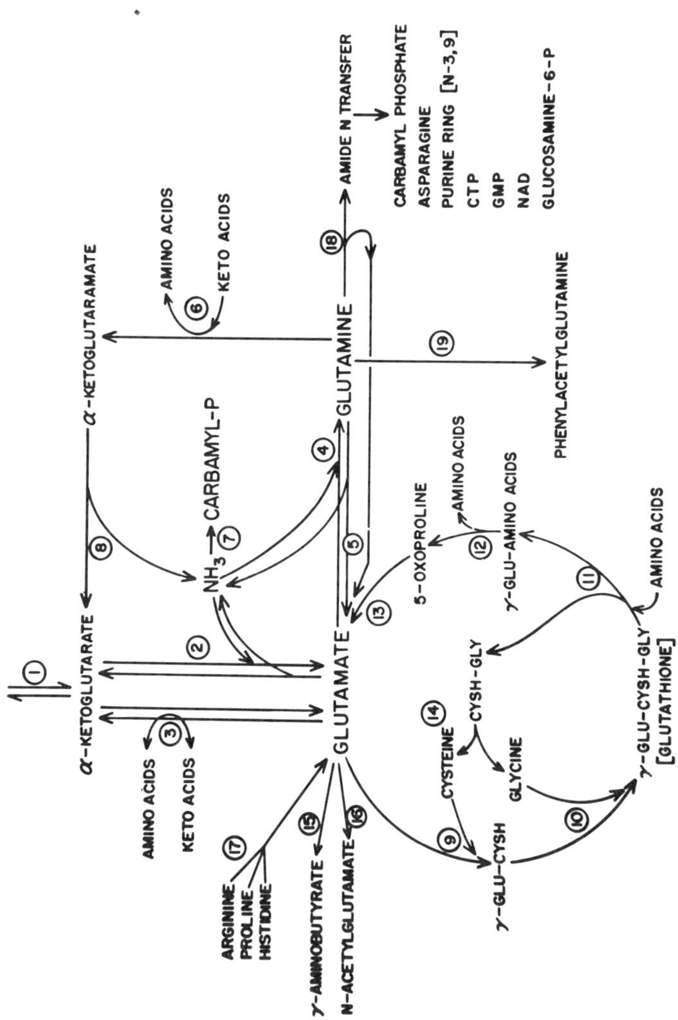

Fig. 10. Overall metabolic scheme for glutamine and glutamate. (95)

TABLE III

Glutamine amidotransferases found only in plants and microorganisms

Glutamate Synthase

α-Ketoglutarate + glutamine + TPNH + H^+ → 2 glutamate + TPN^+

Anthranilate Synthase

Chorismate + glutamine → anthranilate + glutamate + pyruvate

p-Aminobenzoate Synthase

Chorismate + glutamine → p-aminobenzoate + glutamate + pyruvate

Glutamyl-tRNA^{gln} Synthase

Glu-tRNA^{gln} + glutamine + ATP + H_2O → Gln-tRNA^{gln} + ADP + P_i + glutamate

Imidazole Glycerolphosphate Synthase

Phosphoribosyl-formimino-aminoimidazole carboxamide ribonucleotide + glutamine → imidazole glycerol phosphate + aminoimidazole carboxamide ribonucleotide + glutamate

$$\text{Glutamine} + \alpha\text{-keto acid} \rightleftharpoons \alpha\text{-amino acid} + \alpha\text{-ketoglutaramate} \quad (11)$$

$$\alpha\text{-ketoglutaramate} + H_2O \rightarrow \alpha\text{-ketoglutarate} + NH_3 \quad (12)$$

mine. In most cells both glutamate and glutamine are separately coded for and used for protein synthesis. However, several microorganisms including *Bacillus subtilis* and *Bacillus megaterium* use the missense intermediate, glutamyl-tRNA^{gln}, which is converted to the corresponding glutaminyl-tRNA derivative (96). The enzyme that catalyzes this reaction, as well as four others that do not take place in animal tissues but which do occur in microorganisms and plants are listed in Table III.

In man and higher apes, glutamine is used in a so-called detoxication reaction in which phenylacetic acid is coupled (via phenylacetyl adenylate and phenylacetyl coenzyme A) with glutamine to form phenylacetylglutamine (97, 98). In man, the normal daily urinary excretion of phenylacetylglutamine is about 300 mg.

Although recent studies have tended to emphasize reactions catalyzed by the glutamine amidotransferases, reactions involving the α-amino group of glutamine are probably also of considerable metabolic and physiological significance. Reactions involving the α-amino group are catalyzed by the coupled system glutamine transaminase and α-keto acid ω-amidase (99-104). (Reactions (*11*) and (*12*)). Several different glutamine transaminases are now known that exhibit high affinity for glutamine and certain α-keto acids. The glutamine transaminase reactions like other transaminations, are freely reversible, but in contrast to reactions such as those catalyzed by glutamate-aspartate transaminase, the transamination of glutamine becomes essentially irreversible *in vivo*. It follows that the metabolic role of this reaction must be associated with the utilization of glutamine, formation of ammonia, and the utilization of certain α-keto acids for synthesis of the corresponding amino acids. The most plausible idea concerning the physiological role of the glutamine transaminases is that they function in the conversion of α-keto acids to amino acids. The concentrations of amino acids in mammalian tissues probably fluctuate depending on nutritional and other factors, and accumulation of amino acids may lead to their transamination to the corresponding α-keto acids catalyzed by α-ketoglutarate-amino acid transaminase activity. Under these circumstances, the carbon chains of amino acids such as phenylalanine, tyrosine, and methionine might be lost by excretion or degradation. It would appear that the glutamine transaminases function in a salvage process that prevents such a loss. The transamination reactions between glutamine and α-keto acids are driven by removal of α-ketoglutaramate, and the metabolic balance tends to be further stabilized by the formation of ammonia, which may be used for glutamine synthesis. One can thus picture a mechanism by which glutamine transaminase serves as part of a homeostatic mechanism for the preservation of amino acid balance; this may be illustrated as follows:

Phenylpyruvate + glutamine → α-ketoglutaramate + phenylalanine (*13*)

α-Ketoglutaramate → α-ketoglutarate + NH_3 (*14*)

Glutamate + NH_3 + ATP → glutamine + ADP + P_i (*15*)

SUM:

Phenylpyruvate + glutamate + ATP → phenylalanine + α-ketoglutarate + ADP + P_i (*16*)

TABLE IV *

Other glutamine amidotransferases

Carbamyl Phosphate Synthetase

$HCO_3^- + 2\ ATP + glutamine + H_2O \rightarrow$ carbamyl phosphate + 2 ADP + P_i + glutamate

Formylglycinamidine Ribonucleotide Synthetase

Formylglycinamide ribonucleotide + glutamine + ATP + $H_2O \rightarrow$ formylglycinamidine ribonucleotide + ADP + P_i + glutamate

5-Phosphoribosylamine Synthase

5-Phosphoribosyl pyrophosphate + glutamine + $H_2O \rightarrow$ 5-phosphoribosylamine + PP_i + glutamate

CTP Synthetase

UTP + ATP + glutamine + $H_2O \rightarrow$ CTP + ADP + P_i + glutamate

Asparagine Synthetase

Aspartate + ATP + glutamine + $H_2O \rightarrow$ asparagine + glutamate + AMP + PP_i

GMP Synthetase

XMP + ATP + glutamine + $H_2O \rightarrow$ GMP + AMP + PP_i + glutamate

DPN Synthetase

Deamido-DPN+ + ATP + glutamine + $H_2O \rightarrow$ DPN+ + AMP + PP_i + glutamate

Glucosamine-6-phosphate Synthase

Fructose-6-P + glutamine $\rightarrow$ glucosamine-6-P + glutamate

* See (105) for a recent review.

In this way a metabolic coupling between the activities of glutamine transaminase, ω-amidase, and glutamine synthetase can drive the conversion of an α-keto acid to the corresponding amino acid since the overall equilibrium is far to the right.

The 13 enzymes that catalyze the utilization of the amide nitrogen atom of glutamine are listed in Tables III and IV, Of the several that seem to occur only in microorganisms and plants, glutamate synthase seems worthy of special comment. This enzyme, which catalyzes the reductive amination of α-ketoglutarate by glutamine amide, leads to formation of two moles of glutamate for each mole of glutamine used.

Metabolic coupling between this reaction and those catalyzed by glutamine synthetase and glutamate transaminases provides a pathway for irreversible synthesis of amino acids from the corresponding α-keto acids (34):

$$\text{Glutamine} + \alpha\text{-ketoglutarate} + \text{TPNH} + \text{H}^+ \rightarrow 2\ \text{glutamate} + \text{TPN}^+ \qquad (17)$$

$$\text{ATP} + \text{glutamate} + \text{NH}_3 \rightleftharpoons \text{glutamine} + \text{ADP} + \text{P}_i \qquad (18)$$

$$\text{Glutamate} + \alpha\text{-keto acid} \rightleftharpoons \alpha\text{-ketoglutarate} + \text{amino acid} \qquad (19)$$

SUM:

$$\text{ATP} + \text{NH}_3 + \alpha\text{-keto acid} + \text{TPNH} + \text{H}^+ \rightarrow \text{amino acid} + \text{ADP} + \text{P}_i + \text{TPN}^+ \qquad (20)$$

The other glutamine amidotransferases listed in Tables III and IV include enzymes that catalyze a number of well known reactions involved in the biosynthesis of amino acids (e. g., histidine, tryptophan (via anthranilate), p-aminobenzoate) , and nitrogen atoms 3 and 9 of the purine ring and of the amino groups of guanine and cytidine. Glutamine amide nitrogen is the precursor of the nitrogen atom of carbamyl phosphate, which is used for synthesis of the pyrimidine ring. Glutamine amide nitrogen is also used for synthesis of the amide nitrogen atoms of DPN and asparagine, and for the amino group of glucosamine.

Each of the glutamine amidotransferases catalyzes a separate reaction and would therefore be expected to exhibit characteristic structural and functional properties. Not all of the glutamine amidotransferases have thus far been highly purified and studied in detail. Nevertheless, the available data indicate that, in general, the glutamine amidotransferases exhibit certain similar features which seem to reflect their common ability to bind and use glutamine. For most of these enzymes, there is evidence that glutamine binds to an enzyme site which is separate from that which catalyzes the synthesis reaction. After binding, the amide nitrogen atom of glutamine is transferred to a site which can also accept ammonia; ammonia bound at this site can be used for the synthesis reaction. In some instances, it appears that the amide nitrogen of glutamine binds close to or at the site which can bind ammonia. Almost all of the glutamine amidotransferases can use ammonia in place of glutamine. Although ammonia is often reported to be much less active than glutamine, this may not always be true when comparisons are made on the basis of un-ionized NH_3. Most of the glutamine amidotransferases exhibit glutaminase activity which is usually of a very low order. Certain glutamine analogs such as L-2-amino-4-oxo-5-chloropentanoate, L-azaserine, and 6-diazo-5-oxo-L-norleucine interact effectively with glutamine amidotransferases and bind irreversibly to the glutamine binding sites of these enzymes. After such binding, the enzyme often still functions with ammonia but can no longer utilize glutamine. In a number of instances there is direct or indirect evidence that an enzyme sulfhydryl group is involved in the binding of glutamine analogs. It is notable that the glutamine binding capacity of a number of glutamine amidotransferases may be decreased by procedures used during enzyme purification thus leading to a modified amidotransferase which may react more effectively with ammonia than with glutamine.

To illustrate some of these points, we may briefly summarize findings on the glutamine amidotransferase, carbamyl phosphate synthetase, which has been isolated from *Escherichia coli* (52, 106-111). This enzyme consists of a monomeric unit having a molecular weight of about 163,000; the monomer is composed of a heavy subunit (mol. wt. ~ 130,000) and a light subunit (mol. wt. ~ 40,000). The monomer itself can undergo reversible self-association to yield a dimer or higher polymers. As indicated in Fig. 11, the heavy subunit contains the binding sites for ammonia, bicarbonate, ATP, and the allosteric effectors UMP, IMP, ornithine, and ammonia. The heavy subunit has two separate ATP binding sites. The light subunit has the binding site for the γ-glutamyl moiety of glutamine and this site is equipped with a sulfhydryl group which is thought to interact with glutamine to form a γ-glutamyl enzyme. Both subunits seem to contribute to the binding of glutamine, and this is consistent with the relatively low affinity of the separated light subunit for glutamine. When the subunits are separated, the heavy subunit is able to catalyze carbamyl phosphate synthetase from ammonia, but not from glutamine; the heavy subunit also catalyzes

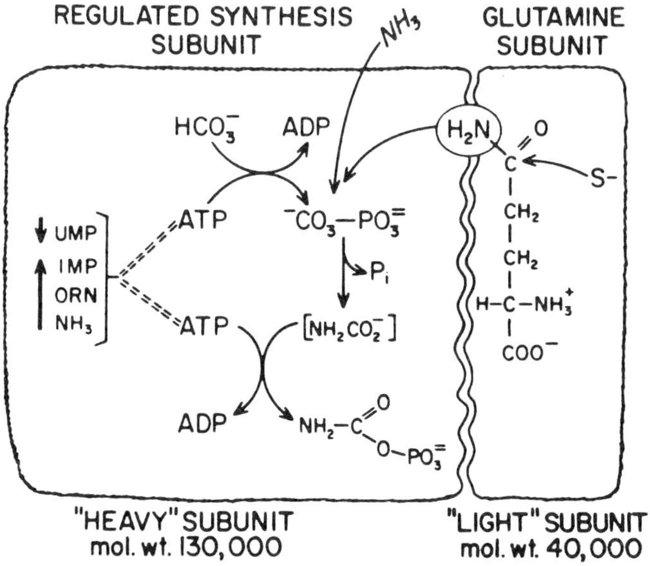

Fig. 11. Subunits of carbamyl phosphate synthetase. (109)

a number of other partial reactions catalyzed by the holoenzyme, and the activities of this subunit are affected by allosteric agents. The only reaction catalyzed by the light subunit is hydrolysis of glutamine. When the enzyme is treated with L-2-amino-4-oxo-5-chloropentanoate, the γ-glutamyl site on the glutamine subunit becomes blocked and the enzyme then reacts only with ammonia and not with glutamine. The two ATP binding sites on the heavy subunit are located in close proximity to the binding site for bicarbonate (and carbamate). A variety of approaches including pulse-labeling studies and trapping experiments of the general type used in elucidating the mechanism of the reaction catalyzed by glutamine synthetase have provided substantial evidence for the formation of enzyme-bound carboxy phosphate as an intermediate in the reaction catalyzed by carbamyl phosphate synthetase. Application of the ^{18}O-scrambling procedure used earlier with glutamine synthetase to the reaction catalyzed by carbamyl phosphate synthetase provided kinetic evidence that carboxy phosphate is an intermediate (112). A mechanism for the reaction catalyzed by carbamyl phosphate synthetase is given in Fig. 12. The enzyme binding sites are shown in A. The binding site for the amide moiety of glutamine lies in the intersubunit space.

It seems probable that ATP, bicarbonate, and glutamine are bound to the enzyme at the start of the normal catalytic cycle and that ATP is bound simultaneously to both ATP binding sites. Carboxy phosphate, rapidly formed by interaction of one of the bound ATP molecules with bicarbonate, reacts with bound NH_3 derived from the amide nitrogen atom of glutamine to form carbamate, which is phosphorylated by the ATP bound at the second ATP site. The reaction thus involves formation of two highly unstable intermediates: carboxy phosphate and carbamate. These compounds are probably stabilized to some extent by virtue of their binding to the enzyme, but the evidence indicates that such stabilization is quite limited. Thus, formation of carbamate by reversal of steps 5 and 4 to yield complex D is associated with prompt formation of ammonia and bicarbonate. Similarly, carboxy phosphate on complex C breaks down readily in the absence of glutamine. Catalysis of the overall reaction appears to be favored by the presence of all of the reactants on the enzyme at the start of the catalytic cycle. Ammonia reacts with carboxy phosphate as soon as it is formed and the formation of carbamate is followed by prompt phosphorylation by the ATP molecule already bound to the enzyme. Although this is probably the only glutamine amidotransferase that catalyzes a reaction involving carboxy phosphate, this intermediate may be formed by other types of enzymes, for example, the biotin enzymes. On the other hand, the structural and functional aspects of this enzyme with respect to its glutamine site and glutamine binding subunit are, in general, characteristic of other enzymes of the glutamine amidotransferase group.

Finally, a few comments may be in order about the regulation of glutamine synthetase and glutamine metabolism in mammalian tissues. Much is known about the regulation of glutamine synthetase in certain microorganisms; in *E. coli* there is substantial evidence for the existence of adenylylated forms of this enzyme which are highly sensitive to feedback inhibition by end-products of glutamine metabolism (34). Presumably in this and in related organisms, there was a selective advantage in not producing more glutamine than is required for various metabolic purposes. In contrast to many microorganisms, mammalian tissues seem to exhibit rather high concentrations of glutamine suggesting that regulation of glutamine synthesis in mammalian tissues may be less important than it is in certain bacteria. Perhaps regulation of the various pathways of glutamine metabolism is more significant in mammalian tissues than is regulation of glutamine synthesis. On the other hand, the regulation of glutamine synthesis in mammals may be adjusted to achieve the

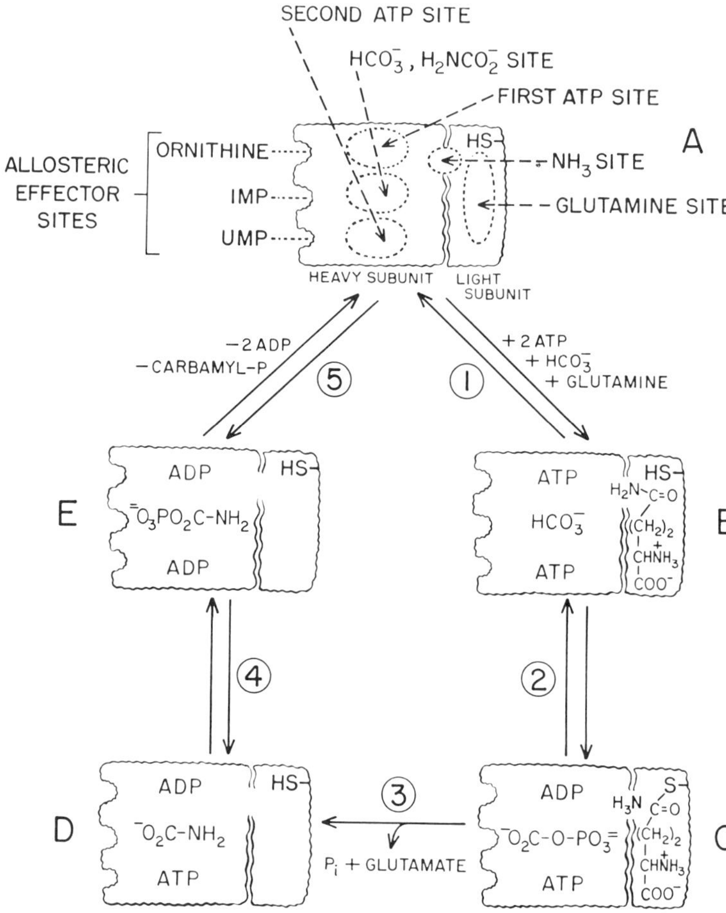

Fig. 12. Proposed mechanism of action of carbamyl phosphate synthetase. (110)

relatively high levels of glutamine apparently needed by mammalian tissues. There are some apparent similarities between the regulation of the glutamine synthetases of *E. coli* and those of mammalian tissues. For example, α-ketoglutarate activates the glutamine synthetases of rat liver and sheep brain. These enzymes are inhibited by glycine, L-alanine and D-alanine, especially in the presence of Mn^{++}. As mentioned above, the studies of Rowe (84) suggest that the concentration of ADP may be a factor in the regulation of glutamine synthetase in muscle. Information now becoming available about the synthesis and utilization of glutamine in synapses suggests that various types of feedback inhibition phenomena may operate to regulate glutamine synthesis and glutaminase (94). The problem of regulation of glutamine synthesis and of inter-organ metabolism of glutamine in mammalian tissues has only begun to be approached, and there is much to be learned in this area.

REFERENCES

1. Hlasiwetz, H., and Habermann, J. (1873), *Ann. 169*, 150.
2. Schulze, E., and Bosshard, E. (1883), *Landwirtsch. Vers.-Sta. 29*, 295.
3. Damodaran, M., Jaaback, G., and Chibnall, A. C. (1932), *Biochem. J. 26*, 1704.
4. Meister, A. (1965), *Biochemistry of the Amino Acids*, 2nd Edition, Vol. 1, Academic Press.
5. Meister, A. (1956), *Physiol. Revs. 36*, 103.
6. Bergmann, M., Zervas, L., and Salzmann, L. (1933),*Ber. 66*, 1288.
7. Krebs, H. A. (1935), *Biochem. J. 29*, 1951.
8. Orstrom, A., Orstrom, M., Krebs, H. A., and Eggleston, L. V. (1939), *Biochem. J. 33*, 995.
9. Elliot, W. H. (1948), *Nature 161*, 128.
10. Elliot, W. H. (1951), *Biochem. J. 49*, 106.
11. Elliott, W. H., and Gale, E. F. (1948), *Nature 161*, 129.
12. Elliott, W. H. (1953), *J. Biol. Chem. 201*, 661.
13. Speck, J. F. (1947), *J. Biol. Chem. 168*, 403.
14. Speck, J. F. (1949), *J. Biol. Chem. 179*, 1387.
15. Speck, J. F. (1949), *J. Biol. Chem. 179*, 1405.
16. Bujard, E., and Leuthardt, F. (1947), *Helv. Physiol. Pharmacol. Acta 5*, C39.
17. Leuthardt, F., and Bujard, E. (1947), *Helv. Med. Acta 14*, 274.
18. Levintow, L., and Meister, A. (1954), *J. Biol. Chem. 209*, 265.
19. Benzinger, T. H., and Hems, R. (1956), *Proc. Natl. Acad. Sci. U. S. A. 42*, 896.
20. Edsall, J. T., and Wyman, J. (1958), *Biophys. Chem. 1*, 210.
21. Pamiljans, V., Krishnaswamy, P. R., Dumville, G., and Meister, A. (1962), *Biochemistry 1*, 153.
22. Ehrenfeld, E., Marble, S. J., and Meister, A. (1963), *J. Biol. Chem. 238*, 3711.
23. Greenberg, J., and Lichtenstein, N. (1959), *J. Biol. Chem. 234*, 2337.
24. Monder, C. (1965), *Biochemistry 4*, 2677.
25. Wellner, V. P., and Meister, A. (1966), *Biochemistry 5*, 872.
26. Meister, A. (1974), *The Enzymes*, Vol. X, 3rd edition, pp. 699-754, Academic Press.
27. Rueppel, M. L., Lundt, S. L., Gass, J. D., and Meister, A. (1972), *Biochemistry 11*, 2839.
28. Haschemeyer, R. H. (1968), *Trans. N. Y. Acad. Sci.* [2] *30*, 875.
29. Haschemeyer, R. H. (1970), *Advan. Enzymol. 33*, 71.
30. Wilk, S., Meister, A., and Haschemeyer, R. H. (1969), *Biochemistry 8*, 3168.
31. Wilk, S., Meister, A., and Haschemeyer, R. H. (1970), *Biochemistry 9*, 2039.
32. Tate, S. S., and Meister, A. (1971), *Proc. Natl. Acad. Sci. U. S. A. 68*, 781.
33. Tiemeier, D. C., and Milman, G. (1972), *J. Biol. Chem. 247*, 2272.

34. Stadtman, E. R., and Ginsburg, A. (1974), *The Enzymes*, Vol. X, 3rd edition, pp. 755-807, Academic Press.
35. Tate, S. S., and Meister, A. (1973), *Symposium on the Enzymes of Glutamine Metabolism* (S. Prusiner and E. R. Stadtman, eds.), p. 77, Academic Press.
36. Buchanan, J. M., and Hartman, S. C. (1959), *Advan. Enzymol. 21*, 199.
37. Buchanan, J. M. (1973), *Advan. Enzymol. 39*, 91.
38. Wedler, F. C., and Boyer, P. D. (1972), *J. Biol. Chem. 247*, 984.
39. Krishnaswamy, P. R., Pamiljans, V., and Meister, A. (1962), *J. Biol. Chem. 237*, 2932.
40. Allison, R. D., Todhunter, J. A., and Purich, D. L. (1977), *J. Biol. Chem. 252*, 6046.
41. Purich, D. L., and Allison, R. D. (1980), *Methods Enzymol. 64*, 41.
42. Vorhaben, J. E., Wong, L., and Campbell, J. W. (1973), *Biochem. J. 135*, 893.
43. Levintow, L., and Meister, A. (1956), *Fed. Proc., Fed. Am. Soc. Exp. Biol. 15*, 299.
44. Boyer, P. D., Koeppe, O. J., and Luchsinger, W. W. (1956), *J. Am. Chem. Soc. 78*, 356.
45. Kowalski, A. Wyttenbach, C., Langer, L., and Koshland, Jr., D. E. (1956), *J. Biol. Chem. 219*, 719.
46. Levintow, L., and Meister, A. (1953), *J. Am. Chem. Soc. 75*, 3039.
47. Rochovansky, O., and Ratner, S. (1967), *J. Biol. Chem. 242*, 3839.
48. Anderson, P. M., and Meister, A. (1965), *Biochemistry 4*, 2803.
49. Nishimura, J. S., Dodd, E. A., and Meister, A. (1963), *J. Biol. Chem. 238*, PC 1179.
50. Nishimura, J. S., Dodd, E. A., and Meister, A. (1964), *J. Biol. Chem. 239*, 2553.
51. Mooz, E. D., and Meister, A. (1967), *Biochemistry 6*, 1722.
52. Powers, S. G., and Meister, A. (1978), *J. Biol. Chem. 253*, 800.
53. Rose, I. A., O'Connell, E. L., Litwin, S., and Tana, J. B. (1974), *J. Biol. Chem. 249*, 5163.
54. Khedouri, E., Wellner, V. P., and Meister, A. (1964), *Biochemistry 3*, 824.
55. Khedouri. E., and Meister, A. (1965), *J. Biol. Chem. 240*, 3357.
56. Tate, S. S., Leu, F. Y., and Meister, A. (1972), *J. Biol. Chem. 247*, 5312.
57. Gass, J. D., and Meister, A. (1970), *Biochemistry 9*, 842.
58. Tsuda, Y., Stephani, R. A., and Meister, A. (1971), *Biochemistry 10*, 3186.
59. Todhunter, J. A., and Purich, D. L. (1975), *J. Biol. Chem. 250*, 3505.
60. Midelfort, C. F., and Rose, A. (1976), *J. Biol. Chem. 254*, 5881.
61. Kagan, H. M., Manning, L. R., and Meister, A. (1965), *Biochemistry 4*, 1063.
62. Meister, A. (1968), *Advan. Enzymol. 31*, 183.
63. Meister, A. (1968), *Fed. Proc., Fed. Am. Soc. Exp. Biol. 27*, 100.
64. Kagan, H. M., and Meister, A (1966), *Biochemistry 5*, 725
65. Kagan, H. M., and Meister, A. (1966), *Biochemistry 5*, 2423.
66. Gass, J. D., and Meister, A. (1970), *Biochemistry 9*, 1380.
67. Gass, J. D. (1970), Doctoral Dissertation, Cornell University Medical College, New York.
68. Ronzio, R. A., and Meister, A. (1968), *Proc. Natl. Acad. Sci. U. S. A. 59*, 164.
69. Ronzio, R. A., Rowe, W. B., and Meister, A. (1969), *Biochemistry 8*, 1066.
70. Rowe, W. B., Ronzio, R. A., and Meister, A. (1969), *Biochemistry 8*, 2674.

71. Manning, J. M., Moore, S., Rowe, W. B., and Meister, A. (1969), *Biochemistry 8*, 2681.
72. Rowe, W. B., and Meister, A. (1973), *Biochemistry 15*, 1578.
73. Christensen, B. W., and Kjaer, A. (1965), *Chem. Commun.*, p. 225.
74. Richman, P. G., Orlowski, M., and Meister, A. (1973), *J. Biol. Chem. 248*, 6684.
75. Orlowski, M., and Meister, A. (1971), *J. Biol. Chem. 246*, 7095.
76. Griffith, O. W., and Meister, A. (1978), *J. Biol. Chem. 253*, 2333.
77. Griffith, O. W., Anderson, M. E., and Meister, A. (1979), *J. Biol. Chem. 254*, 1205.
78. Griffith, O. W., and Meister, A. (1979), *J. Biol. Chem. 254*, 7558.
79. Rowe, W. B., and Meister, A. (1970), *Proc. Natl. Acad. Sci. U. S. A. 66*, 500.
80. Bessman, S. P., Magnes, J. Schwerin, P., and Waelsch, H. (1948), *J. Biol. Chem. 175*, 817.
81. Schwerin, P., Bessman, S. P., and Waelsch, H. (1950), *J. Biol. Chem. 184*, 37.
82. Garber, A J., Karl, I. E., and Kipnis, D. M. (1979), *J. Biol. Chem. 254*, 826.
83. Garber, A J., Karl, I. E, and Kipnis, D. M. (1979), *J. Biol. Chem. 254*, 836.
84. Rowe, W. B. (1978), unpublished data.
85. Pitts, R. F. (1975), *in Symposium on Renal Metabolism, Medical Clinics of North America* (Baruch, S. ed.), *59* (3), 507.
86. Windmueller, H. G., and Spaeth, A. E. (1978), *J. Biol. Chem. 253*, 69.
87. Pitts, R. F., deHaas, J., and Klein, J. (1963), *Am. J. Physiol. 204*, 187.
88. Errera, M., and Greenstein, J. P. (1949), *J. Biol. Chem. 178*, 495.
89. Martínez-Hernández, A., Bell, K. P., and Norenberg, M. D. (1976), *Science 195*, 1356.
90. Berl, S., Takagaki, G., Clarke, D. D., and Waelsch, H. (1962), *J. Biol. Chem. 237*, 2562.
91. Cooper, A. J. L., McDonald, J. M., Gelbard, A. S., Gledhill, R. F., and Duffy, T. E. (1979), *J. Biol. Chem. 254*, 4982.
92. Van den Berg, C. J., and Garfinkel, D. (1971), *Biochem. J. 123*, 211.
93. Benjamin, A. M., and Quastel, J. H. (1972) , *Biochem. J. 128*, 631.
94. Hamberger, A., Chiang, G. H., Sandoval, E., and Cotman, C. W. (1979), *Brain Res. 168*, 531.
95. Meister, A. (1979), *in Advances in Biochemistry and Physiology* (Garattini, *et al.*, eds.), pp. 69-84, Raven Press, New York.
96. Wilcox, M. (1969), *Cold Spring Harbor Symp. Quant. Biol. 34*, 521.
97. Thierfelder, H., and Sherwin, C. P. (1914), *Ber. Chem. Ges. 47*, 2630.
98. Moldave, K., and Meister, A. (1957), *J. Biochem. Chem. 229*, 463.
99. Meister, A. and Tice, S. V. (1950), *J. Biol. Chem. 187*, 173.
100. Meister, A. (1953), *J. Biol. Chem. 200*, 571.
101. Meister, A. (1954), *Science 120*, 43.
102. Cooper, A. J. L., and Meister, A. (1972), *Biochemistry 11*, 661.
103. Cooper, A. J. L., and Meister, A. (1974), *J. Biol. Chem. 249*, 2554.
104. Cooper, A. J. L., and Meister, A. (1977), *Crit. Rev. Biochem. 4* (3), 281, CRC Press.
105. Buchanan, J. M. (1973), *Adv. Enzymol. 39*, 91.
106. Anderson, P. M., and Meister, A. (1965), *Biochemistry 4*, 2803.
107. Powers, S. G., and Meister, A. (1976), *Proc. Natl. Acad. Sci. U. S. A. 73*, 3020.
108. Powers, S. G, Griffith, O W., and Meister, A. (1977), *J. Biol. Chem. 252*, 3558.

109. Powers, S. G., and Meister, A. (1978), *J. Biol. Chem. 253,* 1258.
110. Meister, A., and Powers, S. G. (1978) *Adv. Enz. Reg. 16,* 289.
111. Powers, S. G., Meister, A., and Haschemeyer, R. H. (1980), *J. Biol. Chem.,* in press.
112. Wimmer, M. J., Rose, I. A., Powers, S. G., and Meister, A. (1979), *J. Biol. Chem. 254,* 1854.

2. The Interconvertible Enzyme Cascade that Regulates Glutamine Synthetase Activity

EARL R. STADTMAN, UMBERTO MURA,* P. BOON CHOCK, AND SUE GOO RHEE

Laboratory of Biochemistry
National Heart, Lung, and Blood Institute
National Institutes of Health
Bethesda, Maryland 20205

SUMMARY

Glutamine synthetase (GS) activity in *Escherichia coli* is regulated by a bicyclic cascade system. One cycle involves the adenylylation and deadenylylation of GS subunits. The other cycle involves the uridylylation and deuridylylation of Shapiro's P_{II} regulatory protein. The adenylylation and deadenylylation reactions are catalyzed at different sites on a single adenylyltransferase (AT), whereas the uridylylation and deuridylylation reactions are catalyzed at separate uridylyltransferase (UT) and uridylyl-removing (UR) centers, respectively, on the same enzyme or enzyme complex. The two cycles are linked because *in vivo* the unmodified form of the regulatory protein (P_{IIA}) is required for the adenylylation activity of AT, whereas the uridylylated form (P_{IID}) is required for the deadenylylation activity of AT. Through dynamic coupling of the nucleotidylation and denucleotidylation steps in each cycle, a steady-state is established in which the rates of adenylylation and deadenylylation

* Laboratorio di Biochimica, Facolta di Scienze Matematiche, Fisiche e Naturali, Pisa, Italy. Currently a Visiting Fellow supported by the Fogarty International Center, Bethesda, Maryland.

ISBN: 0-12-506040-8

are equal, and the specific activity of GS is inversely proportional to the fraction of subunits that are adenylylated in the steady-state. By means of allosteric and substrate interactions, the several cascade enzymes can collectively sense simultaneous changes in the concentrations of at least 40 different metabolites. These interactions lead to modulation of the values of reaction constants that govern activities of the cascade enzymes, and as a consequence, the steady-state distribution of adenylylated and unadenylylated GS subunits is shifted accordingly. The dependence of the steady-state level of adenylylation on metabolite control is illustrated by the results of *in vitro* studies in which the mole fraction, $[P_{IIA}] / ([P_{IIA}] + [P_{IID}])$, was varied to simulate the effects of allosteric interactions on the uridylation cycle and in which the concentrations of α-ketoglutarate and glutamine were varied to study the effects of allosteric modifiers on the AT catalyzed reactions. The results demonstrate the remarkable flexibility of the cascade system to metabolite control, and in agreement with theorical predictions, they show that such cascades are capable of enormous signal amplification with respect to primary allosteric stimulii and can generate a cooperative type of response to increasing concentrations of allosteric effectors.

After subjecting intact *E. coli* cells to freezing and thawing, followed by incubations in 0.1% Lubrol W-X, they become permeable to small molecules, but not to proteins of 9 000 daltons or greater. Preliminary studies show that the permeabilized cells contain an active GS cascade system and that the state of adenylylation in such cells is a function of the concentrations of metabolites in the suspending medium which are known to affect activities of the cascade enzymes *in vitro*. The permeabilized cells thus appear to offer a unique opportunity to determine the extent to which results obtained with purified enzymes *in vitro* apply to the regulation of GS *in situ*.

INTRODUCTION

The synthesis of glutamine is the first step in a highly branched biosynthetic pathway leading to the formation of amino acids, purine and pyrimidine nucleotides, glucosamine-6-P, and p-aminobenzoic acid (1). Thus, glutamine synthetase is a key enzyme in metabolic control. Studies of this enzyme in *Escherichia coli* and other gram negative bacteria have shown that the catalytic activity and the level of this enzyme is subject to a highly complex and rigorous control system (2-4). In addition to feedback inhibition by numerous end products of glutamine metabolism and to the modulation by the divalent metal ion level (2), gram negative bacteria have evolved a complex bicyclic cascade system for the regulation of glutamine synthetase activity (4). By means of allosteric and subtrate interactions with the several cascade enzymes, the concentrations of numerous metabolites are monitored continuously, and the specific activity of glutamine synthetase is adjusted automatically to meet cellular requirements for glutamine. In this presentation, we shall sumarize the current status of our studies on the regulation of glutamine synthetase via the interconvertible enzyme cascade.

The glutamine synthetase cascade consists of two tightly linked interconvertible enzyme (protein) cycles. In one cycle, glutamine synthetase (GS) is continuously adenylylated and deadenylylated. In the other cycle, Shapiro's regulatory protein, P_{II}, is continuously uridylylated and deuridylylated. The interrelationship between these two cycles is illustrated in Fig. 1.

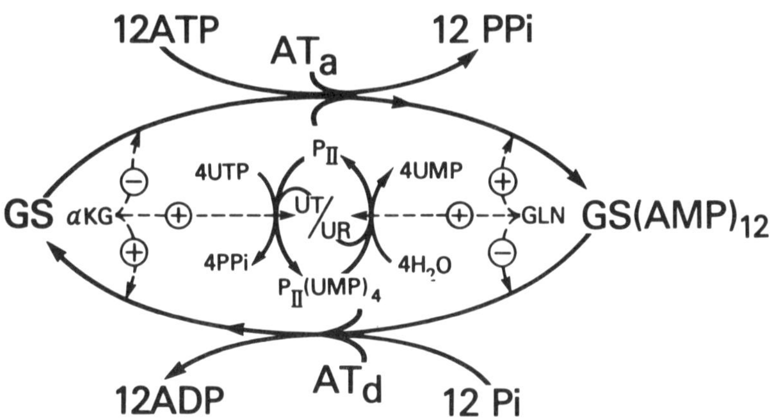

Fig. 1. The relationship between the uridylylation and deuridylylation of P_{II} (inner cycle) and the adenylylation and deadenylylation of glutamine synthetase, GS (outer cycle), and the roles of glutamine (GLN) and α-ketoglutarate (α-KG) in these two interconvertible enzyme (protein) cycles. + indicates stimulation; — indicates inhibition.

The outer cycle depicts the adenylylation and deadenylylation of GS. In the adenylylation reaction, the AMP moiety of ATP is transferred to the hydroxyl group of a unique tyrosyl group in each subunit of GS (5-7). This reaction is opposed by the deadenylylation reaction which involves a transfer of the AMP group of adenylylated GS to Pi with the formation of ADP and regeneration of the unmodified form of GS (8). The adenylylation and deadenylylation reactions are catalyzed at separate AT_a and AT_d sites, respectively, on the same adenylyltransferase, AT, which consists of a single polypeptide of 115,000 daltons (9-11). Because GS is composed of 12 identical subunits, up to 12 equivalents of AMP can be bound (in phosphodiester linkage) to a single GS molecule (2). From the standpoint of regulation, it is significant that under physiological conditions, adenylylation of a given subunit leads to inactivation of that subunit only. Therefore, the specific catalytic activity of GS is inversely proportional to the average number, n, of adenylylated subunits per GS molecule. (For convenience, the value of n, which can vary from 0 to 12, is often referred to as the "state of adenylylation".)

Because the adenylylation and deadenylylation reactions are catalyzed by one and the same enzyme (AT), the two processes are coupled. Therefore, in the absence of appropriate controls, GS will undergo cyclc interconversion between adenylylated and unadenylylated forms, accompanied by the net conversion of ATP and Pi to ADP and PPi. Aimless coupling of the adenylylation and deadenylylation reactions is prevented by the action of Shapiro's protein, P_{II}, which regulates the capacity of AT to catalyze adenylylation of GS, and also by the differential effects of various metabolites on these processes.

As is shown in Fig. 1 (inner cycle) the P_{II} protein also exists in two interconvertible forms: an unmodified form (sometimes referred to as P_{IIA}) and a uridylylated form (sometimes referred to as P_{IID}) (12, 13). The uridylylation of P_{II} involves the transfer of the UMP moiety of UTP to a tyrosyl residue in each subunit and is catalyzed by a specific uridylyltransferase (UT) (14, 15). Because P_{II} is composed of 4 identical subunits, up to 4 equivalents of UMP can be bound to one P_{II} molecule (16). Conversion of the uridylylated P_{II} back to the unmodified form involves the hydrolytic cleavage of the UMP-tyrosyl bond and is catalyzed by a uridylyl-removing activity, UR, which probably exists together with UT in a single multifunctional protein or protein complex, UR/UT (17, 18). The adenylylation cycle and the uridylylation cycle are linked due to the fact that P_{IIA} is required for the adenylylation of GS at the AT_a site, whereas P_{IID} is required for catalysis of the deadenylylation at the AT_d site (11).

In the last analysis (4, 19), however, the state of adenylylation of GS is determined by the concentrations of many metabolites that affect the activities of the converter enzymes, AT, UR, and UT. Most important of these are α-ketoglutarate (α-Kg) and glutamine (Gln) which, in opposition to one another, exert reciprocal effects on the modification and demodification reactions. Thus, as shown in Fig. 1, α-ketoglutarate inhibits the adenylylation but stimulates the deadenylylation of GS, whereas glutamine has opposite effects on these reactions. Also, α-ketoglutarate stimulates the deuridylylation of P_{IID}. It is therefore evident that the ratio of α-ketoglutarate to glutamine is of primary importance in regulating the state of GS adenylylation.

RESULTS

The Bicyclic Cascade

The consequences of linking the uridylylation of P_{II} with the adenylylation of GS are more readily appreciated, if the regulatory system is visualized as a closed type bicyclic cascade as is illustrated in Fig. 2. It follows from theoretical considerations that for any given metabolic condition, a steady-state will be attained in which the rates of covalent modification and demodification of an interconvertible enzyme are equal, and that the distribution of modified and unmodified enzyme in the steady-state is determined by the relative catalytic potentials of the converter enzymes that prevail

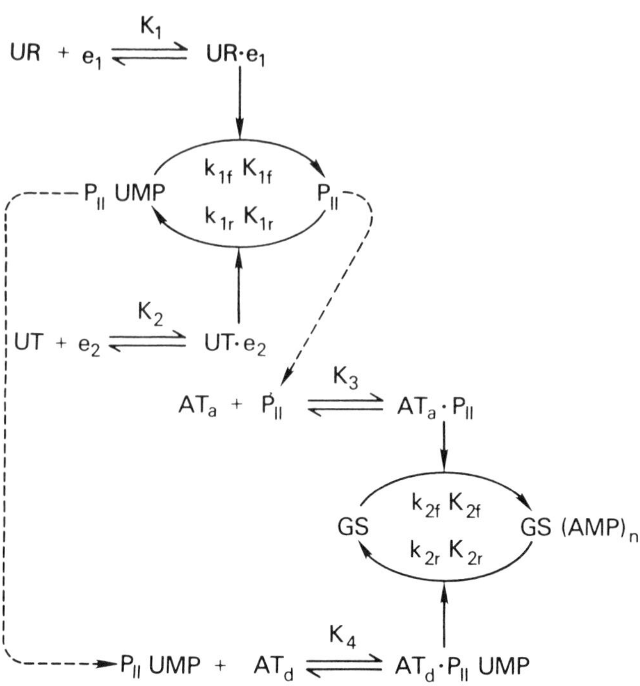

Fig. 2. The bicyclic cascade of glutamine synthetase. In addition to the notations defined in the text, e_1 and e_2 represent an allosteric effector which activates the enzymic activity of UR and UT, respectively.

in that metabolic condition. A theoretical analysis (19, 20) of the steady-state function which describes the GS cascade model depicted in Fig. 2 shows that the state of adenylylation, n, is defined by the equation:

$$n = 12\left[\left(\frac{k_{1r}}{k_{1f}}\right)\left(\frac{k_{2r}}{k_{2f}}\right)\left(\frac{K_{1r}}{K_{1f}}\right)\left(\frac{K_{2r}}{K_{2f}}\right)\left(\frac{K_4}{K_3}\right)\left(\frac{K_2}{K_1}\right)\frac{[UT]}{[UR]}\;\frac{[AT_d]}{[AT_a]}\;\frac{[e_2]}{[e_1]}\times \frac{1+K_1[e_1]}{1+K_2[e_2]} + 1\right]^{-1} \qquad (1)$$

in which K_1, K_2, K_3, and K_4 are the association constants for the reactions leading to the activated converter enzyme complexes, k_{1f}, k_{1r}, k_{2f}, and k_{2r} are specific rate constants for the designated reactions, K_{1f}, K_{1r}, K_{2f}, and K_{2r} are the association constants for the activated converter enzyme-interconvertible enzyme complexes (see Fig. 2). The equation shows that the value of n is a function of 18 different parameters. More significantly, n is a function of 9 different parameter ratios; *i. e.*, the ratio of a parameter that governs the forward step in a given cycle to the corresponding parameter that governs the reverse step in that cycle.

It is noteworthy that the value of each one of the 18 parameters in the steady-state equation is susceptible to modulation by changes in the concentrations of various metabolites (allosteric effectors and substrates) that interact with one or more of the cascade enzymes (4, 20). As a consequence, the GS cascade is endowed with an unusual capacity to respond simultaneously to changes in a large number of different metabolites and to elicit varied responses to a given metabolite depending upon the constraints imposed by others. The GS cascade is thus the basis of an extremely *flexible* regulatory system. Moreover, because the specific activity of GS (as determined by the value of n) is a multiplicative function of 9 different parameter ratios, a small change in several parameters can provoke a relatively large change in the specific activity of GS. In addition, relatively small reciprocal changes in the values of the two parameters that make up a given parameter ratio will result in a relatively large change in that ratio; and small changes in each of several parameter ratios can produce a huge change in the overall product of those ratios (R), which governs the value of n. The data in Fig. 3 show how the value of n will vary in response to the fractional saturation of the first converter enzyme, UR, in the cascade with its allosteric activator, Gln, and to variations in the overall parameter ratio, R. Note that, depending upon the value of R, 50% activation of UR can result in a greater than 50% level of GS adenylylation, and that with values of R less than 0.01, only 1% saturation of UR with Gln is sufficient to establish values of n that are 6 or greater. Thus, it is apparent that substantial levels of GS adenylylation can be provoked by concentrations of allosteric effectors that are well below the dissociation constants of the UR·Gln activation complex. In other words, the GS cascade is endowed with enormous *signal amplification potential* [for comparative purposes signal amplification has been defined as the ratio of the concentration of a primary effector required to produce 50% activation of the first converter enzyme, UR, in the cascade to the concentration of that effector required to sustain a steady-state in which 50% of the target enzyme, GS, is modified (4, 20)].

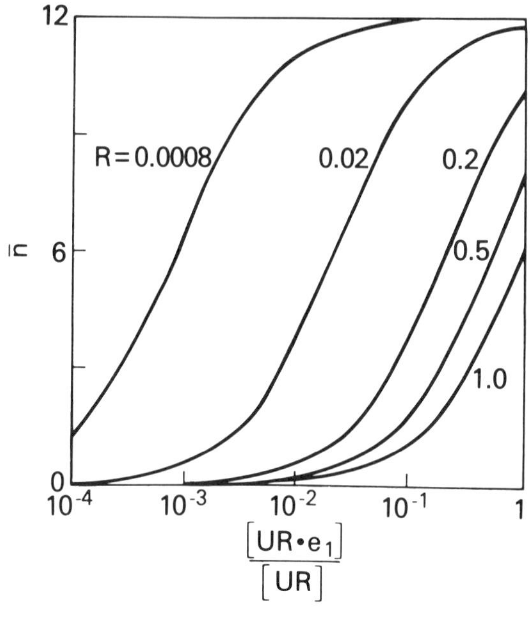

Fig. 3. Theoretical curves simulated with Equation 1 to demonstrate the relationship between the state of adenylylation of glutamine synthetase and the fractional activation of UR as a function of R, where

$$R = \left(\frac{k_{1r}}{k_{1f}}\right)\left(\frac{k_{2r}}{k_{2f}}\right)\left(\frac{K_{1r}}{K_{1f}}\right)\left(\frac{K_{2r}}{K_{2f}}\right)\left(\frac{K_4}{K_3}\right)\left(\frac{K_2}{K_1}\right)\left(\frac{[UT]}{[UR]}\right)\left(\frac{[AT_d]}{[AT_a]}\right)$$

The value of K_1 used is 1.0.

The data in Fig. 3 show also that with saturating concentrations of Gln (*i. e.*, when [Gln·UR] / [UR] = 1.0), the value of n can vary significantly, depending upon the overall parameter ratio R. In other words, the *amplitude* of the adenylylation reaction is subject to modulation by the overall parameter ratio (4, 20).

Although not evident from Equation 1, or from the data in Fig. 3, it follows from other theoretical considerations (4, 20) that when the same effector either stimulates a forward step and inhibits the regeneration step in a given cycle, or stimulates the forward steps in both cycles of the bicyclic cascade, the state of adenylylation will be a sigmoidal function of that effector concentration. Bicyclic cascades are thus able to generate a *cooperative* type of response

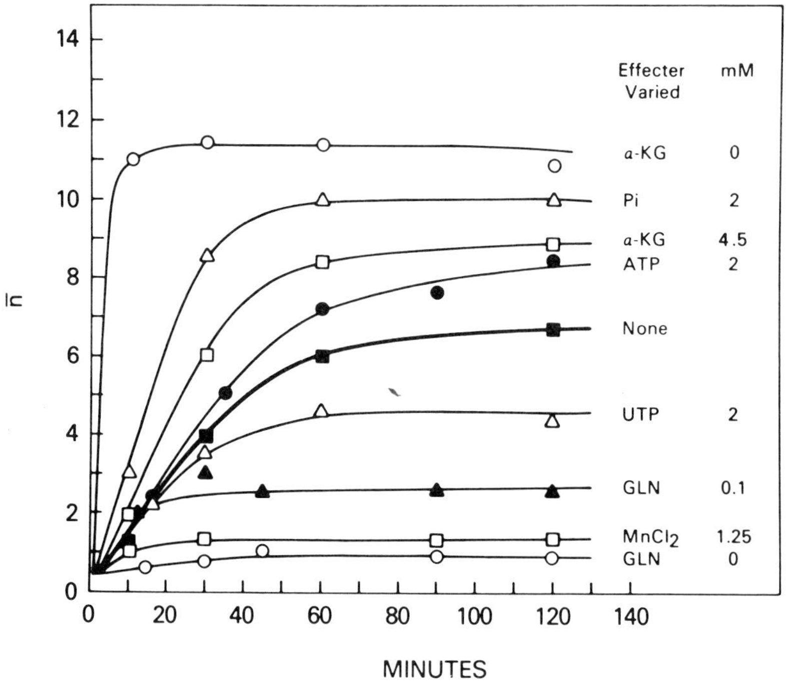

Fig. 4. Effect of metabolite concentrations on the state of adenylylation (n) of glutamine synthetase at steady-state. The heavy line (solid squares) shows the change of n as a function of time when 95 μg. of unadenylylated GS was incubated in a mixture containing 20 mM $MgCl_2$, 20 mM Pi, 1 mM ATP, 1 mM UTP, 15 mM α-ketoglutarate, 0.3 mM glutamine, and partially purified preparations of P_{II}, AT, UR, and UT as previously described (21). When only one metabolite in the mixture was varied as indicated by other curves, a new steady-state was established for n. The data are derived from Reference 21.

to effector concentrations with Hill coefficients up to 4.0. The net effect is to increase the *sensitivity* [for a precise definition see Stadtman and Chock (4)] of the state of adenylylation to changes in effector concentration.

Validity of the Cascade Model

To date over 40 metabolites have been shown to either activate or inhibit one or more of the cascade enzymes (4). It follows, that the interaction of any one of these metabolites with a cascade enzyme will affect the value of one or more of the parameters in Equation 1, and therefore should affect the steady-state level of adenylylation. This prediction is supported by the *in vitro* studies of Segal *et al.* (21) showing that when the unadenylylated GS was incubated in a mixture containing the converter enzymes and arbitrary concentrations of ATP, UTP, Pi, α-ketoglutarate (α-Kg), glutamine (Gln), Mg^{2+}, and Mn^{2+}, the state of GS adenylylation increased with time until a steady-state was reached in which $n \simeq 7.0$ (Fig. 4). Furthermore, a change in the concentration of just any one of the 7 effectors led to a shift in the steady-state

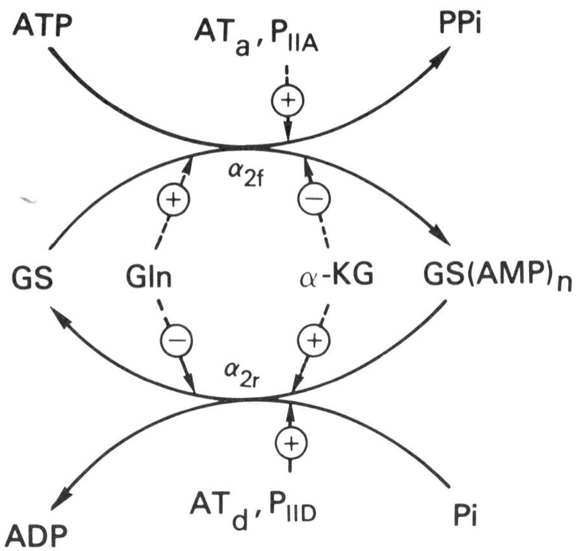

Fig. 5. A simplified monocyclic cascade showing the roles of P_{IIA}, P_{IID}, α-ketoglutarate (α-KG) and Gln on the state of adenylylation of glutamine synthetase. + indicates activation; — indicates inhibition.

value of n. These results demonstrate that the interconversion of GS between adenylylated and unadenylylated forms is a dynamic process, and show also that the steady-state distribution of adenylylated and unadenylylated subunits is governed by the relative concentrations of many different metabolites that affect the activities of the cascade enzymes.

Contributions of the Uridylylation Cycle on GS Adenylylation

It is evident from Fig. 2 that the function of the uridylylation-deuridylylation cycle is to determine the steady-state distribution of P_{IIA} and P_{IID}. Therefore, the contribution of the uridylylation cycle on the adenylylation cycle can be assessed directly by substituting various ratios of P_{IIA} and P_{IID} for the role of UR /UT complex. The bicyclic cascade system is thus converted to the relatively simple monocyclic system illustrated in Fig. 5. A theoretical analysis (11) of this simplified cascade shows that the state of adenylylation of GS is given by the expression:

$$n = 12 \left[1 + \frac{\alpha_{2r}}{\alpha_{2f}} \frac{1}{(P_{IIA})_{mf}} \right]^{-1} \qquad (2)$$

in which $(P_{IIA})_{mf}$ is the mole fraction of P_{IIA} [*i. e.*, $P_{IIA}/(P_{IIA} + P_{IIA})$], and α_{2f} and α_{2r} are each products of 4 different parameters that govern the adenylylation and deadenylylation reactions, respectively. The curves in Fig. 6A show how, according to Equation 2, the value of n will vary as a function of the $(P_{IIA})_{mf}$ and the parameter ratios $\alpha_{2f} / \alpha_{2r}$. Note that as $(P_{IIA})_{mf}$ increases from 0 to 1.0, the value of n will increase in either a linear, hyperbolic, or parabolic manner when the parameter ratio $\alpha_{2f} / \alpha_{2r} = 1.0, > 1.0$, or < 1.0, respectively. In light of the fact that the value of $(P_{IIA})_{mf}$ under physiological conditions is determined by the relative catalytic potentials of UR and UT, and that the parameter ratio $\alpha_{2r} / \alpha_{2f}$ is determined by all allosteric and substrate interactions that affect AT_a and AT_d activities, the curves in Fig. 6A illustrate the importance of the uridylylation cycle in modulating the response of GS adenylylation to changes in metabolite concentrations, and they illustrate also the remarkable flexibility of the cascade system to metabolite control.

It is clear from Fig. 5 that a quantitative evaluation of the contribution of the uridylylation cycle on the adenylylation cascade *in vitro*

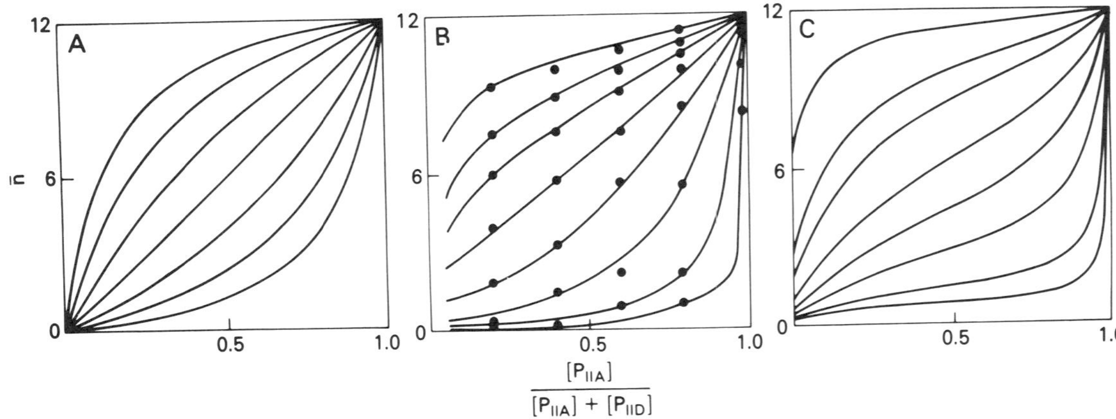

Fig. 6. (A) Steady-state levels of n simulated with Equation 2. The values of α_{2f}/α_{2r} were set at 8, 4, 2, 1, 0.5, 0.25, and 0.125 for the curves from top to bottom, respectively. (B) Steady-state levels of n determined as a function of $(P_{IIA})_{mf}$. The experiments were performed at 25°C in a mixture (50 μl) containing 50 mM 2-methylimidazole (pH 7.2), 20 mM $MgCl_2$, 20 mM Pi, 4 mM ATP, 4 mM phosphoenolpyruvate, 0.1 mM EDTA; 0.5 mM dithiothreitol, 0.5 mM Gln, 0.14 unit of pyruvate kinase, 25 μg of GS, 22 μg of AT, 4.2 μg of P_{II} (a mixture of known amounts of P_{IIA} and P_{IID}); and the concentration of α-ketogutarate was 0, 0.002, 0.004, 0.01, 0.02, 0.04, 0.1 and 0.2 mM for the curves from top to bottom, respectively. (C) Computer simulated curves using the steady-state equation derived for the reaction scheme shown in Fig. 7 and the necessary constants reported by Rhee *et al.* (11) for the corresponding curves in B.

can be obtained by measuring the value of n when GS is incubated with AT in a series of reaction mixtures in which the proportions of pure P_{IIA} and P_{IID} are varied to change the $(P_{IIA})_{mf}$, and in which the ratios of glutamine and α-ketoglutarate are varied to alter the $\alpha_{2r} / \alpha_{2f}$ ratio. A comparison of the results of one such experiment shown in Fig. 6B with the theoretical curves in Fig. 6A shows that the experimentally observed behavior is qualitatively similar to that predicted by Equation 2. In the course of these studies, it became apparent that the discrepancy between the theoretical and observed results is largely attributable to the fact that there are synergistic effects of α-ketoglutarate and glutamine on the interactions of AT with the regulatory protein, and also to the fact that adenylylation and deadenylylation reactions are each catalyzed by three different AT complexes. Therefore, the monocyclic adenylylation cascade is more accurately described by the model shown in Fig. 7. A theoretical analysis of this steady-state model shows that the value of n is described by an equation (not shown) contain-

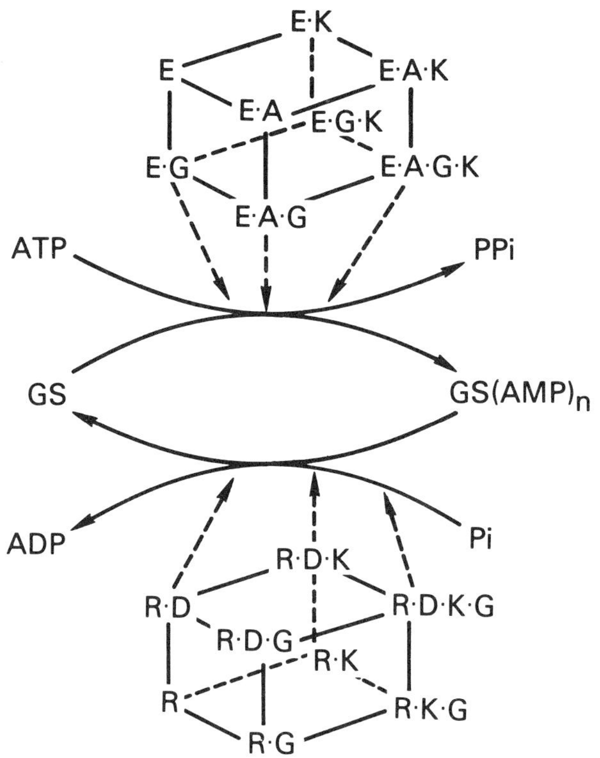

Fig. 7. Mechanistic scheme for the adenylylation-deadenylylation of glutamine synthetase. The notations E, R, A, D, K, and G represent AT_a, AT_d, P_{IIA}, P_{IID}, α-ketoglutarate, and glutamine, respectively.

ing well over a hundred terms, among which are 10 binding constants, 12 synergistic or antagonistic constants, and 6 rate constants (11). Of these 28 constants, 22 were determined directly by means of kinetic measurements, 4 were determined by computer simulation analysis with rigorous constraints imposed by the other constants, and only 2 of which were determined solely by computer simulation (11). By means of the more complex steady-state equation and the 28 constants so derived, the steady-state level of n was calculated for the experimental conditions described in Fig. 6B. A comparison of the curves in Fig. 6B with those in 6C shows that there is relatively good agreement between the experimental observations and theoretical predictions of the cascade model.

Signal Amplification

It is apparent from the theoretical curves in Fig. 3, that only slight activation of the UR converter enzyme by its allosteric activator, glutamine, can elicit a large change in the state of GS adenylylation. Whereas this capacity for signal amplification increases exponentially with the number of cycles in a cascade (4, 20, 22), a theoretical analysis shows that monocyclic cascades are also endowed with considerable amplification potential (4, 20, 23). This is verified by the data in Fig. 8 which are from experiments designed to investigate properties of the modified monocylic adenylylation cascade model depicted in Figs. 5 and 7. In these experiments, the $(P_{IIA})_{mf}$ was fixed at 0.6 and the effect of glutamine concentration on the value of n was examined at various levels of α-ketoglutarate. It is evident that as the parameter ratio, $\alpha_{2r} / \alpha_{2f}$, is altered by decreasing the concentration of α-ketoglutarate from 1000 μM to 3 μM, there is a progressive decrease in the fraction of $AT_a \cdot$ Gln complex that is needed to produce a given state of adenylylation. Indeed, with 170 μM, 30 μM, and 3 μM levels of α-ketoglutarate, only ~ 7%, ~ 0.8%, and 0.1%, respectively, of the AT_a site must be saturated with glutamine in order to achieve 50% adenylylation of the GS (n = 6.0). There is partial agreement between the experimentally derived curves in Fig. 8A and the theoretical curves in Fig. 8B which were calculated using the complex equation (not shown) that described the monocyclic cascade model depicted in Fig. 7 (11).

Apparent Cooperativity

Although it is not obvious from the experimental data as presented in Figure 8A, an in-depth analysis of the these data shows that the

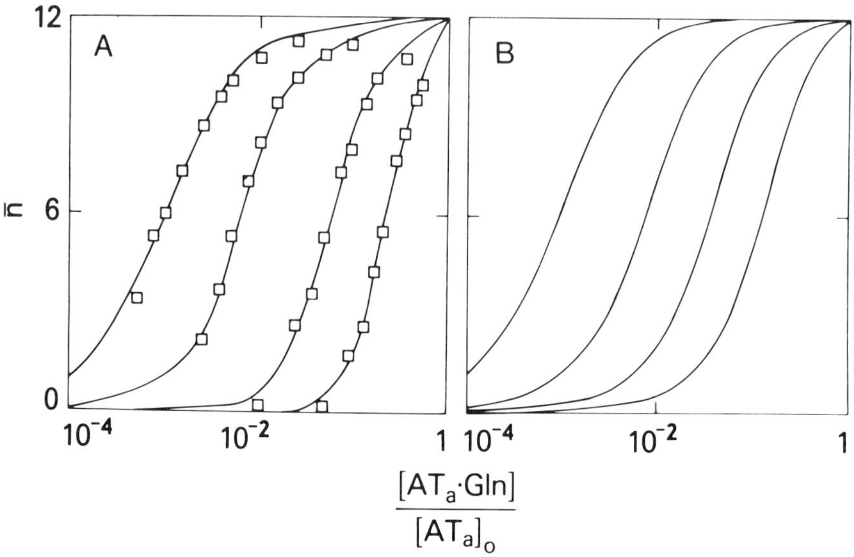

Fig. 8. (A) Steady-state levels of n plotted as a function of fractional activation of AT_a by Gln. $[AT_a \cdot Gln]$ was calculated using the experimentally determined K_d of 80 mM for Gln in the absence of effectors and substrates, $(P_{IIA})_{mf}$ was kept at 0.6, and the R values were varied by changing the concentration of α-ketoglutarate used. For the curves from left to right, [α-Kg] was maintained at 3 μM, 30 μM, 0.17 mM, and 1 mM, respectively. (B) Computer simulated curves for the corresponding curves in A using the steady-state equation derived for the scheme shown in Fig. 7 and the constants reported by Rhee *et al.* (11).

state of adenylylation is a sigmoidal function of the glutamine concentration (11). Relative to a value of 1.0 for a normal hyperbolic saturation function, a sensitivity index of 2.1 (see Reference 4 for definition) is calculated from the experimental data. This corresponds to a Hill number of 1.5 (11). This apparent cooperativity is due to the fact that glutamine not only stimulates the adenylylation of GS at the AT_a site, but it also inhibits the deadenylylation of GS at the AT_d site (see Fig. 5).

Modulation of Adenylylation in Permeabilized Cells

In order to assess the significance of multiple metabolites *in vivo*, we have initiated some studies with permeabilized cells. In such cells, the concentration of the several cascade enzymes are maintained at their normal intracellular levels and the contributions of the many metabolites that affect their activities can be measured by varying the concentrations of the these metabolites in the buffer solution in which the cells are suspended. Permeabilization was achieved by passing the cells through a freeze-thaw cycle and then incubating them with imidazole buffer containing 0.1% Lubrol W-X for 90 minutes at 37°C. With this technique, which will be described in detail elsewhere, between 80-90% of *E. coli* cells become permeable to small molecules without significant loss (< 5%) of intracellular proteins (Mura and Stadtman, unpublished data).

Data summarized in Figures 9 and 10 illustrate the ability of permeabilized cells to catalyze the adenylylation and deadenylylation of GS, and the dependence of this interconversion on the allosteric

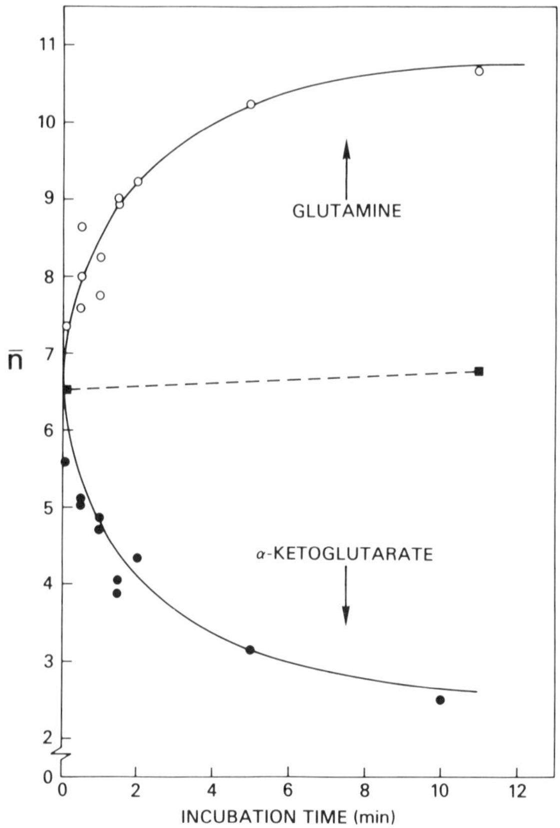

Fig. 9. Effector dependent adenylylation and deadenylylation of GS in permeabilized cells. Reactions were started by mixing in micro test tubes 100μl. of permeabilized cell suspension (160 mg/ml), 50 mM 2-methylimidazole buffer (pH 7.2), with 100μl. of buffer solution containing 2.5 mM ATP, 2.5 mM UTP, 9.5 mM Pi, 2.5 mM $MgCl_2$, and either 4 mM L-glutamine (open circles) or 10 mM α-ketoglutarate (closed circles). After incubation at 37°C for the times indicated, the reactions were stopped by placing the test tubes in liquid nitrogen. 0.8 ml of ice cold 50 mM 2-methylimidazole buffer (pH 7.2) was added, and the samples were sonicated for three 30 second intervals in an ice bath. The resulting suspension was centrifuged 15 minutes at 4°C, and 50 μl of the supernatant was assayed for the state of adenylylation as previously described (2).

effectors α-ketoglutarate and glutamine. Initially, the state of adenylylation of GS in the permeabilized cells used for these studies was about 6.5. As shown by the broken line in Fig. 9, the value of n did not change when the permeabilized cells were incubated in buffer alone. However, when incubated in a buffer mixture containing all the substrates for the cascade enzymes (UTP, ATP, Pi, Mg^{2+}) and the allosteric effector, glutamine, the value of n increased rapidly to a steady-state value of about 10.5. Moreover, when the permeabilized cells were incubated in a comparable mixture containing α-ketoglutarate in place of glutamine, the value of n declined rapidly to a value of about 2.5. The dependence of the steady-state level of adenylylation on the concentration of α-ketoglutarate and glutamine is shown in Fig. 10. It is evident from these results and the results in Fig. 9, that permeabilized cells have the capacity to catalyze the adenylylation and deadenylylation of endogenous GS and that the state of adenylylation in these cells is a function of the concentration of α-ketoglutarate and glutamine, as was predicted from studies with isolated enzymes. It therefore seems likely that permeabilized cells may be exploited in further experiments to examine in detail the role of multiple metabolites in the regulation of GS activity *in situ*.

DISCUSSION

Covalent modifications of interconvertible enzymes have often been thought to serve as biological switches that are used to turn ON or OFF key enzyme activities. This concept is not supported by the results of *in vivo* experiments (24) showing that the state of adenylylation of GS varies over a wide range, depending upon the nutritional state of the cell, nor by results of the *in vitro* experiments summarized here, showing that the interconversion of GS between adenylylated and unadenylylated forms is a dynamic process which leads to steady-states in which the rates of adenylylation and deadenylylation are equal. Furthermore, the specific activity of GS in a given steady-state is inversely proportional to the fraction of adenylylated subunits in that steady-state. This in turn is determined by the relative catalytic potentials of the converter enzymes that prevail in that metabolic condition. The results of *in vitro* studies with purified cascade enzymes are in full agrement with predictions based on a theoretical analysis of GS cascade models. They show that the GS cascade is highly *flexible* with respect to regulation by multiple metabolites; it is capable of enormous *signal*

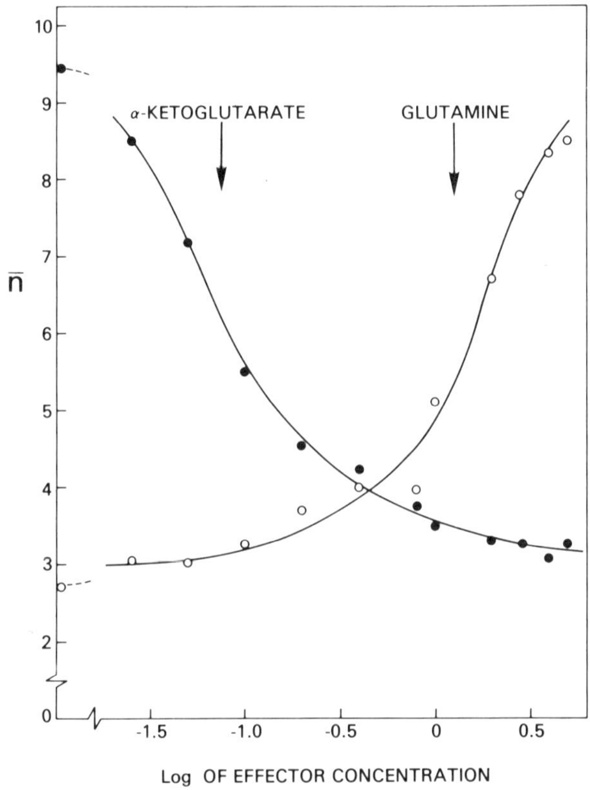

Fig. 10. The effect of α-ketoglutarate and glutamine concentration on the steady-state level of GS adenylylation in permeabilized cells. To study the effect of α-ketoglutarate concentration (closed circles), each reaction mixture (200 μl) contained 100 μl of permeabilized cell suspension (160 mg/ml), 5 mM ATP, 1 mM UTP, 10 mM $MgCl_2$, 10 mM Pi, 1 mM $MnCl_2$, 0.4 mM L-glutamine, 50 mM 2-methylimidazole buffer (pH 7.2), and the amounts (0 to 5 mM) of α-ketoglutarate, as indicated. To study the effect of L-glutamine concentration (open circles), the reaction mixtures contained the same components as above, except the fixed level of glutamine (0.4 mM) was replaced with 1 mM α-ketoglutarate, and the concentration of L-glutamine was varied (0-5 mM) as indicated. After incubation for 20 minutes at 37°C, the state of adenylylation was assayed as in Fig. 9.

amplification; it is able to govern the *amplitude* of the response to a given allosteric effector; it can generate a *cooperative* type of response to increasing allosteric effector concentrations. The results of studies in which the mole fraction, $P_{IIA}/(P_{IIA} + P_{IID})$, was varied to mimic effects of the UR/UT complex on the state of adenylylation, emphasize the importance of the uridylylation cycle in increasing the flexibility of the GS cascade to allosteric control.

It would seem from the studies summarized here that interconvertible enzyme cascades are in reality metabolic integration systems.

Since each of the several cascade enzymes is a separate target for allosteric and substrate interaction, collectively they can sense changes in the concentrations of a large number of different metabolites. By causing changes in the catalytic constants of the converter enzymes, these multiple interactions lead continuosly to changes in the steady-state levels of adenylylation of GS. Thus, manifold effects of changes in the levels of diverse metabolites are integrated, and the specific activity of GS is modulated accordingly.

It is noteworthy that each nucleotidylation-denucleotidylation cycle in the GS cascade is accompanied by the decomposition of one equivalent of nucleoside triphosphate (either UTP or ATP). However, this consumption of "energy-rich" bonds is not a wasteful process, because in the absence of a continual supply of nucleoside triphosphates, the interconvertible enzyme would exist almost exclusively in the unmodified state. The decomposition of ATP is therefore the price the cell must pay in order to maintain steady-state levels of a covalently modified enzyme that are greater than those allowed at true thermodynamic equilibrium. It is the price the cell must pay to support an unusually effective mechanism of cellular regulation.

REFERENCES

1. Stadtman, E. R. (1973) in *The Enzymes of Glutamine Metabolism* (Prusiner, S. and Stadtman, E. R., eds.), pp. 1-6, Academic Press, New York.
2. Stadtman, E. R. and Ginsburg, A. (1974) in *The Enzymes* (Boyer, P. D., ed.) 3rd ed., Vol. 10, pp. 755-807, Academic Press, New York.
3. Magasanik, B., Prival, M. J., Brenchley, J. E., Tyler, B. M., DeLeo, A. B., Streicher, S. L., Bender, R. A., and Paris, C. G. (1974) in *Current Topics in Cellular Regulation* (Horecker, B. L., and Stadtman, E. R., eds.) Vol. 8, pp. 119-138, Academic Press, New York.
4. Stadtman, E. R. and Chock, P. B. (1978) in *Current Topics in Cellular Regulation* (Horecker, B. L., and Stadtman, E. R., eds.) Vol. 13, pp. 53-95, Academic Press, New York.
5. Shapiro, B. M., Kingdon, H. S., and Stadtman, E. R. (1967) *Proc. Nat. Acad. Sci. U.S.A. 58,* 642-649.
6. Wulff, K., Mecke, D., and Holzer, H. (1967) *Biochem. Biophys. Res. Commun. 28,* 740-745.
7. Shapiro, B. M. and Stadtman, E. R. (1968) *J. Biol. Chem. 243,* 3769-3771.
8. Anderson, W. B. and Stadtman, E. R. (1970) *Biochem. Biophys. Res. Commun. 41,* 704-709.
9. Anderson, W. B. Hennig, S. B., Ginsburg, A., and Stadtman, E. R. (1970) *Proc. Natl. Acad. Sci. U.S.A. 67,* 1417-1424.
10. Caban, C. E., and Ginsburg, A. (1976) *Biochemistry 15,* 1569-1580.

11. Rhee, S. G., Park, R., Chock, P. B., and Stadtman, E. R. (1978) *Proc. Natl. Acad. Sci. U.S.A. 75,* 3138-3142
12. Shapiro, B. M. (1969) *Biochemistry 8,* 659-670.
13. Brown, M. S., Segal, A. and Stadtman, E. R. (1971) *Proc. Natl. Acad. Sci. U.S.A. 68,* 2949-2953
14. Mangum, J. H., Magni, G., and Stadtman, E. R. (1973) *Arch. Biochem. Biophys. 158,* 514-525.
15. Adler, S. P., Mangum, J. H., Magni, and Stadtman, E. R. (1973) in *Third International Symposium on Metabolic Interconversion of Enzymes* (Fischer, E. H., Krebs, E. G., and Stadtman, E. R. eds.) pp. 221-233, Springer-Verlag, Berlin and New York.
16. Adler, S. P., Purich, D., and Stadtman, E. R. (1975) *J. Biol. Chem. 250,* 6264-6272.
17. Bancroft, S., Rhee, S. G., Neumann, C., and Kustu, S. (1978) *J. Bacteriol. 134,* 1046-1055.
18. Francis, S. H., and Engleman, E. G. (1978) *Arch. Biochem. Biophys. 191,* 590-601.
19. Stadtman, E. R., Chock, P. B., and Rhee, S. G. (1979) in *From Gene to Protein: Information Transfer in Normal and Abnormal Cells* (Russell, T. R., Brew, K., Shultz, J., and Faber, H. eds.) pp. 521-544, Academic Press, New York.
20. Chock, P. B. and Stadtman, E. R. (1979) in *Proc. 1979 ICN-UCLA Symp. Mol. Cell Biol.* (Atkinson, D. E., ed.) pp. 185-202, Academic Press, New York.
21. Segal, A., Brown, M. S., and Stadtman, E. R. (1974) *Arch. Biochem. Biophys. 161,* 319-327.
22. Chock, P. B. and Stadtman, E. R. (1977) *Proc. Natl. Acad. Sci. U.S.A. 74,* 2766-2770.
23. Stadtman, E. R. and Chock, P. B. (1977) *Proc. Natl. Acad. Sci. U.S.A. 74,* 2761-2765.
24. Senior P. (1975) *J. Bacteriol. 123,* 407-418.

3. The Role of Glutamine Synthetase in the Regulation of Bacterial Nitrogen Metabolism

BORIS MAGASANIK AND DAVID M. ROTHSTEIN

Department of Biology
Massachusetts Institute of Technology
Cambridge, Massachusetts 02139
U.S.A.

SUMMARY

In many microorganisms, glutamine synthetase (GS) plays an essential role in the assimilation of ammonia present at low concentration in the medium. In enteric bacteria both the intracellular level and the activity of GS are determined by the nature and concentration of nitrogen compounds of the medium. Cells grown with an excess of ammonia contain highly adenylylated, and therefore inactive GS at a low level; growth on limiting ammonia or on a poor source of nitrogen results in a high level of non-adenylylated GS. We have found previously that GS activates the synthesis of enzymes capable of providing the cell with ammonia or glutamate by the degradation of amino acids, the hydrolysis of urea and the reduction of nitrate or of dinitrogen. This activation requires a high level of non-adenylylated GS. Our current work deals with the elucidation of the regulation of the synthesis of GS, a mechanism ultimately responsible for the regulation exerted by GS on the enzyme of nitrogen metabolism. We have identified the structural gene for GS, *glnA*, in enteric bacteria, as well as several genes not linked to *glnA*, whose products participate in the regulation of GS formation or of GS activity. The product of the *glnB* gene is P_{II}, a protein shown by Stadtman and his coworkers to stimulate the adenylylation and deadenylylation of GS

ISBN: 0-12-506040-8

by the enzyme adenylyltransferase (ATase). Mutations leading to the inability of P_{II} to stimulate deadenylylation of GS result in the inability of the cell to produce GS at a high level; mutations causing a loss of P_{II} result in production of GS at a high level irrespective of the presence of ammonia in the medium. Thus, formally P_{II} plays the role of a repressor of GS. In addition, formation of GS requires the product of the *glnF* gene, which does not play a role in the regulation of GS activity. On the other hand, the product of the *glnE* gene, ATase, is essential for the regulation of GS activity but plays no role in the regulation of GS production. Finally, the product of the *glnG* gene, a gene very closely linked to *glnA*, coordinates the repression exerted by the *glnB* product and the activation exerted by the *glnF* product. It has become apparent that we are dealing with a complex and subtle system of regulation.

The regulation of the synthesis of enzymes capable of supplying bacterial cells with glutamate and ammonia has been discussed in several recent reviews (1, 2). In enteric bacteria such enzymes, for example those responsible for degradation of histidine and proline in *Klebsiella aerogenes*, are not produced in media containing glucose and ammonia, the preferred energy and carbon sources of the organism, even if the appropriate inducers, histidine or proline are provided. Removal of either glucose or ammonia from the medium allows these enzymes to be produced. In the former case, their synthesis is activated by the catabolite activator protein (CAP) charged with cyclic AMP, and in the latter case by glutamine synthetase (GS). Enzymes able to provide the cell with energy but not with nitrogen, such as β-galactosidase, are regulated by CAP and cyclic AMP, but not by GS, while conversely enzymes capable of providing the cell with nitrogen but not with energy, such as urease, are regulated by GS, but not by cyclic AMP.

The activation of the synthesis of the enzymes of nitrogen metabolism appears to require an increase in the cellular level of GS and its conversion to the unadenylylated form. These ideas were based on the results of experiments which showed that cells starved for ammonia brought about the conversion of GS from the adenylylated to the nonadenylylated form and caused its level to rise. This rise preceded the rise in the level of enzymes of nitrogen metabolism; conversely, addition of ammonia to cells grown on a limiting nitrogen source led to the immediate adenylylation of GS and the immediate arrest of the synthesis of enzymes of nitrogen metabolism (Foor, F. F. and Magasanik, B., unpublished results. In addition, mutants unable to produce GS (Gln^-) or defective in their ability to deadenylylate GS were also unable to produce the enzymes of nitrogen metabolism in response to ammonia starvation. An exception were the mutants producing enzymatically inactive GS antigen: those retained their ability to produce the enzymes of nitrogen metabolism. Mutants with an alteration in the regulation of GS synthesis resulting in high levels of GS in cells growing with an excess of ammonia (GlnC) also produced the enzymes of nitrogen metabolism readily in the presence of ammonia.

The genetic and biochemical analysis of mutants with Gln^- phenotypes allowed a site on the bacterial chromosome, *glnA*, located at a position corresponding to 85 min on the *E. coli* map, to be identified as the structural gene for GS. Mutants with the GlnC phenotype were found at a site closely linked to *glnA*. On the basis of mapping and of complementation experiments it was proposed that the mutations responsible for the GlnC character were actually in the structural gene *glnA*. Thus, it appeared that the GS formation was subject to autogenous regulation (3). The recent discovery in *E. coli* and in *Salmonella typhimurium* of another gene, *glnG*, closely linked to *glnA* whose product appears to regulate the expression of the *glnA* gene, makes it necessary to reevaluate the evidence for the autogenous regulation of GS formation (4. 5).

As originally proposed, the hypothesis postulated that adenylylated GS could repress synthesis of GS (1). Accordingly, the intracellular ratio of glutamine to 2-ketoglutarate would determine not only the biosynthetic activity but also the rate of synthesis of GS. The discovery of a number of genes at different sites of the chromosome whose products affect the formation of GS has shown that in reality the regulation of GS formation is a much more complicated process.

One of these genes is *glnF*, located at a site corresponding to 68 min on the *E. coli* chromosome. Mutants in this site, first discovered in *S. typhimurium* (6) and then in *K. aerogenes* (7) and *E. coli* (8) result in a glutamine requirement, reflecting the inability of the mutant to produce GS at a level higher than the fully repressed level. In *K. aerogenes* this level, 15 units per mg of protein, corresponds to that found in cells of the wild strain growing in a medium containing a poor carbon source such as histidine, as well as ammonia and glutamine. Cells of the wild strain growing on glucose, ammonia, glutamine, on glucose, ammonia, and on glucose, glutamine, contain respectively, 150, 350, and 1,400 units of enzyme per mg of protein. In the case of *E. coli*, it is possible to prevent expression of the *glnF* gene by insertion of the transposon Tn10 (4).

Consequently it is clear that this gene produces a product necessary for the derepression of GS.

The phenotype of the *glnF* mutants can be suppressed by mutations closely linked to *glnA* (7). These mutations result in the same phenotype, GlnR, in a *glnF* mutant or *glnF*+ background: production of GS at a level intermediate between fully repressed and derepressed, almost irrespective of the composition of the medium. It has now been shown in the case of all three organisms that mutations at this site and with this phenotype result from insertions of prophage Mu in the case of *E. coli* (4), of transposon Tn10 in the case of *S. typhimurium* (5), and of transposon Tn5 in the case of *K. aerogenes* (Foor F. F. and Magasanik, B. unpublished results). Thus the GlnR phenotype results from the loss of the product of *glnG*, a gene closely linked to *glnA*.

It appears that as previously proposed for GS (7), the product of the *glnG* gene is both a repressor and an activator of GS synthesis. Accordingly, the product of the *glnF* gene is necessary for the activation exerted by the *glnG* product. In the absence of the *glnF* product, the *glnG* product, represses GS. This is shown by the observation that introduction of a functional *glnG* gene on an episome into a strain with mutations in *glnG* and *glnF*, reduces the GS level to that found in strains carrying a functional *glnG* gene and a mutant *glnF* gene (4, 7). Additional support for the dual role of the *glnG* product as repressor and activator of GS synthesis comes from the isolation of operator constitutive *gln* mutants of *K. aerogenes* (9).

These mutations only affect expression of the contiguously linked *glnA* gene; the GS determined by this gene cannot be repressed below the level characteristic for a mutant lacking the *glnG* product, but the full activation of GS formation still requires functional products of the *glnF* and, presumably, *glnG* genes.

The fact that in cells grown under the most repressing conditions the level of GS corresponds to that found in *glnF* mutants suggests that under these repressing conditions the *glnG* product fails to bring about activation of GS production. It appears that the product of still another gene, *glnB*, is involved at this stage. This gene is located on the chromosome at a position corresponding to 55 min on the *E. coli* map. It has been identified in *K. aerogenes* as the gene for P_{II}, the protein that activates the adenylyltransferase (ATase) responsible for the adenylylation and deadenylylation of GS (10). One mutation in this gene, *glnB3* results in a variant of P_{II} capable of stimulating adenylylation, but not deadenylylation of GS by

ATase. This mutant has the Gln^- phenotype, its GS is highly adenylylated and its ability to increase the level of GS in response to nitrogen starvation is greatly diminished. A similar phenotype results from a mutation in *glnD*, a gene located on the chromosome in a position corresponding to 4 min on the *E. coli* map. This mutation has resulted in the loss of uridylyltransferase (UTase), the enzyme required for the uridylylation of P_{II} responsible for its conversion from an activator of adenylylation to an activator of deadenylylation (11, 12, 13). Another class of mutants results from the insertion of transposon Tn5 into the *glnB* gene; these mutants contain no P_{II} and have the GlnC phenotype, that is, they contain a high level of GS even when grown with an excess of ammonia and glutamine (14). Thus, it appears that the presence of P_{II} results in a repression of GS, which can be relieved by the conversion of P_{II} to P_{II}–UMP.

These observations explain the effect of ammonia on the rate of GS synthesis. In cells starved of ammonia the ratio of glutamine to 2-ketoglutarate is low, a condition that results in the activation of UTase, conversion of P_{II} to P_{II}–UMP and consequently derepression of GS. Conversely, in an excess of ammonia, the ratio of glutamine to 2-ketoglutarate is high, a condition that results in the inhibition of UTase, conversion of P_{II}–UMP to P_{II} and consequently repression of GS.

We originally considered the adenylylation of GS stimulated by P_{II} to be responsible for the repression of GS (10). It has now been shown that this is not the case: the loss of ATase, the product of the *glnE* gene whose location on the chromosome is not yet known, results in the inability to adenylylate GS, but has no effect on GS formation (13, Foor, F., Reuveny, Z., and Magasanik, B., unpublished results). P_{II} is not essential for the repression, since in a double mutant, lacking P_{II} as well as the product of the *glnF* gene, GS is fully repressed. This repression is due to the product of the *glnG* gene, since in a double mutant lacking the *glnG* product, as well as the ability to convert P_{II} to P_{II}–UMP, GS is not fully repressed but is made at the intermediate level characteristic of *glnG* mutants (4, 7). These results can be explained by the assumption that P_{II} antagonizes the *glnF* product in its conversion of the *glnG* product from repressor to activator of GS synthesis. This view is supported by experiments using a *gln*-operator constitutive, *glnB3* double mutant, with functional *glnF* and *glnG* genes. In this strain, the inability to convert P_{II} to P_{II}–UMP did not result in repression of GS, but prevented the activation of GS synthesis (9).

A series of recent experiments has shown that GS itself is required neither for full activation of GS synthesis nor for full repression. In these experiments the promoter of the *glnA* gene of *E. coli* was fused to the structural gene for β-galactosidase, *lacZ* (Unpublished results). Such a strain contains no functional GS, and produces β-galactosidase controlled by the products of the *glnD, glnF* and *glnG* genes. This could be demonstrated by isolating strains carrying the *gln-lac* fusion, as well as a mutation in one of these genes. The mutation in *glnF* resulted in the almost complete loss of ability to produce β-galactosidase, and the mutation in *glnD* greatly lowered the level of β-galactosidase. A mutation in *glnG* resulted in the formation of β-galactosidase at an intermediate level. The results of other experiments with the *gln-lac* fusion strain, nevertheless, suggest a role for GS in the regulation of GS synthesis. It was found that in a merodiploid strain carrying the *gln-lac* fusion as well as a functional *glnA* gene, the level of β-galactosidase was considerably lower than in the haploid *gln-lac* fusion strain, and was regulated in parallel with the level of GS. It is, however, possible that this response of β-galactosidase formation to the presence of a functional GS, reflects an effect of GS on the intracellular level of glutamine.

Of particular interest is the fact that in the *gln-lac* fusion strain the synthesis of histidase is not activated by ammonia starvation. This result clearly militates against the hypothesis that both GS and the enzymes of nitrogen metabolism, such as histidase, respond in parallel to a single activator. The result is in keeping with the idea that GS is essential for the activation of the synthesis of these enzymes, though a role for the *glnG* product cannot be excluded. In this regard, a recent observation is of particular interest; it was shown that in an *E. coli* strain carrying a single functional *glnG* gene and a single functional *glnA* gene in *trans* position, GS synthesis was normally controlled, but the synthesis of the enzymes of nitrogen metabolism, such as histidase, was not activated by nitrogen starvation. Completely normal regulation of GS and of histidase synthesis was only observed in strains carrying functional contiguous *glnA* and *glnG* genes on the chromosome or episome (4).

These results suggest the possibility that a complex of the products of the *glnA* and *glnG* genes is the actual activator of the synthesis of the enzymes of nitrogen metabolism. The formation of this complex would require close proximity of the newly formed subunits, which might result from simultaneous transcription of their linked genes.

The results are also compatible with the hypothesis that *glnA* and *glnG* belong to a single operon, whose transcription could start either at P_1 or at P_2 as shown, with P_1 being the stronger promoter.

P_1 *glnA* P_2 *glnG*

Transcription starting at P_1 would be repressed or activated by the product of the *glnG* gene. Accordingly, the *glnG* product could activate its synthesis, but not completely repress it. Let us assume further that the level of *glnG* product arising by transcription from P_2 would be sufficient for full activation of transcription at P_1, but not for activation of transcription of the *hut* genes or the genes for other enzymes of nitrogen metabolism. Only full activation of transcription of P_1 would bring the *glnG* product to the level required for the activation of transcription of those genes. In cells carrying a functional *glnA* gene and an insertion in *glnG* on the chromosome, as well as a functional *glnG* gene and an insertion in *glnA* on the episome, the *glnG* product could only arise by transcription starting at P_2; its level would suffice for activation of transcription of the chromosomal *glnA* gene, but not for the activation of the transcription of the *hut* genes. This hypothesis would assign all the regulatory functions, previously assigned to GS, to the product of the *glnG* gene, a member of the GS operon.

It is at present not clear how certain mutations resulting simultaneously in the GlnC phenotype, in activation of the synthesis of the enzymes of nitrogen metabolism in excess ammonia, and in over-adenylylation of GS fit into this scheme (1, 2). These mutations are closely linked to mutations in *glnA* and *glnG*. Their definite assignment to one or the other of these genes should sharpen our understanding of the molecular basis of this fascinating regulatory network.

REFERENCES

1. Magasanik, B. (1977) *Trends in Biochem. Sci.* 2, 9-12.
2. Tyler, B. (1978) *Ann. Rev. Biochem.* *47,* 1127-1162.
3. Bender, R. A., and Magasanik, B. (1977) *J. Bacteriol.* *132,* 106-112.
4. Pahel, G., and Tyler, B. (1979) *Proc. Natl. Acad. Sci. U.S.A.* *76,* 4544-4548.
5. Kustu, S., Burton, D., Garcia, E. McCarter, L., and McFarland, N. (1979) *Proc. Natl. Acad. Sci. U.S.A.* *76,* 4576-4580.
6. Garcia, E. Bancroft, S., Rhee, S. G., and Kustu, S. (1977) *Proc. Natl. Acad. Sci. U.S.A.* *74,* 1662-1666.

7. Gaillardin, C. M., and Magasanik, B. (1978) *J. Bacteriol. 133*, 1329-1338.
8. Pahel, G., Zelenetz, A. D., and Tyler, B. (1978) *J. Bacteriol. 133*, 139-148.
9. Rothstein, D. M., and Magasanik, B. (1980) *J. Bacteriol., 141*, 671-679.
10. Foor, F., Janssen, K. A., and Magasanik, B. (1975) *Proc. Natl. Acad. Sci. U.S.A. 72*, 4844-4848.
11. Foor, F., Cedergren, R. J., Streicher, S. L., Rhee, S. G., and Magasanik, B. (1978) *J. Bacteriol. 134*, 562-568.
12. Bloom, F. R., Levin, M. S., Foor, F., and Tyler, B. (1978) *J. Bacteriol. 134*, 569-577.
13. Bancroft, S., Rhee, S. G., Neumann, C., and Kustu, S. (1978) *J. Bacteriol. 134*, 1046-1055.
14. Foor, F., Reuveny, Z., and Magasanik, B. (1980) *Proc. Natl. Acad. Sci. U. S. A. 77*, 2636-2640.

4. Regulation of Nitrogen Metabolism in Escherichia coli

BONNIE TYLER, FREDRIC BLOOM AND GREG PAHEL

Merck, Sharp and Dohme Research Laboratories
Rahway, New Jersey, USA

and

Department of Biology
Massachusetts Institute of Technology
Cambridge, Mass., USA

SUMMARY

In *Escherichia coli* the formation of glutamine synthetase (GS) from the *glnA* gene increases when growth is limited by the availability of nitrogen. This regulation is effected at the level of transcription. In addition, the enzymatic activity of GS is inversely proportional to the supply of nitrogen due to a reversible covalent attachment of AMP groups to each of its twelve subunits. Mutants with lesions in an unlinked gene *glnF* synthesize extremely low levels of GS when grown on glucose in the presence of excess or growth-rate limiting quantities of nitrogen; consequently they are glutamine auxotrophs. A rare revertant of a GlnF mutant always synthesizes a high level of GS in *glnF* strains. The mutation (*gln-501*) responsible for this phenotype maps in the *glnA* region. However, when the *gln-501* mutation is transferred into a GlnF+ strain, the regulation of formation of GS is very similar to that observed in wild-type strains. The high level of GS produced by strains carrying the *gln-501* mutation is independent of the *glnF* allele and is expressed in cells devoid of the *glnF*-product. Since this mutation results in high levels of GS regardless of the presence of the *glnF*-product, we were able to examine the role of the *glnF*-product in nitrogen-regulation of gene expression in

ISBN: 0-12-506040-8

both GlnF+ and GlnF− strains which contained high levels of GS during nitrogen-limited growth. The results strongly suggest that the *gln-F*-product is necessary for nitrogen regulation of gene expression in *E. coli*. Experiments with a *glnF*^{ts} mutant of *Klebsiella pneumoniae* also support this notion.

We have isolated two classes of *E. coli* mutants altered in GS regulation due to insertion of phage Mu in the *glnA* region. One class is Gln− and devoid of GS polypeptide. The other class is Gln+ but does not utilize a variety of compounds as the sole source of nitrogen (Reg− phenotype). This latter class produces, in either a GlnF+ or GlnF− background, a low level of GS regardless of the availability of nitrogen and therefore suppresses the Gln− phenotype of GlnF mutants. Biochemical analysis of the GS in the Reg− strains together with complementation analysis between the Mu-insertions in the Reg and *glnA*::Mu strains strongly suggests that another gene, *glnG*, is tightly linked to *glnA* and is necessary for nitrogen regulation of GS synthesis in *E. coli*.

In this organism, synthesis of GS is also stimulated by carbon/energy limitation by a mechanism which is independent of *glnF*, *glnG*, and the adenylylation state of GS.

We have been using *Escherichia coli* to study the regulation of synthesis of glutamine synthetase (GS, the product of the *glnA* gene) and the GS associated regulation of other enzymes of nitrogen metabolism. The reason for our choice was the refined genetic techniques available for use in *E. coli*.

As stated in the overview by B. Magasanik and D. Rothstein (see chapter 3), the discovery of the *glnF* locus necessitated a revision in the previous model of GS regulation, that *glnA* expression was regulated solely by the GS protein, whose regulatory properties were determined by the adenylylation state of the GS molecule.

In general, strains with mutations at the *glnF* locus are glutamine auxotrophs. Such mutations were initially characterized in *Salmonella typhimurium* (1) and subsequently in *E. coli* (2), *Klebsiella aerogenes* (3) and recently, in this laboratory, in *Klebsiella pneumoniae*. In all these organisms the level of GS is barely detectable when the *glnF* mutants are grown with a good carbon source under nitrogen-excess or nitrogen-limiting conditions. However, we have observed that when *E. coli* GlnF$^-$ strains are grown on carbon and energy-limiting medium, the level of GS increases five-fold and the strains no longer require glutamine for growth. We did not observe this "energy effect" reversal of the Gln$^-$ character of *glnF* mutants with the other enteric bacteria. In addition we found that intergeneric hybrids of *E. coli* and *K. aerogenes glnF* mutants containing the heterologous *glnA* regions did not exhibit this energy effect indicating that it is a property of the *glnA* region of *E. coli*. From the evidence presented below on the *glnA*-linked *glnG*-gene, it seems reasonable that this difference in the regulation of GS synthesis in *glnF* mutants of the enteric bacteria results from variations in the promoter-operator regions of the *glnA* genes.

In *K. aerogenes*, the *glnF* and *gltB* loci are reported to be on opposite sides of *argG* (2). However, our data show that in *E. coli* and *K. pneumoniae* these genes are tightly linked on the same side of *argG*.

GlnF$^-$ mutants have a pleiotropic phenotype. They are not only Gln$^-$ but also are Reg$^-$, meaning that they fail to derepress other enzymes (such as histidase) which normally increase during nitrogen-limited growth. Since the *glnF* gene had not been predicted by the simple model of autogenous regulation of *glnA* proposed by Magasanik and co-workers, we sought information on the role of the *glnF*-product in the regulation of synthesis of GS. In addition it was of interest to ask whether the Reg$^-$ phenotype of *glnF* mutants is merely due to the lack of GS or whether the *glnF*-product is essential for nitrogen regulation of histidase. Among the Gln$^+$ Reg$^+$ revertants of the *glnF* mutant of *K. pneumoniae* we found one which has a temperature sensitive Reg$^+$ phenotype. The mutation causing this temperature sensitive phenotype is 100% linked to the original *glnF* mutation by Pl transduction. Both GS and histidase synthesis are rapidly reduced upon shifting this mutant from permissive to non-permissive temperature indicating that the *glnF* gene product may be directly involved in both GlnA and *hut* regulation (Fig. 1). At 30°C this *glnF*^{ts} strain is Gln$^+$ Reg$^+$, while at 42°C this strain is Gln$^+$ Reg$^-$ and produces a low level of GS regardless of the growth medium. When this *glnF*^{ts} strain is grown at a variety of temperatures increasing from 30° to 40°C the ability to synthesize derepressed levels of histidase is lost before the ability to make high levels of GS (Fig. 2). Therefore at intermediate temperatures the regulation of GS and histidase synthesis is uncoupled in this *glnF* mutant, possibly due to different conformational states of the *glnF* product which act at various sites in regulating GS and histidase synthesis.

We obtained more definitive evidence that the *glnF*-gene product is necessary for nitrogen regulation of histidase expression from studies on a rare revertant of an *E. coli* GlnF$^-$ strain. The unusual *glnA*-linked lesion (*gln-501*) present in this double mutant results in the constitutive production of a high level of GS when present in a GlnF$^-$ background. The high level of GS produced by strains carrying the *gln-501* mutation is independent of the *glnF* allele and

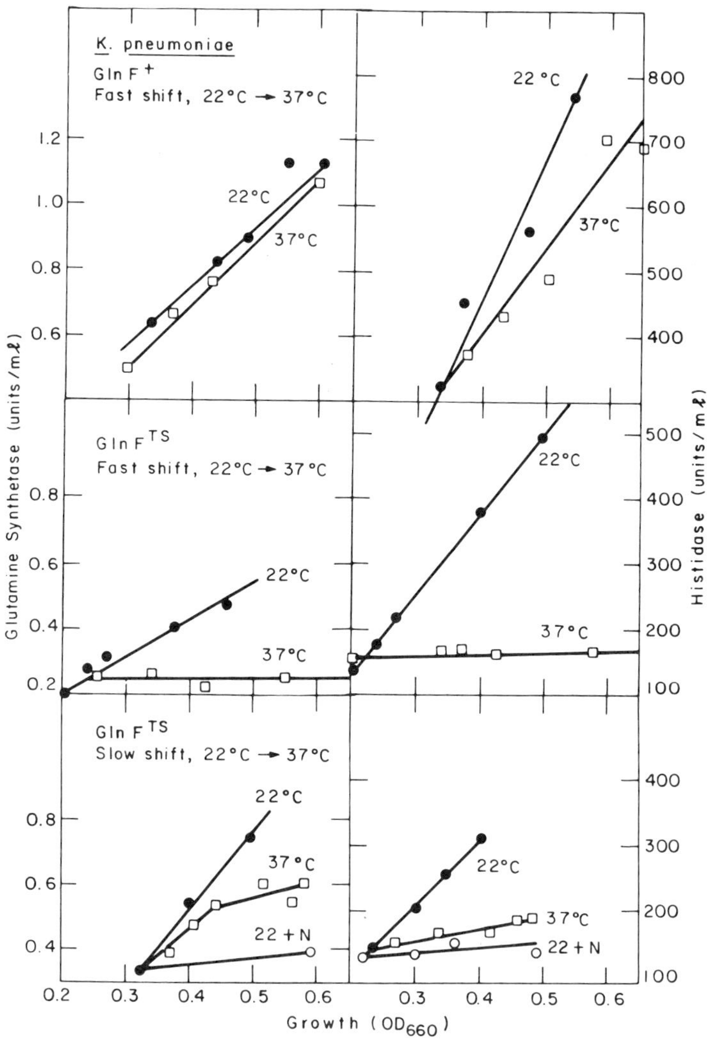

Fig. 1. Effect of temperature shifts on the activity of glutamine synthetase and histidase in GlnF+ and GlnF^{TS} strains.

is expressed in *glnF*::Tn 10 cells which should be devoid of the *glnF*-gene product. However, when the *gln-501* mutation is transferred into a GlnF$^+$ strain, the regulation of formation of GS is very similar to that observed in wild-type strains.

Thus when strains containing the *gln-501* mutation are grown on nitrogen limiting medium, equivalent high levels of non-adenylylated GS are produced in both a Gln$^+$ or GlnF$^-$ background. However physiological studies with these *gln-501* mutants showed that histidase derepression only occurred in the GlnF$^+$ strain; the level of histidase did not increase during nitrogen limited growth of the GlnF$^-$ strain. These data confirm and extend observations on the *K. pneumoniae glnF*^{ts} strain, suggesting that the *glnF*-product is directly involved in histidase regulation as well as GS synthesis.

In the enteric bacteria, selection for Gln$^+$ revertants of *glnF* mutants generally yields, at a high frequency, *glnA*-linked suppressor mutations which result in the production of a low level of GS regardless of the availability of nitrogen in the growth medium (GlnR phenotype). These *glnF*-suppressor mutations result in the Reg$^-$

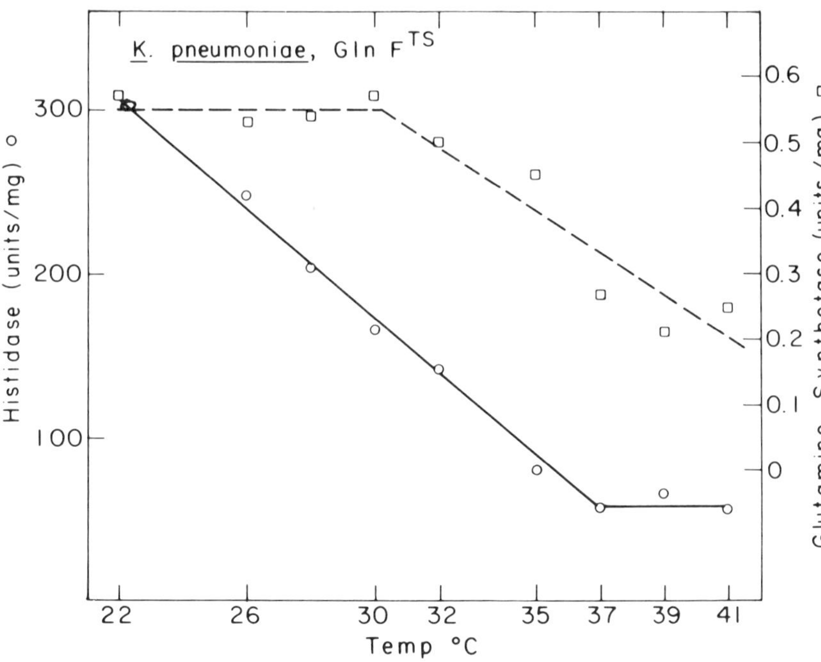

Fig. 2. Effect of different temperatures on the activity of glutamine synthetase and histidase in GlnF^{TS} strain.

phenotype in either a GlnF+ or GlnF− background. In both *K. aerogenes* (2) and *E. coli* (unpublished observations of this laboratory), mutations in the adenylylation system genes have no effect on GS levels even though the adenylylation state of the GS reflects the biochemical lesion of the particular mutation (2, unpublished observation of this laboratory).

In *K. aerogenes* mutations resulting in the GlnR Reg− phenotype were ordered with respect to *glnA*-linked mutations resulting in the Gln− phenotype by three-point mapping using *rha* as the outside marker. The results suggested that these *glnF*-suppressor mutations were clustered on the *rha* distal side of all but one of the mutations causing the Gln− phenotype (2). Since these suppressor mutations were not complemented by three *glnA* mutations for restoration of either GS or histidase regulation, Gaillardin and Magasanik concluded that the *glnF*-suppressor mutations were alterations in the *glnA* gene which result in formation of a GS protein which does not act as either a repressor or activator of GS synthesis (2). However, we have recently shown that in *E. coli,* mutations such as those just described reside in a gene distinct from *glnA*, called *glnG* (4).

The *glnG* locus is defined by Mu insertion mutations isolated after penicillin enrichment for cells which can not utilize arginine as sole nitrogen source (Reg−). Three independent Mu insertions, isolated for their Reg− phenotype, also suppressed a mutation in *glnF.* These cells synthesize about 0.06 units of GS activity/mg of cell protein on glucose-glutamine-minimal medium whether grown under conditions of nitrogen-excess or nitrogen-limitation and, apparently due to the energy effect, about 0.13 units of GS activity/mg of cell protein under carbon and energy limiting conditions. These levels of GS are identical to those found in strains carrying spontaneous *glnA*-linked *glnF*-suppressors in *E. coli*; they all have the GlnR Reg− phenotype.

The *glnG*::Mu mutations are recessive to the wild-type *glnA*-region with respect to regulation of GS and histidase in both a GlnF+ and GlnF− background. Thus they do not appear to be in a control region for *glnA*. The GS produced by the *glnG*::Mu mutants appears to have the same molecular weight as wild-type GS. Finally we showed that the *glnG*::Mu and *glnA*::Mu insertions complement to restore GS regulation. Since Mu insertions are polar, the initiation of transcription of the *glnA* and *glnG* genes must be from separate promoters in these merodiploids:

glnA::Mu	*glnG*+
glnA+	*glnG*::Mu

Thus the *glnG*::Mu mutations must be in a separate gene from the structural gene for GS.

The role assigned to the *glnG* product is analogous to that given to GS in *K. aerogenes* GlnR mutants (2). Thus, the product of *glnG* is most likely involved in the expression of derepressed levels of GS *glnF* positive strains, and in the expression of very repressed levels in *glnF* mutant strains. Due to the fact that the *glnG* mutations are the result of Mu insertion it seems that this phenotype is due to the loss, rather than alteration, of *glnG* product. The spontaneous *glnF* suppressors in *E. coli* have not been positively assigned to *glnG*. Whether or not the *K. aerogenes* mutations with the GlnR phenotype, interpreted as mutations in *glnA*, might actually be in *glnG* is also not known. A strain containing a *glnA*-linked Tn5 insertion resulting in the GlnR phenotype has been isolated in *K. aerogenes* (F. Foor, personal communication).

It has been proposed that GS regulates both its own synthesis and the synthesis of many other enzymes, such as histidase, which are involved in nitrogen assimilation. Since there is a regulatory gene closely linked to *glnA* the question arises whether all regulation previously attributed to GS is in fact due to the *glnG* product. If the *glnG* product alone were responsible for this regulation, one would expect complementation of a *glnA*::Mu mutant by a *glnG*::Mu mutant to restore normal regulation under all conditions. However, we have observed two anomalies in such complementation tests. Such merodiploids synthesize an abnormally high level of GS when grown under extreme nitrogen starvation. In addition histidase synthesis is not derepressed in response to nitrogen-limited growth, that is the *glnG*::Mu/*glnA*::Mu merodiploids still have the Reg^- phenotype (4).

We also find that *glnA*::Mu strains are Reg^-. This result could reflect a change in the intra-cellular level of certain key nitrogenous compounds, but many *glnA* mutants are actually derepressed for histidase at all times (Reg^c phenotypes). Therefore it seems probable that either GS or some other product in the *glnA* operon is involved in nitrogen regulation of histidase. The results with *glnA-lac*-Mu reveal a high rate of transcription of the *glnA* operon, at least up to the Mu DNA.

It appears therefore that regulation of expression of *glnA* and the *hut* operons is not always coupled and that a product of the *glnA* operon may be involved in nitrogen-control of histidase formation. In summary our results indicate that three conditions must be met for the wild-type Gln+ Reg+ phenotype: a wild-type *glnG*-product, an intact *glnA*-operon and, most curiously, the *glnA* and *glnG* genes must be located in *cis* to each other. This last requirement is most unusual considering that the Reg+ phenotype is determined by monitoring expressions of genes such as those of the *hut* operons, which are not linked to the *glnA*-*glnG* region.

What explanations could account for the lack of a *trans* effect in restoring the Reg+ phenotype in the merodiploids comprised of a *glnA*::Mu mutation on one piece of DNA and a *glnG*::Mu mutation on the other:

glnA+	*glnG*::Mu	?
glnA::Mu	*glnG*+	

There are several very general mechanisms which might account for these observations. First, one or both of the *glnA* or *glnG* products may only interact with another protein when it is synthesized in the *cis* position and this complex may be required for normal regulation, particularly with respect to the activation of unlinked genes such as the *hut* operons. Second, it may be that the Gln+ Reg+ phenotype requires a particular ratio of the products from genes in the *glnA*-region and that this balance is not achieved in these merodiploids. This notion suggests that there may be, in the *glnA*-region, at least one other gene associated with nitrogen metabolism, and that this gene may be proximal to the Mu insertions in the *glnA* or *glnG* operons. Third, the organization and regulation of transcription of the *glnA* and *glnG* genes could lead to the Reg− phenotype of these merodiploids. In particular if we assume that these genes are contiguous and that the *glnG* product can arise from transcription originating at either of two promoters, P1 or P2, then the following model would explain the Reg− phenotype of the merodiploids used in the complementation analysis of the *glnA*::Mu and *glnG*::Mu mutations.

P1 *glnA*::Mu	P2 *glnG*+
P1 *glnA*+	P2 *glnG*::Mu

Let us assume that transcription from the P2 promoter is not regulated but occurs at a lower frequency than that originating at P1 in cells fully derepressed for GS synthesis. If we further assume that transcription from the P1 promoter normally requires an activating form of the *glnG* product then the Reg⁻ phenotype of our *glnA*::Mu/*glnG*::Mu merodiploids can be explained. In this case we suggest that *glnG*-product is produced at a low level by transcription from the P2 promoter and potentially, at a much higher level, from transcripts originating at P1 but extending through P2. We would then speculate that, in wild-type cells, the low level of *glnG*-product resulting from transcripts starting at P2 is sufficient for full derepression of *glnA* transcription from P1. However, this low level of *glnG* is not sufficient to stimulate transcription of the *hut* genes. Thus since the insertion of Mu in *glnA* eliminates read-through transcription from P1 in the merodiploid described above, the *glnA*::Mu/*glnG*::Mu merodiploids have the Reg⁻ phenotypes. Whether any of these three models is correct remains to be tested. However, one thing is clear: the regulation of GS and the other enzymes involved in nitrogen assimilation is extremely complex. As indicated from the data presented above, the *glnG* product functions as either an activator or repressor of *glnA* transcription. Since very little GS is produced in the absence of the *glnF* product, then the *glnG*-product or some other protein must function as a repressor in *glnF* mutants. Magasanik (see chapter 3) has reviewed the evidence indicating that this protein is likely to be the *glnG*-product and not PII. Therefore we propose that it may be that the *glnG*-product functions as a repressor of transcription from P1 unless the *glnF*-product is present, in which case the *glnG*-product can be converted to an activator of transcription. However, if this model is correct, then the data of Foor, Reuveny and Magasanik (personal communication) suggest that this conversion can be blocked by the PII protein. This notion remains to be tested.

If, as proposed by Magasanik and Rothstein (see chapter 3), the unmodified product of the *glnB* gene (PII) functions as an anti-activator of *glnA* transcription, then the modification of PII by uridylylation contributes significantly to regulation of *glnA* expression. The placement of uridylyl-transferase and uridylyl-removing enzyme activities in such a central role in *glnA* regulation is interesting in the light of our observations pertaining to strains of *E. coli* containing *glnD*::Tn10 mutations which eliminate these activities. These observations depend on the characteristics of two classes of *glnA*-linked mutations with the GlnC phenotype. In both of these classes high levels of GS are produced under the normally repressing conditions of growth. However, when these strains are grown under conditions of extreme repression, one class (termed "classical" GlnC) continues to synthesize a derepressed level of GS while the other (termed "repressible" GlnC) synthesizes a very low level of GS. The interesting result is that introduction of a *glnD*::Tn10 mutation into a strain containing a mutation resulting in "repressible" GlnC phenotype completely reverses the "repressable" character of this mutation and results in the formation of high levels of GS in cells grown under conditions of extreme repression. Similar observations were made with GlnF$^+$ strains carrying the *gln-501* mutation. Since these GlnD$^+$ and GlnD$^-$ strains, grown under conditions which are normally extremely repressing, synthesize GS having, apparently, the same adenylylation state, these observations on the *gln-501* mutants and the strains with the "repressible" GlnC phenotype raise the possibility that the *glnD*-product, UTase, is itself involved in regulation of GS, adding another dimension of complexity to this system.

In any case, since we must now consider the possibility that the *glnG*-product is responsible for aspects of regulation of nitrogen metabolism which were previously assigned to GS, one of the most interesting questions is the site or sites of the *glnA* linked mutations resulting in these two classes of GlnC phenotype. Are these mutations in *glnA, glnG,* or some other gene in the *glnA*-region?

REFERENCES

1. Garcia, E., Bancroft, S., Rhee, S. G., and Kustu, S. (1977) *Proc. Natl. Acad. Sci. U.S.A. 74*:1662-1666.
2. Pahel, G., Zelenetz, A. D., and Tyler, B. (1978) *J. Bacteriol. 133*:139-148.
3. Gaillardin, C. M. and Magasanik, B. (1978) *J. Bacteriol. 133*:1329-1338.
4. Pahel, G. and Tyler, B. (1979) *Proc. Natl. Acad. Sci. U.S.A. 76*:4544-4548.

5. Analysis of Mutations Affecting the Regulation of Nitrogen Utilization in *Salmonella typhimurium*

JEAN E. BRENCHLEY, DAVID M. BEDWELL,
SUSAN M. DENDINGER AND JOHN M. KUCHTA *

Department of Biological Sciences
Purdue University
W. Lafayette, IN. 47907

SUMMARY

We have examined the ability of *Salmonella typhimurium* LT2 to regulate a number of enzymes in response to different nitrogen sources in the growth media. Studies of proline utilization show that even though *S. typhimurium* has a slow growth rate in glucose-proline medium, it has the ability to increase the activity of the proline catabolic enzyme, proline oxidase. This nitrogen control response is also reflected by increases in the activities of glutamine synthetase and several amino acid transport systems in cells grown with limiting nitrogen sources. Mutants were selected in which the ability to grow on a variety of nitrogen sources, such as arginine and proline, was altered. One class had mutations linked to the glutamine synthetase structural gene, *glnA*, and had elevated levels of glutamine synthetase, and glutamine, arginine, aspartate and lysine transport, but reduced proline oxidase activity. Strains with these mutations grow more rapidly on glucose-arginine medium, more slowly on glucose-proline medium, and are more sensitive to D-glutamine than a *glnA*+ control. A revertant selected for D-glutamine resistance has a new mutation, also closely linked to *glnA*, that causes lower levels of glutamine

* Present Address: Department of Biology, Duquesne University, Pittsburgh, PA. 15219.

ISBN: 0-12-506040-8

synthetase, reduced transport for several amino acids, and the inability to grow on arginine and glutamate as nitrogen sources. The characterization of one of these revertants shows that it is similar to another class of mutants selected for their ability to produce glutamine synthetase in the absence of the *glnF* factor. We have initiated genetic studies to locate the mutations causing altered nitrogen utilization relative to those causing glutamine auxotrophy and to a Tn10 element about 30% linked to *glnA*.

INTRODUCTION

Although *Salmonella typhimurium* has been used extensively for studies of biosynthetic pathways, little is known about the enzymes needed for the degradation of organic nitrogen sources. We are interested in the overall question of how nitrogen utilization is controlled and are examining these processes in *S. typhimurium*. The results of studies of proline degradation and amino acid transport show that a regulatory mechanism for nitrogen control does exist for *S. typhimurium*. To further study this control, we have isolated mutants altered in their ability to utilize various nitrogen sources and have examined the effects of the mutations on proline utilization, glutamine synthetase expression and amino acid transport regulation. We have found that a number of mutations altering growth on nitrogen sources such as arginine and proline are linked to *glnA*, the structural gene for glutamine synthetase. This paper describes the characterization of the physiological and genetic properties of some of these nitrogen utilization mutants.

MATERIALS AND METHODS

A description of the general growth conditions and glutamine synthetase and transport assays is contained in the following references: Brenchley (1, 2) ; Brenchley, *et al.* (3) ; Funanage, *et al.* (4) and Stadtman, *et al.* (5) . Proline oxidase and β-galactosidase were measured in whole cells using the procedures of Dendinger and Brill (6) and Prival and Magasanik (7) respectively.

The derivation of the strains used is given in Table I. The Tn10 transposable element was introduced into *S. typhimurium* using a P22 mutant as a vector (8) .

TABLE I

Derivation of strains

Strain	Relevant Characteristic	Derivation
MK53	*K. aerogenes*	Prival and Magasanik (1971)
15-59	*S. typhimurium* AR65 *(put C3)*	Dendinger and Brill (1970)
JL907	*gln+*	J. Ingraham
JB653	*gln-1003*	Selection for glutamine auxotrophy
JB667	*gln-1010*	Selection for glutamine auxotrophy
JB673	*gln-1007*	Selection for glutamine auxotrophy
JB741	F′ *lac*	F′ *lac* transferred from *E. coli* into JL907
JB756	*gln+*	Funanage and Brenchley (1977)
JB757	*gln-1000*	Funanage and Brenchley (1977)
JB1108	*gln-982*	Selection for methionine sulfoximine resistance
JB1110	*gln-983*	Selection for methionine sulfoximine resistance
JB1064	*gln-981*	Funanage *et al.,* (1978)
JB1065	*gln-1000, gln-1001*	Selection for D-glutamine resistance
JB1136	*glnF986::Tn10*	J. Roth

RESULTS AND DISCUSSION

Regulation of Proline Utilization

We are interested in examining the regulation of some enzymes needed for the degradation of amino acids as nitrogen sources. For these initial studies we monitored the regulation of proline oxidase, an enzyme of the proline utilization (Put) pathway. The levels and regulation of proline oxidase in *S. typhimurium* strain JL907 and *K. aerogenes* strain MK53 grown in various growth media were compared. The results (Table II) show that both strains have aproximately the same proline oxidase activity when the cells are

TABLE II

Proline oxidase activities for S. typhimurium and K. aerogenes cultures

Strains	Growth Media [a]			
	Glu-Amm-Pro	Glu-Pro	Gly-Amm-Pro	Gly-Pro-
S. typhimurium LT2 JL907	7.8	14.3 (1.8) [b]	19.2	59.3 (3.1)
K. aerogenes MK53	10.4	46.6 (4.5)	32.5	ND [c]
S. typhimurium 15-59 AR65	13.7	23.0 (1.7)	32.2	75.5 (2.3)

[a] The abbreviations for the growth media are: Glu, glucose; Amm, ammonia; Pro, proline; Gly, glycerol.

[b] The numbers in parenthesis represent the ratio of the activity obtained for cells grown in media with proline as a nitrogen source relative to that obtained for cells with ammonia and proline. The values reported for strain JL907 are averages of at least 5 separate experiments with a range of ± 2 units from those reported. The activities are nmoles of product formed per min. per mg protein.

[c] ND; Not Determined.

grown in glucose-ammonia-proline media; however, the levels are higher for the *K. aerogenes* strain when proline is the sole nitrogen source. Interestingly, the levels in *S. typhimurium* also increase about two fold in the glucose-proline medium despite a decrease in generation time from about 1 to 8-12 h. A more pronounced increase in proline oxidase is found when the results with the glycerol-ammonia-proline and glycerol-proline cultures are compared. Measurements of the differential rate of proline oxidase synthesis for cells growing in glucose-proline medium demonstrated that this increase is linear for at least three generations (data not shown). Thus, the slow growth of *S. typhimurium* in glucose-proline medium does not reflect the absence of a nitrogen control response by proline oxidase in this strain. As reported by Prival and Magasanik (7), the proline oxidase levels of *K. aerogenes* increase approximately four-fold for cells grown in glucose-proline medium and the growth rate for *K. aerogenes* was significantly faster in this medium (about 2 h generation time) than that of *S. typhimurium* strain JL907 (8-12 h).

Because proline is necessary as an inducer of proline oxidase, it was possible that the smaller increase in proline oxidase for strain JL907 could be caused by a partial exclusion of the inducer when cells are growing slowly in the glucose-proline medium. To eliminate this possibility, these experiments were repeated using a *S. typhimurium* 15-59 mutant, AR65, which produces proline oxidase constitutively (Table II). This strain has slightly higher proline oxidase activity than strain JL907 for all media, but the relative increases for strain 15-59 grown with proline as a nitrogen source are similar. The proline oxidase activities of both strains are subject to carbon catabolite repression by glucose and show a clear response to the nitrogen-limited growth conditions. This increase is in contrast to the further decrease found for an enzyme such as β-galactosidase that is not subject to nitrogen control. Strain JB741 was constructed by transferring an F'*lac* from *Escherichia coli* into JL907, and the activities of β-galactosidase and proline oxidase were measured in cells grown in different media (Table III). Both enzymes are subject to carbon catabolite repression as demonstrated by the lower

TABLE III

β-galactosidase and proline oxidase activities for S. typhimurium strain JB741 carrying an F' lac plasmid

Growth Media [a]	Enzyme Activities [b] β-Galactosidase	Proline Oxidase
Glycerol-ammonia-proline	22,269 (2.1)	14.1 (2.3)
Glucose-ammonia-proline	10,555 (1.0)	6.1 (1.0)
Glucose-p.oline	4,458 (0.4)	12.6 (2.1)
Glucose-proline + cAMP[a]	42,078 (4.0)	29.7 (4.9)

[a] cAMP; cyclic adenosine 3', 5'-monophosphate. cAMP was added at 10 mM. IPTG (isopropyl-β-D-thiogalactoside) was added to all cultures at 0.5 mM to induce the β-galactosidase activity.

[b] The numbers in parentheses represent the activities of β-galactosidase and proline oxidase relative to the specific activities found for cultures grown in glucose-ammonia-proline media.

activities when glucose is the carbon source as compared with glycerol. Both enzyme activities also increase when cyclic AMP is added to the media. The striking difference between the regulation of these enzymes is found for cells grown in a glucose-proline medium where the β-galactosidase levels are further reduced whereas the proline oxidase levels are increased. Thus, the nitrogen control response, as illustrated by the ability of the cells to increase enzyme activities during a nitrogen-limited growth condition, is evoked for proline oxidase and not for β-galactosidase.

Despite the ability of *S. typhimurium* to increase the proline oxidase levels, the cells grow very slowly in a glucose-proline medium. This aspect may reflect the lack of nitrogen control for the primary permease for proline transport (Funanage and Brenchley, unpublished data). In addition, the regulation of proline catabolism in *S. typhimurium* does appear to have some interesting features that make it different from other nitrogen utilization pathways. Some of these differences will be discussed further with regard to the unusual phenotypes of the mutants with altered nitrogen utilization properties.

Nitrogen Control of Glutamine Transport

The regulation of glutamine transport has been studied for *E. coli* by Willis *et al.* (9) and for *S. typhimurium* by Betteridge and Ayling (10). Their results show that glutamine uptake normally increases when cells are grown with a limiting nitrogen source and suggest that the glutamine transport system responds to nitrogen control in a manner similar to that found for proline oxidase. We confirmed these results for glutamine transport (4) and extended the observations for the transport of other amino acids. Kustu *et al.* (11) have shown that these increases in uptake rates correspond to elevated periplasmic binding proteins for glutamine, arginine and histidine. Because little is known about the catabolic pathways for some amino acids, such as arginine, that are used as nitrogen sources by *S. typhimurium*, the response of transport systems to nitrogen control makes transport a useful model for understanding the regulation of nitrogen utilization. The transport of amino acids can be considered as the first step in their catabolism, and we have monitored transport in mutants having altered growth on amino acids as nitrogen sources.

Selection of Mutants with Elevated Glutamine Synthetase Activities

Mutants with elevated glutamine synthetase activities and altered growth on arginine and proline as nitrogen sources were selected in two ways. The first approach used a glutamine auxotroph similar to the *glnB* mutants of *K. aerogenes* (12). These strains are glutamine auxotrophs because the loss of the *glnB* product prevents the synthesis of glutamine synthetase even though its structural gene is unaltered. Some revertants selected for glutamine prototrophy have second site mutations genetically linked to the *glnA* gene and produce elevated levels of glutamine synthetase. Strain JB757 was constructed in which the new mutation, *gln-1000*, was transduced into a *glnB*+ background and JB756 was constructed as an isogenic *glnA*+ control with normal glutamine synthetase activities (13).

The second selection used a mutant, JB664, that is unable to grow on a number of nitrogen sources (such as arginine) due to a mutation causing low glutamate synthase activity. Revertants of this Asm− phenotype were selected for growth on glucose-arginine medium. Revertants with elevated glutamine synthetase activities were purified and the mutations shown to be linked to *glnA*. Again, a transductant, JB1064, carrying only the *glnA* linked mutation, *gln-981*, was constructed and used in further experiments (4).

Characterization of Strains JB757 and JB1064

Strains JB757 and JB1064 have about five-fold and two-fold higher glutamine synthetase activities, respectively, than the control JB756 when grown in glucose-ammonia media (Table IV). These activities can be further increased in cells grown in a glucose-glutamate medium (Table IV) showing that the enzyme can be controlled normally even though the activities are elevated.

The growth phenotype of strains JB757 and JB1064 is unusual. These strains grow more rapidly than the control on glucose-arginine medium, but grow more slowly on glucose-proline medium (4). The inability to grow on glucose-proline is of special interest considering the data presented in Table II showing the slight increase of proline oxidase in wild-type cells grown in glucose-proline, a growth condition in which cells have very high gluta-

TABLE IV

Glutamine synthetase activities for nitrogen utilization mutants

	Glutamine Synthetase Activities		
		Strain	
Growth Medium	JB756 (*gln*+)	JB757 (*gln-1000*)	JB1064 (*gln-981*)
Glucose-Ammonia	158	830	300
Glucose-Glutamate	562	1712	1172

mine synthetase activities. Strain JB757 grows very slowly on glucose-proline, and the proline oxidase activities of cells grown in this medium are lower, rather than higher, than the activities for cells grown in a glucose-ammonia-proline medium. Thus, the regulation of proline oxidase appears to be reversed by the presence of the *gln-1000* mutation. Although it is not clear how the *gln-1000* mutation causes the slower growth on glucose-proline medium, there is a clear connection between this mutation and the altered nitrogen control response of JB757.

Because increased levels of glutamine synthetase are correlated with increased transport, we wanted to determine if the mutations in strains JB757 and JB1064 affected glutamine uptake. Measurements of the high affinity glutamine uptake for cells grown in a glucose-ammonia medium showed a significant increase for the mutants over that of the control, JB756 (Fig. 1). To determine whether the phenotype, as demostrated by the more rapid growth on glucose-arginine and slower growth on glucose-proline, is correlated with changes in transport, the rates of arginine and proline uptake were assayed (Table V). Arginine uptake is increased, especially for strain JB757 which also has the higher glutamine synthetase activity. In addition, aspartate and lysine uptake rates are increased in these strains. Proline transport, however, actually decreases. Thus, there is a correlation between the better growth on glucose-arginine and poor growth on glucose-proline and the regulation of arginine and proline transport in strains JB757 and JB1064.

Selection and Characterization of Mutants with Reduced Glutamine Synthetase Activities

The above nitrogen utilization mutants have elevated glutamine synthetase levels. To determine the effect of mutations causing low glutamine synthetase activities, use was made of the observation that D-glutamine inhibits the growth of strain JB757, but not the $glnA^+$ control in glucose-ammonia medium (14). A revertant, JB1065, selected for resistance to D-glutamine has a phenotype quite different from its parent JB757. As opposed to the ability of JB757 to grow more rapidly on glucose-arginine medium than a gln^+ control, the D-glutamine resistant mutant fails to grow on glucose-arginine (or glucose-glutamate) medium. The poor growth of JB757 on glucose-proline medium is unchanged in JB1065 (Table VI).

The unusual growth of strain JB1065 suggested that it had not reverted to wild type, but rather had obtained a new mutation in addition to the original *gln-1000* lesion. Transduction experiments

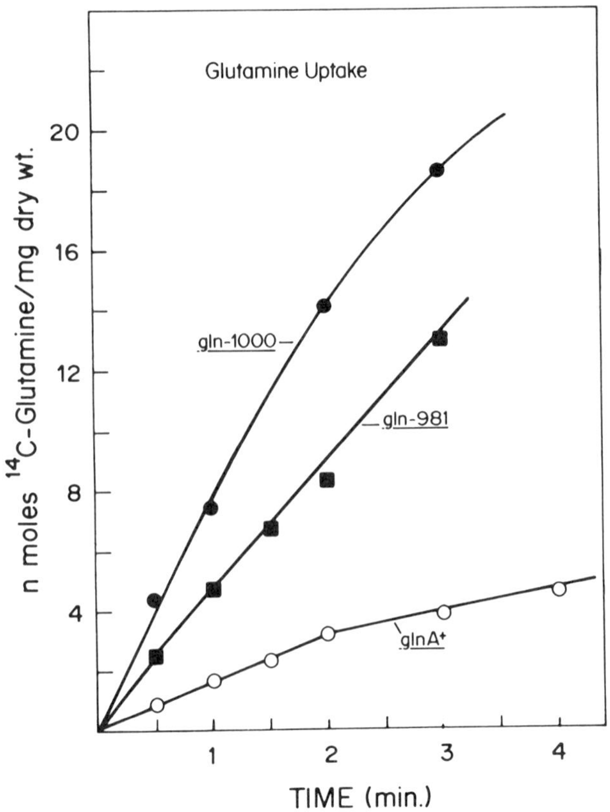

Fig. 1. Rates of L-[14C]glutamine uptake for *S. typhimurium* mutants. Symbols represent: O, JB756; ●, JB757; ■, JB1064. (Data taken from (4)).

demonstrated that strain JB1065 retained the *gln-1000* mutation and had a new, closely linked mutation, *gln-1001*. The presence of the *gln-1001* mutation serves to reduce the elevated glutamine synthetase levels caused originally by *gln-1000* (Table VII). In addition, the transport rates for glutamine, glutamate and arginine are reduced for strain JB1065 (Table VII).

The inability of JB1065 to use arginine as a nitrogen source is similar to the phenotype observed by mutations suppressing the requirement for the *glnF* factor. The *glnF* gene is unlinked to *glnA* and produces a factor necessary for glutamine synthetase production (15, 16). A *glnF* mutant is a glutamine auxotroph because of the reduced glutamine synthetase levels. The majority of Gln$^+$ revertants have suppressor mutations closely linked to *glnA* that cause low glutamine synthetase levels. To determine whether the *gln-1000* and/or *gln-1001* mutations suppress a *glnF* mutation, use was made

TABLE V

Amino acid uptake by strains JB756, JB757 and JB1064

Amino Acid Uptake [a]

Strain	Arginine	Aspartate	Lysine	Proline
JB756	3.38	0.06	2.55	3.28
JB757	7.72	0.20	7.96	1.83
JB1064	4.62	ND [b]	4.31	2.12

[a] Cells were grown in glucose-ammonia medium where the growth rates were comparable. The ^{14}C-labeled amino acids were added to the reaction mixture at the following concentrations: lysine and arginine, 1 μM; proline and aspartate, 10 μM. Values are given as specific activities in nanomoles of amino acid incorporated per minute per milligram (dry weight) of cells.
[b] ND, Not Determined.

(Data taken from (4)).

of a Tn10 transposon conferring tetracycline resistance that had inserted in the *glnF* gene to construct strains with these mutations. Phage grown on the *glnF*::Tn10 mutant were used to transduce strains JB756, JB757 and JB1065 to tetracycline resistance. The addition of the *glnF::Tn10* mutation to both JB756 and JB757 produced glutamine auxotrophs showing that the *glnF* factor is needed for glutamine synthetase production. However, the *gln-1000* and *gln-1001* mutations in strain JB1065 suppressed the need for the *glnF* factor and the new strain with the *glnF::Tn10* remained a prototroph (Table VI).

Recently, mutations causing suppression of *glnF* lesions in *E. coli* (17) and *S. typhimurium* (18) were shown to lie in a gene adjacent to, but distinct from *glnA*. This gene (designated either *glnG* or *glnR*) may encode a protein that becomes an activator of glutamine synthetase production when the *glnF* product is available. One explanation for the phenotype of strain JB1065 is that the *gln-1001* mutation converts this protein into an activator independent of the need for the *glnF* product. The *gln-1000* mutation that initially causes high glutamine synthetase levels could either be in this same

TABLE VI

Comparison of the phenotypes of strains JB756, JB757 and JB1065

	Phenotype Characteristic [a]			
Strain	Glu-Arg Growth	Glu-Pro Growth	D-Gln Sensitivity	*glnF* Suppression
JB756 (*glnA*+)	++	+	R	No
JB757 (*gln-1000*)	+++	—	S	No
JB1065 (*gln-1000* *gln-1001*)	—	—	R	Yes

[a] The abbreviations are: Glu, glucose; Arg, arginine; Pro, proline; and D-Gln, D-glutamine.

The symbols for the Glu-Arg and Glu-Pro growth represent: —, no growth; +, growth within 72 h; ++, growth within 24 h; +++, growth within 12 h.

The designations for D-Gln sensitivity are R for resistance to inhibition at 1 mM and S for sensitive to inhibition at 1 mM.

TABLE VII

Comparison of enzyme activities for strains JB756, JB757 and JB1065

	Enzyme Activities [a]			
Strain	Glutamine Synthetase	Glutamine Transport	Glutamate Transport	Arginine Transport
JB756 (*glnA*+)	236	2.9	7.4	2.6
JB757 (*gln-1000*)	974	9.9	9.4	5.5
JB1065 (*gln-1000*, *gln-1001*)	187	1.2	3.9	1.7

[a] The amino acids were used at 1 μM concentrations for the transport assays. Cells were grown in glucose-ammonia media where rates of the strains are comparable.

control gene or in a separate, closely linked gene. However, it is clear that the *gln-1000* mutation alone is not sufficient to suppress *glnF* mutations (Table VI) even though it does affect nitrogen control. The *gln-1001* mutation serves not only to suppress *glnF* lesions, but it also reverses the elevated levels of glutamine synthetase, high transport rates and rapid growth on glucose-arginine medium that is characteristic of the *gln-1000* containing strain.

To examine the phenotypes and genetics of other *glnF* suppressors (without the additional *gln-1000* mutation) several revertants of the *glnF::Tn10* strain were selected. These strains resemble strain JB1065, except that the glutamine synthetase levels are lower. The suppressor mutations are closely linked to *glnA* and their positions relative to other mutations affecting glutamine synthetase levels is given below.

Genetic Characterization of Nitrogen Utilization Mutants

Because of the interesting effects of these and other *glnA* linked mutations altering glutamine synthetase and other nitrogen control enzyme levels, we are constructing a fine structure map of the *glnA* region using cotransductions, three factor crosses, and deletion mapping. A summary of the arrangement of some of these mutations based on cotransduction frequencies with a Tn10 transposable element in this region is presented (Fig. 2) .The mutations causing *glnF* suppression are the most distal to the Tn10 insertion. The *gln-1000* and *gln-981* mutations that cause high glutamine synthetase levels appear to be between the *glnF* suppressors and a series of mutations for glutamine auxotrophy. Two mutations, *gln-982* and *gln-983*, lie near this region. These lesions alter the glutamine synthetase activity and render the enzyme activity resistant to methionine sulfoximine inhibition (19). Although these mutations also prevent growth on glucose-arginine (the Asm$^-$ phenotype), they do not suppress *glnF* mutations. The mutations *gln-990, gln-1003, gln-1007*, and *gln-1010* all cause glutamine auxotrophy and the loss of glutamine synthetase activity.

The characterization of these and several other mutants in which nitrogen utilization has been altered has shown that these mutations are clustered near the glutamine synthetase structural gene and correspond to one or more control genes for glutamine synthetase and/or a regulator for nitrogen utilization. Strains with the Tn10 or Tn5 transposable elements linked to *glnA* are now being used in a series of three-factor crosses with the mutations affecting

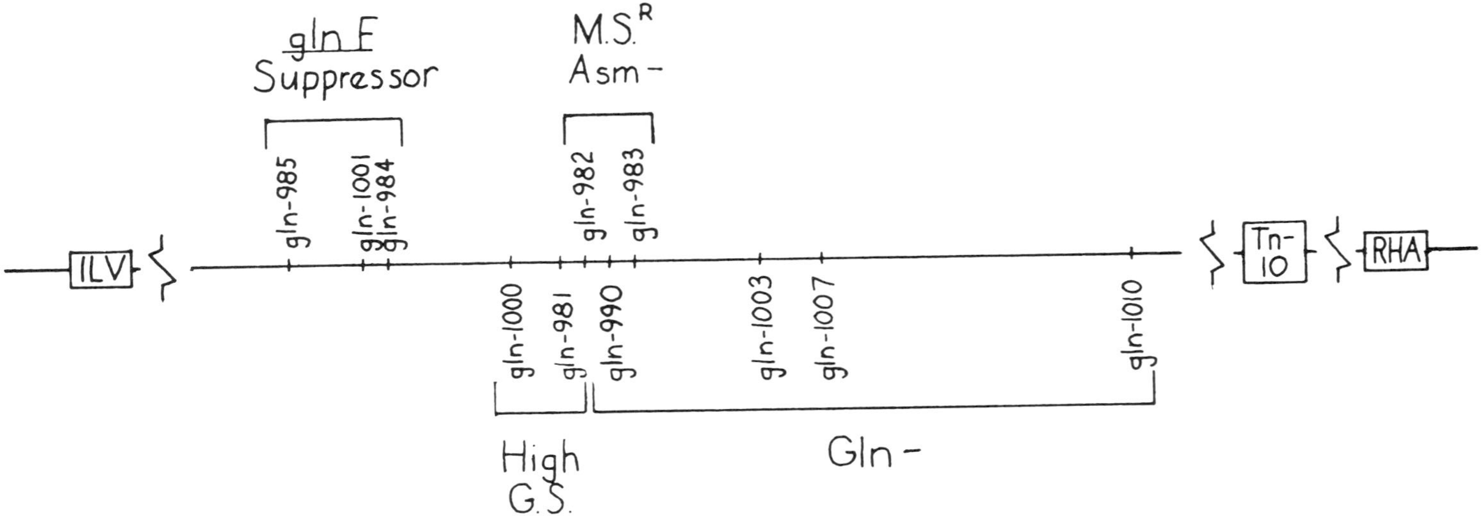

Fig. 2. A diagrammatic representation of the mutations in the *glnA* region of the *S. typhimurium* chromosome. The mutations within parentheses are placed together to indicate their similar phenotypes. The mutations within a cluster are closely linked, but their location relative to one another has not been determined. Three factor crosses have confirmed the order for the *gln-1000, gln-1003, gln-1007* and *gln-1010* mutations.

nitrogen utilization to construct a detailed map of the region. These data are being combined with that obtained using deletions of this region generated by the excision of the transposable elements. In addition, to obtain a detailed understanding of this region, we have cloned a portion of the *S. typhimurium* chromosome complementing *glnA* mutations onto pBR322 and are using the material to correlate the genetic and physical maps (Koduri and Brenchley, unpublished results). The goal is to combine the information from the physiological studies presented here with the detailed knowledge of the DNA structure and sequence to understand the regulation of glutamine synthetase and other nitrogen utilization enzymes at the molecular level.

It is clear from the preliminary studies presented here that *S. typhimurium* has a regulatory mechanism for controlling growth on various nitrogen sources and that many mutations altering this response lie near or within *glnA*. It is not clear how the *gln-1000* and *gln-981* mutations cause differential changes, such as more rapid growth on glucose-arginine and poor growth on glucose-proline media. If these mutations alter a nitrogen control factor, then the change has the opposite effect on the expression of glutamine synthetase and proline oxidase. As further information on the orientation of the mutations for *glnF* suppression and the mutations for increased glutamine synthetase levels is obtained, it will be interesting to determine whether they lie in separate domains of the same gene or in distinct genes. The use of various cloned fragments for this region in complementation tests and physical studies of this DNA should help resolve the questions remaining about the number and functions of control genes near *glnA*.

Acknowledgments. This investigation was supported by Public Health Service grant GM 25251 and grant PCM 23248 from the National Science Foundation. J. E. B. is a recipient of a Research Career Development Award from the National Institute of General Medical Sciences.

REFERENCES

1. Brenchley, J. (1973) *J. Bacteriol. 114*, 666-673.
2. Brenchley, J., Prival, M. and Magasanik, B. (1973) *J. Biol. Chem. 248*, 6122-6128.
3. Brenchley, J. Baker, C. and Patil, L. (1975) *J. Bacteriol. 124*, 182-189.
4. Funanage, V., Ayling, P., Dendinger, S. and Brenchley, J. (1978) *J. Bacteriol. 136*, 588-596.

5. Stadtman, E., Ginsburg, A., Ciardi, J., Yeh, J., Hennig, S. and Shapiro, B. (1970) *Adv. Enzyme Regul. 8,* 99-118.
6. Dendinger, S. and Brill, W. (1970) *J. Bacteriol. 103,* 144-152.
7. Prival, M. and Magasanik, B. (1971) *J. Biol. Chem. 246,* 6288-6296.
8. Kleckner, N., Roth, J. and Botstein, D. (1977). *J. Mol. Biol. 116,* 125-159.
9. Willis, R., Iwata, K., and Furlong, C. (1975) *J. Bacteriol. 122,* 1032-1037.
10. Betteridge, P. and Ayling, P. (1976) *J. Gen. Microbiol. 95,* 324-334.
11. Kustu, S., McFarland, N., Hui, S., Esmon, B., and Ames, G. (1979) *J. Bacteriol. 138,* 218-234.
12. Prival, M., Brenchley, J. and Magasanik, B. (1973) *J. Biol. Chem. 248,* 4334-4344.
13. Funanage, V. and Brenchley, J. (1977) *Genetics 86,* 513-526.
14. Kuchta, J. and Brenchley, J. (1978) *Abstr. Ann. Mtg. Am. Soc. Microbiol.* K13.
15. Garcia, E., Bancroft, S., Rhee, S. and Kustu, S. (1977) *Proc. Nat. Acad. Sci. U. S. A., 74,* 1662-1666.
16. Gaillardin, C. and Magasanik, B. (1978) *J. Bacteriol. 133,* 1329-1338.
17. Pahel, G. and Tyler, B. (1979) *Proc. Nat. Acad. Sci. U.S.A., 76,* 4544-4548.
18. Kustu, S., Burton, D., Garcia, E., McCarter, L., and McFarland, N. (1979) *Proc. Nat. Acad. Sci. U. S. A., 76,* 4576-4580.
19. Miller, E. and Brenchley, J. (1979) *Abstr. Ann. Mtg. Am. Soc. Microbiol.* K118.

6. Nitrogen Regulation in *Salmonella typhimurium*

SYDNEY KUSTU, DORIS BURTON, EMILIO GARCÍA,
LINDA McCARTER AND NANCY McFARLAND

Department of Bacteriology
University of California
Davis, California 95616

SUMMARY

Two positive regulatory elements, the *glnF* and *glnR* gene products, are required for nitrogen control in *Salmonella*. These gene products have a pleiotropic role in controlling synthesis of glutamine synthetase and amino acid transport components in response to availability of nitrogen. Based on genetic studies we have formulated a working model for the functions of the *glnF* and *glnR* products in mediating nitrogen control.

INTRODUCTION

In bacteria synthesis of a number of proteins is subject to general control by availability of nitrogen in the growth medium (1, 2, 3, 4, 5). Synthesis of these proteins is increased under nitrogen limiting conditions. This phenomenon was discovered by Neidhardt and Magasanik, who demonstrated that the histidine degradative enzymes of *Klebsiella aerogenes* could be synthesized at high levels under nitrogen-limiting conditions even in the presence of glucose as carbon source (1). The elegant work of Prival and Magasanik demonstrated that nitrogen control in *Klebsiella* is mediated by a cytoplasmic factor (s) different from cyclic AMP and its receptor protein (6).

Proteins whose synthesis is subject to nitrogen control vary with the bacterial genus (6, 7). In *Salmonella* they include the enzyme glutamine synthetase and a set of amino acid transport components,

ISBN: 0-12-506040-8

among them periplasmic binding proteins for glutamine, lysine/arginine/ornithine, histidine, and glutamate/aspartate (8). (Periplasmic binding proteins lie between the cytoplasmic and outer membranes and appear to function in an early step in the transport process.) We have focussed on periplasmic transport components because they are water soluble and their amino acid binding activities can be quantitated by equilibrium dialysis. Synthesis of glutamine synthetase and the amino acid binding protein is elevated under nitrogen-limiting conditions. Since the structural genes encoding glutamine synthetase (*glnA*) and the histidine- and glutamine- binding proteins are not clustered on the *Salmonella* chromosome (8), nitrogen control affects the expression of several unlinked genes.

We have found that the products of two positive regulatory genes, *glnF* and *glnR*, are required for nitrogen control in *Salmonella* (9, 10). These gene products have a pleiotropic role in controlling synthesis of glutamine synthetase and amino acid transport components. Based on genetic studies we have formulated a simple working model for the functions of the *glnF* and *glnR* products in mediating nitrogen control (10).

Magasanik, Tyler and their colleagues have proposed models for nitrogen control in which the enzyme glutamine synthetase (reviewed in 11 and 12) or proteins involved in its covalent modification (13) are central genetic regulatory elements. Our results suggest that regulatory characteristics which they attributed to glutamine synthetase can be explained as characteristics of the *glnR* (and *glnF*) products. Whether the actions of the *glnR* and *glnF* products alone can account for nitrogen control or additional regulatory elements are also required (13, 14, 15, 16) remains to be resolved.

MATERIALS AND METHODS

Materials. The frameshift mutagen ICR191E (2-chloro-6-methoxy-9-[3-(2-chloroethyl) aminopropylamino] acridine dihydrochloride) was kindly donated by H. J. Creech, Chemotherapy Laboratory of the Institute for Cancer Research, Philadelphia, PA.

Isolation of Strains. All strains were derived from *Salmonella typhimurium* LT2. Strains with *glnF* mutations (SK99 and SK100) were isolated from an ICR-mutagenized culture of strain TA831 (*his*Δ*F645*) by penicillin enrichment for glutamine auxotrophs. Strain SK294, carrying the *glnR138* mutation was isolated from an ICR-mutagenized culture of strain SK214 (8) by penicillin enrichment for loss of ability to use D-histidine as histidine source, a selection for decreased histidine transport. Strain SK293 (*glnR137*::Tn10 *hisF645*) was isolated by transducing strain SK35 [*gln*Δ*(AR)60 hisF 645*] to glutamine independence with pooled tetracycline-resistant clones containing random insertions of the Tn10 transposon in the *Salmonella* chromosome and screening transductants for tetracycline resistance. Strain SK398 (*glnR137*::Tn10 *hisF645*) was obtained by transducing strain TA831 to tetracycline resistance with phage grown on strain SK293.

Assays. Assays for glutamine synthetase and amino acid binding protein activities and for levels of the corresponding antigens were done as described (8, 17). Synthetase activities were corrected for "blanks" in the absence of adenosine diphosphate and arsenate.

Characterization of Glutamine Synthetase. A sequence of 6 residues at the amino terminus of glutamine synthetase was determined by automated Edman degradation on a Beckman 890C sequencer (kindly performed by Alan J. Smith) (18). A sequence of 7 residues at the carboxyl terminus of glutamine synthetase was determined by digesting it with carboxy-peptidase A (CPA), and identifying the amino acids released on a Durrum D 500 amino acid analyzer (19). (Amino acid analyses were kindly performed by K. Kanagaki.) Glutamine synthetase was treated with snake venom phosphodiesterase (SVP) to remove adenylyl groups (17, 20). Glutamine synthetase (before and after SVP treatment) was subjected to cleavage with trypsin under mild conditions similar to those described by Eisenberg and co-workers (21).

RESULTS AND DISCUSSION

As indicated in Fig. 1, the *glnF* gene is well separated on the chromosome from *glnA*, the structural gene for glutamine synthetase (9), whereas the *glnR* gene (called *glnG* in *E. coli* by Pahel and Tyler (22)) is closely linked to *glnA* (approximately 75% by P22-mediated transduction (10)). (We have identified *glnA* as the structural gene for glutamine synthetase because some strains with spontaneous *glnA* mutations produce glutamine synthetase antigen which is catalytically inactive and other strains produce glutamine synthetase with altered catalytic and physical properties (20).)

Strains with mutations to loss of function of *glnF* (induced with the frameshift mutagen ICR191E or by insertion of the tetracycline-resistance transposon Tn10) are glutamine auxotrophs (Table I).

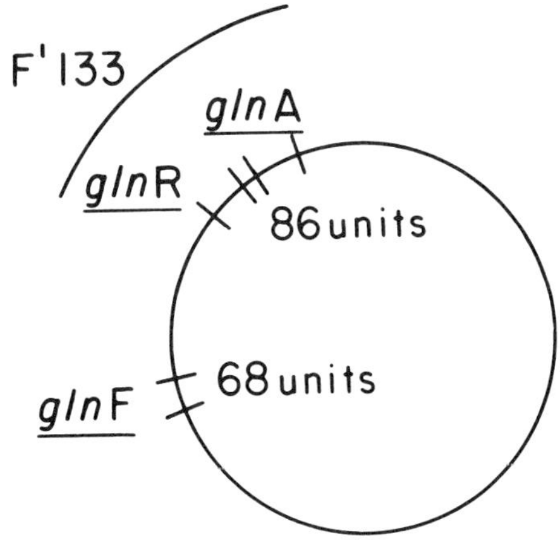

Fig. 1. Locations of the *glnF* and *glnR* genes on the *Salmonella* chromosome: *E. coli* F′ factor used for complementation analysis. The *glnF* gene is linked to the *argG* (9) and *cod* (E. Garcia, unpublished results) genes by P22-mediated transduction. Although we have not determined the order of genes in the *glnA* region, Dr. Tyler's group has determined that *glnR* (*glnG*) is counterclockwise of *glnA* in *E. coli*. In this figure we have assumed that the same is true in *Salmonella.*

They have very low glutamine synthetase activity (<1% the maximal activity of wild type) and low activities of the periplasmic glutamine- and lysine/arginine/ornithine-binding proteins. Unlike the wild type strain, they are unable to increase synthesis of these proteins under nitrogen-limiting conditions. For both glutamine synthetase and binding proteins low levels of activity in *gln F* strains are correlated with low levels of antigen (data not shown).

Strains with mutations to loss of function of *glnR* (induced with ICR191 or Tn10) resemble *glnF* strains in that they have low levels of glutamine synthetase activity and antigen and are unable to increase synthesis of this enzyme or periplasmic transport proteins under nitrogen-limiting conditions (Table I). The *glnR* strains differ from *glnF* strains in that they produce detectable glutamine synthetase (approximately 3% the maximal level of wild type) and are able to grow in the absence of glutamine (Table I).

Mutations to loss of function of *glnF* or *glnR* affect the expression of several unlinked genes; thus, *glnF* and *glnR* appear to encode diffusible products. Consistent with this, mutant *glnF* or *glnR* alleles are recessive to the corresponding wild type allele with regard to expression of *glnA* (9, 10). The fact that mutations to loss of function of either *glnF* or *glnR* cause a pleiotropic decrease in ability to synthesize proteins under nitrogen control indicates that the *glnF* and *glnR* products are required for synthesis of these proteins and

TABLE I

Doubling times, glutamine synthetase and binding protein activities in regulatory mutant strains

Strain[a]	Doubling time[b] (min) −gln	Doubling time[b] (min) +gln	Doubling time[b] (min) +gln, arg, ura, hypo	GS (μmoles/min-mg) N-excess[c]	GS (μmoles/min-mg) N-limiting[c]	Glutamine-binding protein N-excess (pmoles/mg dry weight cells)	Glutamine-binding protein N-limiting (pmoles/mg dry weight cells)	LAO-binding[d] protein N-excess (pmoles/mg dry weight cells)	LAO-binding[d] protein N-limiting (pmoles/mg dry weight cells)
TA831 (wild type)	44	43	38	0.15	1.25	4.2	14.3	2.4	28.6
SK99 (*glnF75*)	≧320	57	43	<0.01	<0.01	1.4	1.1	0.5	0.8
SK100 (*glnF76*)	≧320	58	43	<0.01	<0.01	1.3	0.9	1.6	0
SK398 (*glnR137*::Tn10)	60	48	40	0.03	0.03	1.6	1.0	0	1.0
SK294 (*glnR138*)	55	45	40	0.03	0.04	1.2	0.9	1.4	0
SK464 (*glnR137*::Tn10 *glnF75*)	57	47	41	0.03	0.03	1.2	0.9	0.3	0.4
SK465 (*glnR138 glnF75*)	59	47	40	0.03	0.04	1.6	0.9	0.5	0
SK214 (*gln-128*)	+[e]	+[e]		2.4	2.5	45.9	47.1	54.3	131
SK278 (*gln-128 glnF75*)	—[f]	+[f]		0.01	0.01	1.5	0.9	0.7	1.5

[a] All strains contain the hisΔF645 mutation. Strains SK294 and SK465 also carry the gln-128 mutation (8).

[b] Minimal medium (14) without glutamine; with glutamine; or with glutamine and the additional supplements described in (c). Media contained glucose (0.4%) as carbon source and 20 mM NH_4^+.

[c] N-excess: Minimal glucose medium with 20 mM NH_4^+ and 3 mM glutamine as N-sources. N-limiting: 3 mM glutamine as sole N-source. Both media were supplemented with 0.2 mM arginine, uracil, and hypoxanthine (and 0.002 mM thiamine), major end products that contain a nitrogen atom derived from glutamine, to stimulate the growth rate of mutant strains.

[d] Lysine/arginine/ornithine. Calculated as described previously (8).

[e] The doubling time of strain SK214 is several minutes longer than that of strain TA831 in the presence or absence of glutamine (8).

[f] Like other *glnF* strains, strain SK278 is a glutamine auxotroph. Doubling times have not been determined.

(Data taken from references (8) and (10)).

therefore are positive regulatory factors. (Neither product is involved in reversible covalent modification of glutamine synthetase (9, 14).)

The *glnR* gene is closely linked on the chromosome to *glnA*, but we have established that it is distinct from *glnA*. Unlike *glnA* strains, strains with mutations to complete loss of function of *glnR* produce detectable glutamine synthetase and are able to grow in the absence of glutamine (Table I). In addition, the properties of glutamine synthetase from a mutant strain with insertion of the Tn10 transposon in *glnR* are the same as those of glutamine synthetase from wild type (10). The rationale for studying properties of glutamine synthetase from the insertion mutant is diagrammed in Fig. 2. When the Tn10 transposon, indicated by the heavy line in Fig. 2, is inserted into a bacterial gene, it disrupts transcription and translation of that gene, causing complete loss of function of the gene product (23). If the Tn10 transposon was inserted in the *glnA* gene, even at the end of the gene, the carboxy terminal sequence of amino acids of glutamine synthetase would be altered because transcription and translation of a portion of the *glnA* gene would be followed at some point by transcription and translation into the Tn10 transposon. By contrast, if the Tn10 transposon was inserted in a regulatory gene distinct from *glnA*, the properties of glutamine synthetase would be normal.

We have purified glutamine synthetase from the *glnR*::Tn10 strain and have found that its specific activity and stability to heat, acetone, and acid used during the purification were the same as those of glutamine synthetase from wild type (10). Mobilities on sodium dodecyl sulfate/polyacrylamide gel electrophoresis (Fig. 3) indicated that the subunit molecular weight of glutamine synthetase from the insertion mutant was the same as that from wild type as were the molecular weights of fragments produced by mild proteolytic cleavage of the synthetase with trypsin (10). In addition, glutamine synthetase from the insertion mutant and the wild type could not be distinguished immunologically by microcomplement fixation (10; kindly performed by Linda Baumann), which is very sensitive to small sequence differences between homologous proteins (24). These lines of evidence indicated that the Tn10 transposon in the mutant strain was not inserted in the middle of the *glnA* gene.

A sequence of six amino acids at the amino terminus of glutamine synthetase from the mutant was the same as that of enzyme from wild type (NH_2Ser-Ala-Glu-His-Val-Leu-) (10). The carboxy-terminal set of seven amino acids released by digestion with 10 μg of carboxypeptidase A and the pattern of release of the first five residues with limiting amounts of carboxypeptidase A were the same for

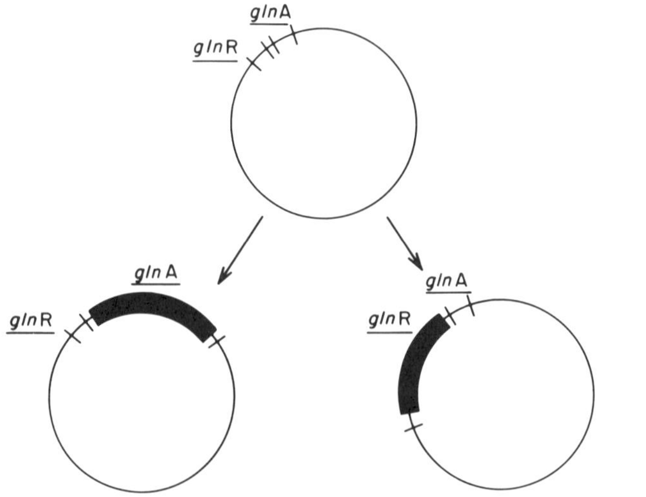

Fig. 2. Consequences of insertion of the Tn10 transposon into *glnA* or *glnR*. The Tn10 transposon is indicated by a heavy line. If the transposon is inserted in *glnA*, the properties of glutamine synthetase—in particular the carboxy terminal sequence of amino acids—must be altered. By contrast, if the transposon is inserted in a regulatory gene distinct from *glnA*, the properties of glutamine synthetase will be normal (see text).

glutamine synthetase from mutant and wild type (Table II). The carboxy terminal sequence inferred was Phe-Glu-Leu-Tyr-Tyr-Ser-Val-COO^- (10). Since the properties of glutamine synthetase from the insertion mutant—including the carboxy terminal sequence of amino acids—were normal, the Tn10 transposon is not inserted in *glnA*. Rather it must be inserted in a regulatory gene closely linked to *glnA*, which we have called *glnR*. (Pahel and Tyler have called the homologous gene in *E. coli glnG* (22)).

As indicated above, the *glnR* and *glnF* products are positive regulatory factors. Loss of function of either results in a pleiotropic decrease in ability to synthesize proteins under nitrogen control. However, *glnR* strains differ from *glnF* strains in that they produce detectable glutamine synthetase and are able to grow in the absence of exogenous glutamine. In addition, mutations to complete loss of function of *glnR* suppress mutations to loss of function of *glnF*—that is, *glnR glnF* strains have exactly the same properties as *glnR* single mutant strains (Table I; reference 10). This observation indicates

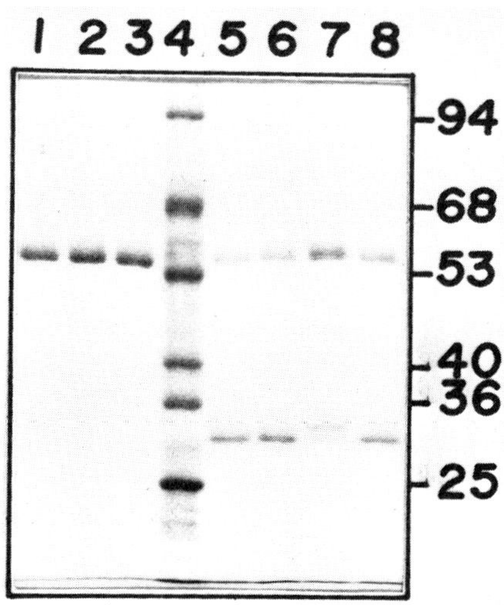

Fig. 3. Glutamine synthetase and trypsin-cleaved glutamine synthetase from strains TA831 and SK398 on SDS-PAGE (9%). Molecular weights of standards (well 4) are indicated in kilodaltons at the right. All other wells (numbered at the top) contain ~0.3 μg protein. Wells 1 and 2 contain SVP-treated glutamine synthetase from strains TA831 and SK398, respectively; well 3 contains a mixture of SVP-treated glutamine synthetase from the two sources; wells 5-8 contain trypsin-cleaved glutamine synthetase from TA831 (SVP-treated), SK398 (SVPtreated), TA831, and SK398, respectively. The upper band is uncleaved glutamine synthetase. The lower band (apparent mol. wt. ~31,000) is the larger trypsin fragment. The smaller fragment ran to the dye front. The upper bands of the doublets observed for glutamine synthetase and the larger trypsin fragment of glutamine synthetase from TA831 (well 7) are adenylylated subunit and fragment, respectively (21). (Data from reference (10)).

that the *glnF* and *glnR* products function together to mediate nitrogen control and is consistent with a model in which the *glnF* product (directly or indirectly) modifies the *glnR* product to convert it from a form with negative regulatory character to a form with positive regulatory character (Fig. 4). In the absence of any *glnF* product (in a *glnF* mutant strain) the *glnR* product apparently has exclusively negative regulatory character (at least with regard to *glnA* expression) because a *glnR glnF* strain, which lacks both the *glnR* and *glnF* products, synthesizes more glutamine synthetase and grows better than a *glnF* strain. The dual positive and negative character proposed for the *glnR* product has a precedent in similar dual character of the *araC* product (25). Gaillardin and Magasanik previously proposed a model for nitrogen regulation involving positive and negative control by glutamine synthetase as the genetic regulatory element (26).

Apart from the partial negative character of the *glnR* product, which can be detected in a *glnF* background, the *glnR-glnF* system

TABLE II

Digestion of glutamine synthetase with carboxypeptidase A [a]

	nmoles amino acid released / nmole GS subunit [b]						
	.01	.02	.04	.08	.16	1.0	10.0 [c] μg CPA
TA831							
val	.60	.77	.75	.80	.77	.81	.91
ser	—	.06	.15	.29	.54	.79	.95
tyr	—	.06	.23	.58	1.0	1.6	1.8
leu	—	—	—	.07	.22	.76	.93
glu	—	—	—	—	—	—	.15
phe	—	—	—	—	—	—	.11
SK398							
val	.52	.72	.80	.80	.80	.81	.87
ser	—	.07	.14	.27	.49	.78	.92
tyr	—	.07	.25	.55	.97	1.6	1.71
leu	—	—	—	.07	.17	.77	.88
glu	—	—	—	—	—	—	.18
phe	—	—	—	—	—	—	.12

[a] Digestion at 37°C for 60 min. Each digestion mixture (0.07 ml final volume, pH 7.4) contained 0.25 mg GS, 21 mM $NaHCO_3$, 7 mM EDTA, 0.085% SDS, and 0.1% LiCl. GS was pre-incubated with SDS and EDTA for 15 min at 45°C before addition of CPA. No free amino acids were found in untreated GS controls or in CPA controls.

[b] Based on a molecular weight of 50,000 daltons (19).

[c] 0.5 mg GS from TA831 and 0.65 mg from SK398 in a digestion mixture of 0.14 ml. The sequence inferred-phe-glu-leu-tyr-tyr-ser-valCOOH, is based, in part, on the fact that CPA releases tyr rapidly, ser slowly, glu very slowly and phe rapidly (30). (Data taken from reference (10)).

is a positive regulatory system. Thus, there is a striking analogy between this system for nitrogen control and the *crp-cya* system for carbon and energy regulation (27). If the analogy extends to mechanism, *glnR* would encode a macromolecular regulator of transcription and *glnF* would encode an enzyme that catalyzes synthesis of a low molecular weight coregulator of transcription—a signal of nitrogen deficiency (Fig. 4). When bound to the low molecular weight signal, the *glnR* product would function as an activator of transcription for genes subject to nitrogen control; in the absence of the nitrogen deficiency signal the *glnR* product would function as an inhibitor (repressor) of transcription, at least for *glnA*. (Note that a model in which *glnR* encodes the macromolecular regulator

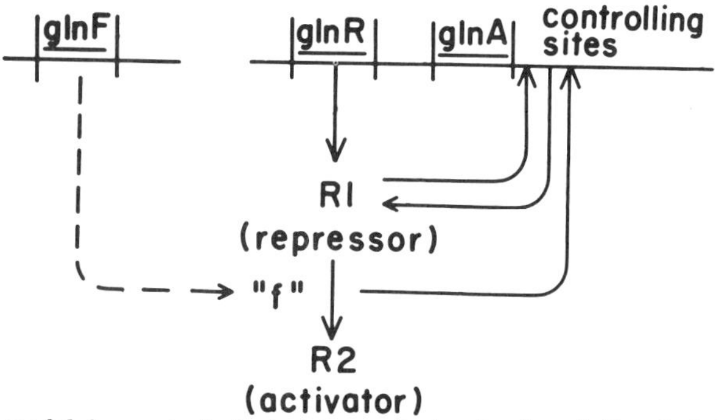

Fig. 4. Working model for control of *glnA* transcription by the *glnR* and *glnF* products. The product of the *glnR* gene exists in two conformations: R1, which represses transcription of *glnA* and R2, which activates transcription of *glnA* (and other genes under nitrogen control). The *glnF* product, which may catalyze formation of a low molecular weight signal of nitrogen deficiency "f", leads to formation of R2, the activator form. (Data from reference (10)).

of transcription accounts for the fact that the properties of *glnR glnF* strains are the same as those of *glnR* strains—in the absence of the macromolecular regulator, the presence or absence of the coregulator is of no consequence). We are presently trying to identify the *glnR* product and the postulated nitrogen limitation signal produced by the *glnF* product on the basis of their ability to activate synthesis of glutamine synthetase in a coupled *in vitro* transcription-translation system.

We would like to add two comments with regard to function of the *glnR* product. First, mutations that result in high level synthesis of nitrogen controlled proteins under all growth conditions (8; strain SK214 in Table I)—that is, in the "GlnC phenotype" (28, 29), lie under a *glnAR* deletion. If, in fact, these mutations lie within *glnR*, they may have altered the *glnR* product so that it activates transcription at unusually low concentrations of signal. (Expression of the GlnC phenotype depends on some level of *glnF* function since this phenotype cannot be expressed in a *glnF*⁻ background (Strain SK278 in Table I)).

Finally, the extent of nitrogen control varies among the enteric bacteria—for example, nitrogen control of enzyme synthesis is much more extensive in *Klebsiella* than in *Salmonella*. Since Prival and Magasanik have demonstrated that the difference between *Klebsiella* and *Salmonella* is due, at least in part, to a difference in a cytoplasmic factor(s), it will be of interest to determine whether it can be accounted for by intergeneric differences in the *glnR* product.

Acknowledgements. We thank L. N. Csonka for helpful criticisms of the manuscript. This work was supported by U. S. Public Health Service Grant GM 21307.

REFERENCES

1. Neidhardt, F. C., and Magasanik, B. (1957) *J. Bacteriol. 73*, 253-259.
2. Pengra, R. M., and Wilson, P. W. (1957) *J. Bacteriol. 75*, 21-25.
3. Mortenson, L. F., Morris, J., and Jeng, D. Y. (1967) *Biochem. Biophys. Acta 141*, 516-522.
4. Woolfolk, C. A., Shapiro, B. M., and Stadtman, E. R. (1966) *Arch. Biochem. Biophys. 116*, 177-192.
5. Mecke, D., and Holzer, H. (1966) *Biochim. Biophys. Acta 122*, 341-351.
6. Prival, M. J., and Magasanik, B. (1971) *J. Biol. Chem. 246*, 6288-6296.
7. Brill, W., and Magasanik, B. (1969) *J. Biol. Chem. 244*, 5392-5402.
8. Kustu, S. G., McFarland, N. C., Hui, S. P., Esmon, B., and Ames, G. F. L. (1979) *J. Bacteriol. 138*, 218-234.
9. Garcia, E., Bancroft, S., Rhee, S. G., and Kustu, S. (1977) *Proc. Natl. Acad. Sci. U.S.A. 74*, 1662-1666.
10. Kustu, S., Burton, D., Garcia, E., McCarter, L., and McFarland, N. (1979) *Proc. Natl. Acad. Sci. U.S.A. 76*, 4576-4580.
11. Magasanik, B. (1976) *Prog. Nucleic Acid Res. Mol. Biol. 17*, 99-115.
12. Tyler, B. (1978) *Annu. Rev. Biochem. 47*, 1127-1162.
13. Foor, F., Reuveny, Z., and Magasanik, B. (1979) *Abstracts of the Annual Meeting of the American Society for Microbiology*, 176.
14. Bancroft, S., Rhee, S. G., Neumann, C., and Kustu, S. (1978) *J. Bacteriol. 134*, 1046-1055.
15. Foor, F., Cedergren, R. J., Streicher, S. L., Rhee, S. G., and Magasanik, B. (1978) *J. Bacteriol. 134*, 562-568.
16. Bloom, F. R., Levin, M. S., Foor, F., and Tyler, B. (1978) *J. Bacteriol. 134*, 569-577.
17. Stadtman, E. R., Ginsburg, A., Ciardi, J. E., Yeh, J., Hennig, S. B., and Shapiro, B. M. (1970) *Adv. Enzyme Regul. 8*, 99-118.
18. Niall, H. D. (1973) *Methods Enzymol. 27*, 942-1010.
19. Ginsburg, A. (1972) *Adv. Protein Chem. 26*, 1-79.
20. Kustu, S. G. and McKeregham, K. (1975) *J. Bacteriol. 122*, 1006-1016.
21. Lei, M., Aebi, U., Heidner, E. G., and Eisenberg, D. (1979) *J. Biol. Chem. 254*, 3129-3134.
22. Pahel, G. and Tyler, B. (1979) *Proc. Natl. Acad. Sci. U. S. A. 76*, 4544-4548.
23. Kleckner, N., Roth, J., and Botstein, D. (1977) *J. Mol. Biol. 116*, 125-159.
24. White, T. J., Ibrahimi, I. M., and Wilson, A. C. (1978) *Nature (London) 274*, 92-94.
25. Englesberg, E., and Wilcox, G. (1974) *Annu. Rev. Genet 8*, 219-242.
26. Gaillardin, C. M., and Magasanik, B. (1978) *J. Bacteriol. 133*, 1329-1338.
27. Pastan, I., and Adhya, S. (1976) *Bacteriol. Rev. 40*, 527-551.
28. Brenchley, J. E., Prival, M. J., and B. Magasanik. (1973) *J. Biol. Chem. 248*, 6122-6128.
29. Prival, M. J., Brenchley, J. E., and Magasanik, B. (1973) *J. Biol. Chem. 248*, 4334-4344.
30. Ambler, R. P. (1967) *Methods Enzymol. 11*, 155-166.

7. Genetic Characterization of Mutations Affecting Glutamine Biosynthesis and its Regulation in *Escherichia coli K 12*

FERNANDO BASTARRACHEA, SUSANA BROM, ALEJANDRA A. COVARRUBIAS, AURORA OSORIO AND FRANCISCO BOLIVAR

Departamento de Biología Molecular
Instituto de Investigaciones Biomédicas
Universidad Nacional Autónoma de México
Apartado Postal 70228, México 20, D. F.
México.

SUMMARY

Mutations affecting the enzymatic activity and regulation of glutamine synthetase (GS) of *Escherichia coli* K12 are described. One particular missense mutation, *glnA20,* conferred to cells a tight glutamine auxotrophy and the capacity to synthesize an enzymatically inactive GS protein which was normally repressed by ammonium ions. In *glnA+/glnA20* merodiploids, glutamine repressed synthesis of enzymatically active GS. These results are discussed in terms of a regulatory role of the GS oligomer, either directly or indirectly, in the expression of its coding unit, the *glnA* structural gene. Recombination analysis with P1 bacteriophage between *glnA* point mutants or Tn5 insertions in *glnA* and certain deletions entering into the *glnA* region, as well as complementation of the different mutations by newly isolated hybrid plasmids, provided evidence for the following order of genes: *old polA glnG glnA chlB rha.* The *glnG* and the *glnA* genes appear to lie in different cistrons.

ISBN: 0-12-506040-8

INTRODUCTION

In enteric bacteria, glutamine synthetase (GS) is the only enzyme responsible for the synthesis of glutamine. The biosynthetic activity of this enzyme is reversibly modified by the covalent attachment of an adenylyl moiety to each of its 12 subunits. The GS enzymatic activity is inversely proportional to the number of covalently bound AMP groups (1).

GS expression, on the other hand, appears to be regulated by the availability of the nitrogen source present in the medium (2).

In order to carry out the genetic analysis of GS regulation, a large number of mutants have been isolated and characterized in several laboratories. Two classes of mutations in the *glnA* structural gene, located at min 85 on the *E. coli* map, lead to glutamine auxotrophy. One of these does not synthesize a GS polypeptide (GlnA$^-$ phenotype), and another, which synthesizes an enzymatically inactive one (Gln (AC)$^-$ phenotype). A third class of mutations results in the high level constitutive synthesis of GS (GlnC phenotype). Mainly because mutations resulting in the GlnC phenotype apparently mapped between mutations conferring the Gln (AC)$^-$ phenotype, mutations leading to GS constitutivity were considered to lie also in the *glnA* gene (3, 4). From these studies, the autogenous regulation concept emerged, i. e., that expression of *glnA* was subject to regulation by GS itself (for reviews, see 5, 6).

In addition, transcription of genes coding for enzymes of nitrogen metabolism capable of supplying bacterial cells with glutamate, such as the arginine, histidine or proline degrading enzymes, appeared to require derepressed levels of unadenylylated GS. Thus, glutamine auxotrophs unable to synthesize any GS-like protein were also incapable of activating synthesis of enzymes of nitrogen metabolism such as the histidine-degrading enzymes (the Reg$^-$ phenotype) (7). Within this context, Gln(AC)$^-$ mutants which synthesize enzymatically inactive GS antigenic proteins, as well as GlnC mutants are phenotypically Reg$^+$.

At this stage, mutations in the new locus *glnF* were discovered by Kustu and coworkers (8) in *Salmonella typhimurium* and subsequently in *Klebsiella aerogenes* (4) and in *E. coli* (9). *glnF* mutants are glutamine auxotrophs and also Reg$^-$; they also synthesize GS amounts even lower than those found under extreme repressing conditions, regardless of whether the nitrogen source present in the medium is limiting or in excess. It is not known yet if cells carrying *glnF* mutations are Reg$^-$ because they are highly repressed for GS synthesis. *glnF* mutations, which map at min 68 of the *E. coli* chromosome (9) far away from the *glnA* region, are suppressed by secondary mutations originally thought to lie at the *glnA* structural gene (4). More recent findings by Pahel and Tyler (7) and by Kustu and coworkers (10) indicate that these *glnF*$^-$ supressor mutations alter another locus termed *glnG* (or *glnR*) in the vicinity of the *glnA* gene.

The question then arises as to whether all properties previously attributed to GS can now be attributed to the product of the *glnG* gene. Genetic evidence indicates that the *glnF* product has a positive regulatory character and that the *glnG* product has both a positive and a negative regulatory character. According to Kustu *et al.* (10), the product of the *glnF* gene can modify the *glnG* product thus converting it to a form with positive regulatory character. In the absence of any *glnF* product, the *glnG* product apparently only has a negative regulatory character, at least with regard to *glnA* expression. This model, however, does not exclude a possible role of GS in the expression of the Reg$^+$ phenotype which cannot be entirely attributable to *glnG*.

In order to get a deeper insight into this complex regulatory system, it was deemed necessary to isolate and characterize more mutations affecting glutamine biosynthesis and its regulation in *E. coli*. We wish to report here the results we have obtained with the study of certain newly isolated mutations.

MATERIALS AND METHODS

Bacteria and Bacteriophages

All bacterial strains were derivatives of *E. coli* K12. Mutants were isolated from the glutamine prototrophic strains RRI (F$^-$ *proA2 leuB6 thi-1 lacY1 galK2 rpsL20 hsdS* r_B^- m_B^-) or from strain MX614 (F$^-$ *thi ilv-680 (pro-lac)* △). MX615 is a spontaneous (gal-*att*λ-*bio*) △ derivative of MX614. MX752 is an SuIII$^+$ *zif*::λ *kan2* derivative of MX615. F14 merodiploids were constructed by mating AB1206 (F′ 14/*proA 2 his-4 thi-1 rpsL8 lacY1 galK2 tfr-3 supE44* (*ilv-argH*) △ $_{E2}$) as donor with MX705 as recipient, and selecting for Gln$^+$ (His$^+$) transconjugants as described by Miller (11). Bacteriophage P1 *vira* was kindly provided by Neil Willetts and λKan2 was kindly provided by Michel Faelen.

Culture media. The minimal NN medium as well as the complex Luria medium used have been described (12).

Enzyme determination. All determinations were performed according to Bender *et al.* (13) with minor modifications (12).

Other procedures. Purification of GS, preparation of antibodies and the rocket immunoelectrophoresis technique have been previously described (12). Protein was determined by the method of Lowry *et al.* (14) using bovine serum albumine as a standard.

RESULTS

Isolation and characterization of the glnA20 mutation

Glutamine auxotrophs have been isolated by a variety of methods. Point mutants were isolated after chemical mutagenesis with ethyl methanesulfonate (15), followed by penicillin treatment (11) to enrich for glutamine auxotrophs. All of our mutants, including insertions and deletions have been isolated from two different lines of *E. coli* K12, RR1 and MX614. MX705, which was derived from RR1, carries *glnA20*, a typical missense mutation. MX705 reverts to the Gln^+ phenotype at a rate of 5 x 10^{-10} per cell per division cycle. In addition, the F14 merogenote which is known to carry the *glnA*$^+$ allele (3) fully complemented the *glnA20* mutation of MX705 (12).

The total amount of GS ($-Mg^{2+}$) as well as the amount of nonadenylylated GS ($+Mg^{2+}$) was determined using the γ-glutamyl transferase assay in cells of the parental strain RR1 and in those of the MX705 mutant (Table I). MX705 cells did not synthesize any detectable GS activity. Wild type RR1 cells synthesized derepressed levels of half adenylated GS when grown on NN medium with either 0.5 mM ammonium or glutamate as sole nitrogen sources. GS activity was also measured in cells of the merodiploid strain MX735 (F14 *glnA*$^+$/*glnA20*). Under derepressed conditions, MX735 synthesized GS similar in amount and in its level of adenylylation to the GS synthesized by cells of the haploid strain RR1 (Table I). Apparently this is consistent with the fact that bacterial cells normally harbor only one to nearly two copies of any particular F-merogenote per chromosome in the autonomous state (16, 17).

A 15 mM ammonium concentration supported good growth of RR1 and MX735 cells (Table I), even though it caused a strong repression of GS synthesis. This was consistent with the well known pattern of enteric bacterial cells synthesizing small amounts of highly adenylylated GS in response to their culturing on high ammonia concentrations (see 6, for review).

TABLE I

GS specific activities of E. coli strains

Inocula were grown overnight on minimal medium with the indicated nitrogen source. Cultures were diluted 1:10 into the corresponding fresh medium and aereated until growth reached 80 to 100 Klett units. Haploid Gln− segregants were found in merodiploid cultures grown on glutamine but only at frequencies of 0.5 to 5%.

Strain (genotype)	Nitrogen source	GS assay $-Mg^{2+}$	GS assay $+Mg^{2+}$
RR1	0.5 mM NH_4^+	1.03	0.52
(*glnA*+)	15 mM NH_4^+	0.09	0.02
	Glutamate	0.80	0.43
	Glutamine	0.64	0.57
MX705 (*glnA20*)	Glutamine	0.01	0.01
MX731	0.5 mM NH_4^+	1.95	1.07
(ColEl *glnA*+/*glnA20*)	15 mM NH_4^+	0.08	0.01
	Glutamate	1.99	1.26
	Glutamine	0.16	0.06
MX735	0.5 mM NH_4^+	0.64	0.36
(F14 *glnA*+/*glnA20*)	15 mM NH_4^+	0.08	0.01
	Glutamate	0.64	0.34
	Glutamine	0.08	0.05
MX615	15 mM NH_4^+	0.08	0.03
(*glnA*+)	Glutamine	0.52	0.30
MX727 (*glnA21*::Tn5)	Glutamine	< 0.01	<0.01
MX746	0.5 mM NH_4^+	1.96	1.11
(ColEl *glnA*+/*glnA21*::Tn5)	15 mM NH_4^+	0.17	0.04
	Glutamate	1.59	0.81
	Glutamine	1.68	1.46

Growth of enteric bacteria on glutamine as sole nitrogen source, results in a condition of nitrogen limitation (18). Apparently, exogenous glutamine is not transported into the cell at a rate sufficient to cause repression. In keeping with this observation, glutamine allowed derepressed synthesis of GS in the haploid parental strain RR1. Quite unexpectedly, glutamine caused marked repression of GS synthesis in MX735 cells, similar to that caused by 15 mM ammonium. Glutamine repression has not been previously observed in haploid cells of *E. coli* (9) or *Klebsiella aerogenes* (19, 20, 4). The small amount of GS synthesized by MX735 cells in the presence of glutamine as sole nitrogen source was partially adenylylated.

The finding of glutamine repression of GS synthesis in MX735 cells was similarly observed in merodiploids carrying a ColE1 hybrid plasmid (pACR1 *glnA*$^{+}$/*glnA20*). A thorough characterization of the pACR1 plasmid has been reported (12).

The glutamine effect and the glnA20 allele

In order to test the possibility that the repression exerted by glutamine on GS synthesis in the merodiploids was triggered by the *glnA20* defective protein, we made use of strain MX727. These cells carry the *glnA21*::Tn5 mutation due to insertion of the Tn5 translocatable element (21). The Tn5 insertion apparently caused complete loss of *glnA* expression, since cross reacting material against GS antibodies was not detected in cell free extracts of MX727 cells (Fig. 1; 12).

MX727 cells were transformed to glutamine prototrophy and colicin immunity with the pACR1 plasmid. One of the transformants was purified and termed MX746. As shown in Table I, MX746 cells synthesized normal amounts of half adenylylated GS enzyme still repressible by 15 mM ammonium, but no longer repressible by glutamine.

The *glnA20* allele has to be considered as belonging to the GlnA^{-} phenotypic subclass (6). It differs from mutations associated with the Gln (AC)$^{-}$ phenotype in that the *glnA20* allele is repressible by ammonium ions (Fig. 1). Most other mutations associated with the Gln(AC)$^{-}$ phenotype characterized from *Klebsiella pneumoniae* (22) and *Klebsiella aerogenes* (19) give rise to enzimatically inactive GS's which are nonrepressible by ammonia. From the results presented in Table I, it appears that glutamine normally represses GS synthesis, albeit slightly. Thus, wild type cells (RR1) grown on glutamine as only nitrogen source had only a 25 to 40% decrease of

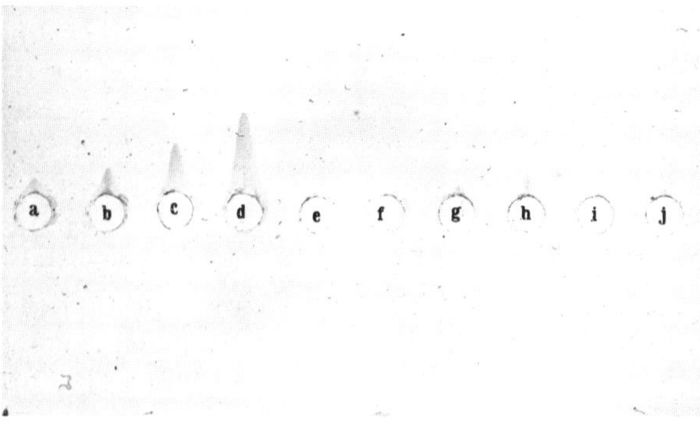

Fig. 1. Rocket immunoelectrophoresis of glutamine synthetase from different cell extracts. Procedures were performed according to Weeke (25). One percent agarose gels containing 0.5% of anti-GS gamma globulin fractions were used. In the wells there were added cell extracts from strain MX615 grown on ammonia (*a* and *b*) or on glutamine (*c* and *d*), MX705 grown on ammonia plus glutamine (*e* and *f*) or on glutamine alone (*g* and *h*) and MX727 grown on glutamine (*i* and *j*). The amount of total protein of the extracts added were 10 and 20 μg/ml, respectively, for every pair of wells such as *a* and *b*, etc. Results similar to those obtained with MX615 were also obtained for strain RR1 (12). Cells were grown on either 15 mM NH_4Cl or on L-glutamine (1 mg/ml).

the GS synthesized under fully derepressing conditions i. e., in the presence of 0.5 mM ammonium. Even the enzymatically inactive GS product made by the *glnA20* allele in MX705 cells was repressed only about 25% in the presence of glutamine (12). In cells of the MX746 (pACR1 *glnA*$^+$/*glnA21*::Tn5) strain, GS synthesis was only 20% repressed when compared to cells grown in a low ammonium concentration. A 15 mM ammonium concentration, by contrast, repressed GS synthesis by 90% or more in most of the strains where testing was allowed. The only cases diverting from the above pattern were the merodiploids carrying the chromosomal *glnA20* mutation. Therefore, the possibility exists that the inactive product of the *glnA20* mutation plays a role in the high degree of repression of GS synthesis caused by glutamine. Since in MX705 cells, the synthesis of the inactive *glnA20* product was only slightly repressible by glutamine, in the absence of the wild type GS protein, the *glnA20* mutant product does not seem to be self-sufficient to promote its own repression by glutamine. Therefore, the most likely possibility to explain glutamine repression in heterozygous *glnA*$^+$/*glnA20* merodiploids is through participation of the GS mixed subunits that are most probably formed.

The finding that mutations associated with the Gln (AC)$^-$ phenotype give rise to enzymatically inactive GS's which are non repressible by ammonia have led Bender and Magasanik (18) to propose that the true repression signal for *glnA* expression may not be ammonium itself but glutamine. Since in MX705 (*glnA20*) cells, ammonium ions are mainly incorporated into glutamate but not into glutamine, either glutamate or a compound derived directly or indirectly from it, or the ammonium ion itself is the signal for repression of the *glnA20* enzymatically inactive product.

Growth of *glnA20 gltB gdh E. coli* triple mutants in a minimal glucose medium supplemented with glutamate and glutamine should lead to derepressed synthesis of the *glnA20* protein, provided the product of the *gltB* and *gdh* genes are not involved in the regulation of GS synthesis. Under these conditions externally added ammonia if transported inside the cells should not be metabolized. By measuring the GS-cross reacting material made from *glnA20* in the absence and in the presence of a high ammonium concentration we should be able to know whether or not the ammonium ion is the repression signal for expression of the *glnA20* allele.

The Reg phenotype of MX705 and MX727

We tried to determine the Reg phenotype of MX705 and MX727 cells by following their growth response in glucose minimal medium with either arginine or proline as main nitrogen source. This phenotype, however, is not easy to score in Gln$^-$ cells. In addition to the limiting nitrogen source such as arginine or proline, one has to add to the medium relatively elevated concentrations of glutamine in order to satisfy the requirement for this amino acid.

Low concentrations of glutamine (75 μg/ml) did not support growth of either MX705 or MX727 cells (Fig. 2). The respective Gln$^+$ parental strains RR1 and MX615, by contrast, grew well with this low amount of glutamine. The inability of MX705 and MX727 cells to grow on low glutamine concentrations may be due to an impairment in their efficiency to transport this amino acid inside the cells under these conditions. In fact, Kustu *et al.* (23) have shown that Reg$^-$ mutants of *S. typhimurium* have diminished ability to transport several amino acids, including glutamine.

As shown in Fig. 2, high concentrations of arginine or proline, in the presence of 75 μg/ml of glutamine, still did not support growth of MX705 or MX727 cells. In fact, these strains only grew well when high glutamine concentrations were added to the minimal medium.

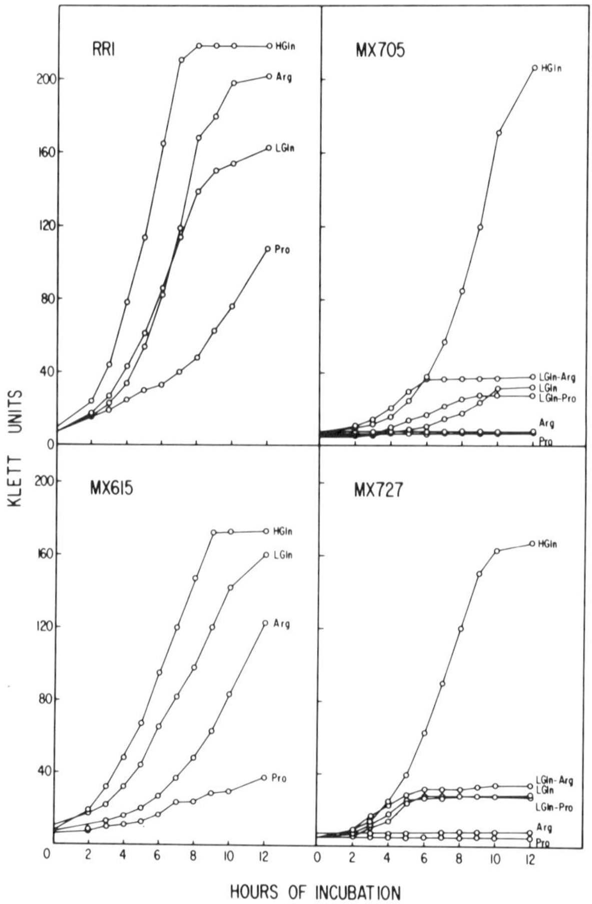

Fig. 2. Growth responses of strains RRI, MX705, MX615 and MX727 to different nitrogen sources in glucose minimal medium. HGln, 1 mg/ml of L-glutamine; LGln, 75 μg/ml of L-glutamine; Arg, 2 mg/ml of L-arginine; Pro, 2 mg/ml of L-proline.

These results seem to suggest that strains MX705 and MX727 are Reg⁻. The following argument, however, allow us to cast doubt on this conclusion, at least with regard to strain MX705. Pahel and Tyler (7) and Kustu *et al.* (10) showed that the Reg⁻ phenotype characteristic of *glnG* mutations confer to cells two discernible manifestations in response to nitrogen limitation. One is the inability to activate operons such as *hut* or of certain amino acid transport systems, and the other is a low level constitutivity of GS synthesis. MX705 cells synthesize almost normal levels of an enzymatically inactive GS antigenic protein when grown on media containing high concentrations of glutamine. Therefore, at least by this criterion, MX705 cells are Reg$^+$. Whether enzymes such as those coded for by the *hut* operon are activated by growing MX705 cells i. e., in the presence of high concentrations of glutamine, as they should, remains to be demonstrated.

MX727 cells on the other hand, by analogy with other Gln$^-$ mutants incapable of synthesizing a GS-like protein, must be Reg$^-$ under all growth conditions.

Merodiploid cells of the genetic constitution *glnA*$^+$/*glnA20* (MX735) and *glnA*$^+$/*glnA21*::Tn5 (MX746), which are Gln$^+$, also grew in the presence of arginine or proline as only nitrogen sources, i. e., they were Reg$^+$ (data not shown). It is not known if expression of the Reg$^+$ phenotype of the *glnA*$^+$/*glnA20* merodiploids is affected by glutamine at the same time it represses GS synthesis.

Ordering of genes in the glnA region.

Pl cotransduction experiments between pairs of *E. coli* strains carrying mutations in certain genes of this genetic region (Tables II and III) have allowed us to confirm that *polA* is the closest known marker on one side of the *glnA* gene and to show that *chlB* is the closest known marker on the *rha* side of *glnA*. The new gross map thus far elaborated is shown in the bottom of Table III.

Deletion mapping of the glnA region.

Deletions of the *glnA* gene have been selected as heat resistant survivors of an *E. coli* strain carrying a thermosensitive λ prophage inserted at the *rha* operon (24). Deletion mapping experiments, using these mutants confirmed the order *polA glnA rha* of this genetic region.

In our laboratory, we have taken advantage of strain MX752 carrying a λKan2 prophage inserted at approximately 0.16 min of the site corresponding to the *glnA20* mutation, to isolate deletions. Pl cotransduction experiments demonstrated that in MX752 the λKan2 bacteriophage is inserted between the *polA* and the *glnA* genes. After brief heat pulses (42°), thermoresistant survivors were isolated from MX752. By determining the Gln and the Reg phenotypes of the survivors and by recombination analysis of Pl transductional pairwise crosses between them and a set of strains carrying point mutations in the *glnA* region, we were able to characterize overlap-

TABLE II

Order of genes in the glnA region of the E. coli chromosome

P1 Donor	Recipient	Selected transductants	Transductants per 10^6 P1	Nº Scored	Percentage of transductants that score as Old+	Pol−	Chl^R	Rha−
C127 (*chlB*)	MX705 (*glnA20*)	Gln+	0.81	200	—	—	20	—
K184 (*old-5, rha4*)	MX705	Gln+	16	344	52	—	—	7
JW164 (*polAl, rha*)	MX705	Gln+	42	208	—	60	—	—
MX232	MX726 (*old*::Tn5, *glnA21*::Tn5)	Gln	1.62	104	44	—	—	—
C127	K184	Rha+	N.D.	98	3	—	45	—

C127 was obtained from N. Murray, K184 from P. Overath and JW164 from M. C. Gómez-Eichelmann. MX232 carries all relevant genes in the wild allelic state.

N.D. Not determined.

TABLE III

Pl Cotransduction frequencies and map intervals of pairs of linked genes in the glnA region of the E. coli K12 chromosome

Gene loci	Cotransduction frequency (%)	Interlocus distance in min (26)
glnA-old	46	0.46
glnA-rha	7	1.15
glnA-chlB	20	0.82
old-rha	2.5	1.41
rha-chlB	45	0.47
glnA-polA	60	0.31

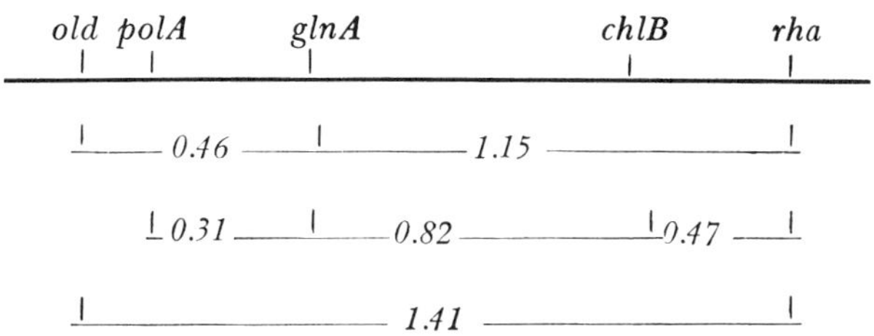

ping deletions extending from the original λKan2 insertion site and ending into the *glnG* and *glnA* genes (S. Brom and F. Bastarrachea, unpublished experiments). We found, in agreement to previous evidence (7), that the regulatory gene, *glnG*, lies close to the *polA* proximal end of the *glnA* gene. Whether *glnG* and *glnA* are contiguous is not yet completely clear. All together, our mapping results are consistent with the following order of genes in this chromosomal region: *old polA glnG glnA chlB rha.*

DISCUSSION

We have presented evidence indicating that the *glnA20* allele confers to MX705 cells the ability to synthesize an enzimatically inactive GS protein capable of reacting with antibodies prepared against purified *E. coli* GS. Moreover, expression of this GS polypeptide is still sensitive to ammonia repression as wild type GS.

Under derepressed conditions, i. e., on glucose minimal medium in the presence of glutamine as only nitrogen source, MX705 cells synthesized normal amounts of the *glnA* inactive product. However in merodiploids carrying the *glnA20* allele on the chromosome and the $glnA^+$ allele either in the F14 merogenote or in a ColE1 hybrid plasmid, glutamine caused marked repression of GS synthesis. Therefore, the conclusion has been reached that glutamine represses GS synthesis in $glnA^+/glnA20$ merodiploids by acting as a negative effector on the mixed GS subunits that are most probably formed. If mixed subunits containing both wild type and mutant GS monomers are involved in the synthesis of enzymatically active GS in the $glnA^+/glnA20$ merodiploids, then a conclusion that cannot be avoided is that GS normally regulates directly or indirectly expression of its *glnA* gene.

Another alternative possibility is that mixed subunits are synthesized by the $glnA^+/glnA20$ merodiploids in normal amounts during growth on glutamine but that the GS oligomers thus formed are less enzymatically active as compared to GS oligomers exclusively made of wild type subunits. This possibility appears unlikely since in the presence of limiting amounts of ammonia, the $glnA^+/glnA20$ merodiploids synthesized high levels of GS (Table I), presumably made of the same mixed subunits that were synthesized in the presence of glutamine.

On the other hand, genetic experiments indicate that the *glnG* product activates *glnA* expression. Two lines of evidence support this contention. First, that mutations in the *glnF* gene, the product of which is necessary to activate GS expression, can be suppressed by mutations in the *glnG* gene (10, 4). Second, that GS regulation is restored in merodiploids presumably containing *glnG*::Mu $glnA^+$ and *glnA*::Mu $glnG^+$ in *trans* configuration. Such merodiploids, however, synthesize normally regulated levels of GS when grown under nitrogen limitation and still retain the Reg^- phenotype of either haploid parent (7). Even though the results of these complementations are not completely understood, they seem to suggest that *glnG* and *glnA* lie in different cistrons.

In our laboratory we have been able to clone small DNA fragments of the *glnA* region into the pBR322 plasmid vehicle (A. A. Covarrubias, M. Rocha, F. Bolívar and F. Bastarrachea, to be published). The polypeptide products that are transcribed from a variety of these plasmids in a minicell system have been analyzed by SDS acrylamide gel electrophoresis and by specific immunoprecipitation with antibodies prepared against purified wild type GS. These studies have revealed that a gene coding for an approximately 65,000 d. polypeptide is contiguous to the *glnA* gene. Moreover, one of the plasmids we have constructed, which carries the gene coding for the 65,000 d. polypeptide but not the *glnA* gene, complements chromosomal deletions of the genotype (*zif-glnG*) $\Delta glnA^+$ to the Reg^+ phenotype. These results provide convincing evidence for the notion that *glnG* and *glnA* lie in separate transcriptional units.

Acknowledgment. This work was supported by grants PNCB 1649 and 1650 from the Consejo Nacional de Ciencia y Tecnología (México).

REFERENCES

1. Ginsburg, A. and Stadtman, E. R. (1973) in *The Enzimes of Glutamine Metabolism* (Prusiner, S. and Stadtman, E. R., eds.) pp. 9-43, Academic Press, New York.
2. Woolfolk, C. A., Shapiro, B. M., and Stadtman, E. R. (1966). *Arch. Biochem. Biophys. 116,* 177-192.
3. Streicher, S. L., Bender, R. A. and Magasanik, B. (1975). *J. Bacteriol. 121,* 320-331.
4. Gaillardin, C. M. and Magasanik, B. (1978). *J. Bacteriol. 133,* 1329-1338.
5. Magasanik, B. (1977). *Trends Biochem. Sci. 2,* 9-12.
6. Tyler, B. (1978). *Ann. Rev. Biochem. 47,* 1127-1162.
7. Pahel, G. and Tyler (1979). *Proc. Nat. Acad. Sci. U.S.A. 76,* 4544-4548.
8. García, E., Bancroft, S., Rhee, S. G. and Kustu, S. (1977). *Proc. Nat. Acad. Sci. U.S.A. 73,* 1662-1666.
9. Pahel, G., Zelenetz, A. D. and Tyler, B. M. (1978). *J. Bacteriol. 133,* 139-148.
10. Kustu, S., Burton, D., Garcia, E., McCarter, L., and McFarland, N. (1979). *Proc. Nat. Acad. Sci. U.S.A. 76,* 4576-4580.
11. Miller, J. (1972). *Experiments in Molecular Genetics,* 2nd. ed. pp. 82-107. Cold Spring Harbor, Cold Spring Harbor Laboratory, New York.
12. Covarrubias, A. A., Sánchez-Pescador, R., Osorio, A., Bolívar, F. and Bastarrachea, F. (1980). *Plasmid, 3,* 150-164.
13. Bender, R. A. Janssen, K. A., Resnick, A. D., Blumenberg, M., Foor, F. and Magasanik, B. (1977). *J. Bacteriol. 129,* 1001-1009.

14. Lowry, O. H., Rosebrough, N. J., Farr, A. L., and Randall, R. J. (1951). *J. Biol. Chem. 193,* 265-275.
15. Clark, A. J. (1967). *J. Cell. Physiol. 70* (Suppl. 1), 165-180.
16. Hayes, W. (1968). *Studies of Bacteria and their Viruses* 2nd. ed. pp. 746-808. John Wiley and Sons, New York.
17. Frame, R. and Bishop, J. O. (1971). *Biochem. J. 121,* 93-103.
18. Bender, R. A. and Magasanik, B. (1977). *J. Bacteriol. 132,* 100-105.
19. Streicher, S. L., DeLeo, A. B. and Magasanik, B. (1976). *J. Bacteriol. 127,* 184-192.
20. Bender, R. A. and Magasanik, B. (1977). *J. Bacteriol. 132,* 106-112.
21. Berg, D. E., Davies, J., Allet, B. and Rochaix, J. (1975). *Proc. Nat. Acad. Sci. U.S.A. 72,* 3628-3632.
22. Shanmugam, P. T., Chan, I. and Morandi, C. (1975). *Biochim. Biophys. Acta 408,* 101-111.
23. Kustu, S. G., McFarland, N. C., Hui, S. P., Esmon, B. and Ames, G. F. (1979). *J. Bacteriol. 138,* 218-234.
24. Pahel, G., Bloom, F. R. and Tyler, B. (1979). *J. Bacteriol. 138,* 653-656.
25. Weeke, B. (1973). Rocket immunoelectrophoresis. In: *A manual of quantitative immunoelectrophoresis* (Axelsen, NN., Kroll, J. and Weeke, B. eds.), pp. 37-46. Oslo: Universitetsforlaget.
26. Wu, T. T. (1966). *Genetics 54,* 405-410.

8. Isolation of *E. coli* Chromosomal DNA Regions Carrying *glnA*, *gdh* and *gltB* Structural Genes

ALEJANDRA COVARRUBIAS, RAY SÁNCHEZ-PESCADOR, EDMUNDO LOZOYA, FERNANDO BASTARRACHEA, AND FRANCISCO BOLIVAR

Departamento de Biología Molecular
Instituto de Investigaciones Biomédicas
Universidad Nacional Autónoma de México
Apartado Postal 70228, México 20, D. F.,
México.

SUMMARY

Three hybrid plasmids were isolated from the Clarke-Carbon *E. coli* gene bank. These recombinant DNA molecules, pACR1, pRSP1 and pRSP20 complemented *glnA, gdh* and *gltB* mutations respectively. In every case, the plasmid conferred to otherwise mutant cells the capacity to synthesize the corresponding wild type enzymes: glutamine synthetase, glutamate dehydrogenase and glutamate synthase, respectively. Evidence is presented that indicates the presence of the *glnA* gene in pACR1, the *gdh* gene in pRSP1 and *gltB* gene(s) in pRSP20. Initial characterization of these plasmids using restriction endonucleases, is also presented.

INTRODUCTION

In *Escherichia coli* and other enteric bacteria, three enzymes play a crucial role for the assimilation of ammonia into glutamate and glutamine. Glutamine synthetase (GS), is not only responsible for

ISBN: 0-12-506040-8

the biosynthesis of glutamine but also for the incorporation of ammonia into glutamate through a coupled reaction with glutamate synthase (GOGAT) (1, 2). In addition, glutamate can be synthesized from ammonia and oxoglutarate by a reaction catalyzed by glutamate dehydrogenase (GDH).

Enteric bacterial strains lacking GDH do not require glutamate for growth. Such mutants grow as well as wild type strains on both low and high ammonia concentrations and on other nitrogen sources (1, 2).

Mutants lacking GOGAT can grow without glutamate on minimal medium containing glucose and excess ammonia but are unable to grow on minimal medium containing a low ammonia concentration or other organic compounds as sole sources of nitrogen (2, 3). Mutants lacking both GDH and GOGAT activities can not utilize ammonia as the only nitrogen source and require glutamate for growth (4).

Mutations affecting the synthesis and regulation of GS have been isolated in different enteric bacteria. Genetic studies have shown that these mutations are localized at a number of distinct loci around the chromosome: *glnA, glnB, glnD, glnE, glnF* and *glnG* (*glnR*) (5, 6, 7). From all these loci, the *glnA* gene, which has been mapped at 85 min in the linkage map of *E. coli* (8), is known to be the structural gene for the GS subunit.

To study the physical organization and the expression of their respective genes in different genetic backgrounds, we sought to isolate individual hybrid plasmids which contain genes that complement *E. coli* chromosomal mutations affecting GS, GDH and GOGAT activities, respectively.

We have screened the Clarke and Carbon *E. coli* gene bank (9), for the presence of GS, GDH and GOGAT gene-complementing mutations and have isolated three different plasmids (10). One of these, pACR1, carrying the GS structural gene, *glnA*; a second, pRSP1 carries the *gdh* gene; and a third, pRSP20, complements the *gltB31* mutation that affects GOGAT activity. The isolation and initial characterization of these plasmids are described in the present paper.

MATERIALS AND METHODS

Bacterial Strains. The *E. coli* strains used in this work are listed in Table I, and have been described before (10).

Media and growth conditions. For most experiments the NN minimal-medium was used (10). Glucose at 0.2% was always added as

TABLE I

List of Strains

Strain	Genotype	Source or Reference
CS520	HfrC *trpA58 metB1 glyV50 supF58 relA1*	Clarke and Carbon (9)
RRI	F- *proA2 leuB6 thi-1 lacY1 galK2 rpsL20 hsdS*	Bolivar *et al.* (13)
AB1206	F′ 14/*proA2 his-4 thi-1 rpsL8 lacY1 galK2 tfr-3 supE44* (*ilvE-argH*) ΔE2	A. J. Pittard via A. J. Clark
PA340	F *thr-1 leuB6 gdh-1 hisG1 gltB31 argH1 thi-1 ara-14 lacY1 gal-6 malA1 xyl-7 mtl-2 tonA2* λRλ− *rpsL9 supE44*	F. Jacob via B. J. Bachmann
MX705	F- *glnA20 proA2 leuB6 thi-1 lacY1 galK2 rpsLB0 rsdS*	F. Bastarrachea *et al.* Chapter 7
MX727	F- *thi ilv-680* (*pro-lac*) △ (*gas-pgl-att*λ*bio* △*glnA21::Tn5*	F. Bastarrachea *et al.* Chapter 7
MX729	Same as PA340 but carrying pRSP1	Covarrubias *et al.* (10)
MX730	Same as PA340 but carrying pRSP20	Covarrubias *et al.* (10)
MX731	Same as MX705 but carrying pACR1	Covarrubias *et al.* (10)
MX746	Same as MX727 but carrying pACR1	Covarrubias *et al.* (10)

carbon source. Nitrogen sources were either 0.5mM or 15 mM NH_4Cl, L-glutamate (2mg/ml) or L-glutamine (1mg/ml). Other growth factors were added at predetermined optimal concentrations from 1 x 10^{-4}M to 2 x 10^{-3}M. Luria broth modified by the omission of glucose, was used as the rich medium. For solid media, agar (Difco) at a final concentration of 1.5% was added. Strains were maintained as slants of Luria medium except MX705 which was mantained on Luria supplement with glutamine (1mg/ml) and merodiploid strains which were maintained on glucose-NN medium with 15 mM NH_4Cl.

Plasmid DNA purification and restriction endonuclease analysis. ColEl and ColEl hybrid plasmids were prepared by the method of Betlach *et al.* (11) after chloramphenicol (200 mg/l) amplification. All the restriction enzymes used in this work were prepared by the method of Greene *et al.* (12) with the exception of the enzyme *SmaI* which was obtained from Bethesda Research Labs. The reaction conditions for the various restriction enzymes have been previously reported (13). Agarose gel electrophoresis was carried out in a vertical slab gel apparatus using a Tris-Borate-EDTA System (13).

Enzyme assays. For enzyme assays, culture were grown on glucose-NN medium with the nitrogen sources added as indicated. Culture and harvesting conditions were those described by Bender *et al.* (14). The GS assay used for whole cells and extracts was the glutamyl transferase assay described by Shapiro and Stadtman (15), using the final concentration of reactants recommended by Bender et al. (16), except that the pH was 7.33, the isoactivity point for both the adenylylated and nonadenylylated forms of *E. coli* GS (17). In parallel assays, 60 mM $MgCl_2$ was added to measure the adenynylation state of GS subunits, since Mg^{2+} inhibits the activity of the adenylylated subunits only (18). To determine the average state of adenylylation ($\tilde{n}$), we used the formula suggested by Stadtman *et al.* (18).

For determination of GDH and GOGAT activities, 100 ml cells were grown to a density of 6 x 10^8 cells/ml on glucose NN medium with 30mM NH_4Cl as only nitrogen source, except those of strain PA340 which were grown on the same medium supplement with 2 mg/ml of glutamate.

Cells were centrifuged, resuspended in a 1/100 volume of buffer (50mM Tris, ph7.6 plus 10 mM mercaptoethanol and disrupted in a Branson sonifier. A clear supernatant was obtained after centrifugation for 30 min at 10,000xg. The GDH and GOGAT activities were determined spectrophotometrically by recording the initial rate of NADPH oxidation at 340nm as described by Meers *et al.* (19).

Enzyme purification. To purify GS the bacteria were grown in aereated cultures at 37°C in glucose NN medium containing either 0.2% glutamic acid plus 0.5mM NH_4Cl or glutamine (1 mg/ml) as nitrogen source. The cells were harvested in the late logarithmic growth phase, washed twice with extraction buffer (5mM KH_2PO_4 0.5 mM EDTA, 50 mM K_2HPO_4, pH 7.2), resuspended in the same buffer containing 2.25 mM $MnCl_2$ and disrupted by sonication. The suspension was centrifuged 15 min at 10,000 rpm and the supernatant subjected to a purification procedure based on affinity chromatography on anthranilate-bound Sepharose (20, See also Chapter 11). Glutamate synthase was purified from the same strain as reported to homogeneity, as described by Sakamoto *et al* (21), with some modifications. Instead of the Sepharose 6B chromatography step, we utilized a preparative flat bed electrophocusing apparatus with Sephdex G75 (superfine) as the support (LKB, Application note 198). Glutamate synthase was purified from the same strain as reported by Sakamoto *et al.* (21)

Immunoprecipitation of GS. Immunoprecipitation studies were carried out as previously described by Palacios *et al.* (22). The *E. coli* anti-GS gamma globulin fraction used was kindly donated by S. López and R. Palacios.

Purification and labelling of minicells. Plasmid containing minicells were grown and purified as described by Meagher *et al.* (23). Minicells were labeled with 50 μCi/mmol^{35}S-methionine in the presence of 50μg/ml of each of the unlabeled amino acids.

Other methods. Colicin El was prepared from strain JC411 ColEl as described by Schwartz and Helinski (24) up to the ammonium sulfate precipitation step. Transformants were tested for colicin immunity by cross-streaking on L-broth agar plates previously spread with the colicin El preparation described above. Protein was determined in whole cells or in extracts by the method of Lowry *et al.* (25) using bovine albumin as a standard. Polyacrylamide slab gel electrophoresis in the presence of sodium dodecyl sulfate, was performed according to Laemmli (26).

RESULTS

Isolation of ColE1 hybrid plasmids

One liter batches of M9 medium were inoculated with 100 different *E. coli* strains from the Clarke-Carbon bank, and plasmid DNA was isolated. We were able to obtain 22 DNA pools. With these DNA pools, two recipient strains were transformed: PA340, in order to isolate recombinant plasmids complementing either *gdh* or *gltB31* mutations, and MX705, for plasmids complementing the *glnA20* allele. We were able to isolate a recombinant plasmid, termed pACR1, from one Gln$^+$ MX705 transformed clone and two plasmids, termed pRSP1 and pRSP20, from Glu$^+$ PA340 transformed clones.

GDH and GOGAT activities in cells harboring plasmids pRSP1 or pRSP20

The activities of GDH and GOGAT were determined in cells extracts from strains MX729 (PA340/pRSP1) and MX730 (PA340/pRSP20). Results are shown in Table II. GDH activity was only

TABLE II

Glutamate Dehydrogenase and Glutamate Synthase Specific Activities Determined by Plasmids pRSP1 and pRSP20

Strain	Genotype	GDH	GOGAT
CS520	*gdh+ gltB+*	184.1	51.2
PA340	*gdh-1 gltB31*	2.1	0.6
MX729	PA340/pRSP1	448.6	0.5
MX730	PA340/pRSP20	2.3	128.6

One hundred ml cultures were grown to a density of 6 x 10^8 cells/ml on glucose NN medium with 30 mM NH_4Cl as the only nitrogen source, except those of strain PA340 which were grown on the same medium supplemented with 2 mg/ml glutamate.

detected in strain MX729. In this case, the GDH specific activity was approximately 2.4 fold higher than that observed for the wild type strain CS520. The GOGAT activity of MX729 cells was as low as to that obtained for PA340 cells carrying no plasmid. By contrast, MX730 cells synthesized GOGAT with an activity higher by a factor of 2.5 than the one synthesized by wild type CS520 cells. Consequently, plasmid pRSP1 complemented the *gdh* mutation, and plasmid pRSP20 complemented the *gltB31* mutation from PA340.

GS activity determined by pACR1

The total amount of GS ($-Mg^{+2}$) as well as the amount of nonade nylylated GS ($+Mg^{+2}$) was determined using the transferase assay in cells of strains MX731 (MX705/pACR1), MX746 (MX727/pACR1), MX705(*glnA20*), MX727 (*glnA21*::Tn5), RR1 (the *glnA*$^+$ strain from which MX705 was derived) and, CS520 (*glnA*$^+$) from which the colony bank of ColE1 hybrid plasmids was originally made (9). In addition, the GS biosynthetic activity was measured by the forward reaction assay (14, 16). These results are summarized in Table III.

Wild type *glnA*$^+$ strains RRI and CS520 synthesized derepressed levels of partially adenylylated GS when grown on medium with either 0.5 mM ammonium or glutamate as sole nitrogen sources. Under similar growth conditions, MX731 cells had a two-fold higher GS activity, half of which was also adenylylated. This high GS activity

TABLE III

GS Specific Activities and Adenylylation States of E. coli Strains

Strain (genotype)	Nitrogen Source	GS assay $-Mg^{2+}$	GS assay $+Mg^{2+}$	ñ	Forward reaction assay
CS520	0.5 mM NH_4^+	0.92	0.43	6	0.40
(*glnA*+)	15 mM NH_4^+	0.08	0.01	10	0.01
	Glutamate	0.66	0.34	6	0.35
	Glutamine	0.53	0.40	3	0.32
RR1	0.5 mM NH_4^+	1.03	0.52	6	0.52
(*glnA*+)	15 mM NH_4^+	0.09	0.02	9	0.02
	Glutamate	0.80	0.43	6	0.43
	Glutamine	0.64	0.57	1	0.37
MX705 (*glnA20*)	Glutamine	<0.01	<0.01	—	0.01
MX731	0.5 mM NH_4^+	1.95	1.07	5	0.93
(pACR1 *glnA*+/*glnA20*)	15 mM NH_4^+	0.08	0.01	10	0.01
	Glutamate	1.99	1.26	6	0.88
	Glutamine	0.16	0.06	7	0.06
MX727 (*glnA21*::Tn5)	Glutamine	<0.01	<0.01	—	0.01
MX746	0.5 mM NH_4^+	1.96	1.11	5	0.68
(pACR1 *glnA*+/*gln21*::*Tn5*)	15 mM NH_4^+	0.17	0.04	9	0.01
	Glutamate	1.59	0.81	6	0.45
	Glutamine	1.68	1.46	2	0.72

Inocula were grown overnight in minimal medium with the nitrogen source indicated in the table. Cultures were diluted 1:10 with the corresponding fresh medium and aereated until growth reached approximately 6×10^8 cells/ml.

Enzyme determinations were carried out as described in MATERIALS AND METHODS. Samples of cultures of the merodiploid strains grown on glutamine as sole nitrogen source were removed before harvesting and screened for the presence of Gln− segregants. These were found only at frequencies of 5% for MX731, and 0.5% for MX746.

could be the result of increased gene dosage, since ColE1 hybrid plasmids are known to exist in bacterial cells at 15-20 copies per chromosome (27).

A 15 mM ammonium concentration supported good growth of all Gln$^+$ strains that were tested (table III), eventhough it caused a strong repression of GS synthesis. The well known pattern of enteric bacterial cells synthesizing small amounts of highly adenylylated GS in response to their culturing on high ammonia concentration (5), was strictly followed by all strains that were tested.

Minicells carrying pACR1, pRSP1 and pRSP20

To demonstrate that these plasmids carry the GS, GDH and GOGAT structural genes, the synthesis of these proteins was studied in minicells. Minicells carrying the hybrid plasmids were purified and allowed to synthesize protein in the presence of ^{35}S-methionine.

Labeled proteins from minicells extracts were separated on sodium dodecyl sulfate-polyacrylamide slab gels and the results are shown in Figs. 1 and 2. In the case of minicells carrying pACR1, a plasmid coded protein (Fig. 1A, lane a) comigrates with purified GS monomer. Considering the results presented in the preceding sections, this radioactive polypeptide is very likely to be GS monomer. However, the possibility that the band could also be truncated colicin E1 (23), had to be discarded.

Therefore, radioactive minicells extracts were treated with rabbit anti GS gamma globulin as described (22). The GS specific antibodies precipitated only a 53,000 d. protein band, that again comigrates with purified GS monomer (Fig. 1B lanes h and i). These results clearly indicate the presence of *glnA* in pACR1.

In minicells extracts carrying pRSP1 plasmid, it can also be seen a ^{35}S labeled plasmid coded protein comigrating with purified GDH monomer (Fig. 2A lanes a and c respectively). To demostrate that this protein was effectively GDH monomer, a double dimension electrophoresis was performed using radioactive extracts from minicells carrying pRSP1. In this experiment it was possible to visualize a protein comigrating with purified GDH monomer (Sánchez-Pescador *et al.*, unpublished results).

Plasmid pRSP20 was able to complement the *gltB31* mutation. To demonstrate the presence of the structural gene(s) for the GOGAT enzyme it was necessary to use the minicells system. Fig. 2B lane b, shows the ^{35}S labeled proteins coded by plasmid pRSP20; the position of purified GOGAT subunits are shown by the arrows in lane c. It can be seen that plasmid coded proteins comigrate with purified GOGAT subunits.

Restriction endonuclease analysis of hybrid-plasmids

The three ColE1 hybrid plasmids pRSP1, pRSP20 and pACR1, were purified and treated with a collection of restriction enzymes in order to obtain preliminary restriction maps and molecular weights. The relative position of the endonuclease sites present in each DNA molecule, was determined by agarose gel electrophoresis of single and double digestion of plasmid DNA with different restriction enzyme combinations (10).

As can be seen Fig. 3, a 0.7 md restriction fragment can be derived from all three hybrid plasmids after digestion of DNA with the restriction endonuclease *Pst* I. The same fragment, which includes a *Sma*I site, can be obtained from plasmid ColE1. Therefore, we assume that this fragment is part of the inactivated colicin E1 structural gene in the ColE1-hybrid plasmids (28). These data allowed us to localize the position of the colicin E1 immunity gene and the origin of replication in the hybrid plasmids, assuming that no deletion or insertion of foreing DNA in the ColE1 segment occurred during the bank construction.

The hybrid plasmid, pACR1, has a molecular weight of 10.6 X 10^6 daltons and carried unique restrictions sites for the endonucleases *Bgl* II, and *Hind* III. It carried two substrates sites for the restriction enzymes *Eco* RI, *Sal* I and *Bam* HI, three for *Sma* I and four *Pst* I. There were four *Hpa* I restriction sites in the molecule.

The plasmid pRSP1 has a molecular weight of 8.0 X 10^6 daltons, one substrate site for each of the restriction enzymes *Sal* I, *Bgl* II, and *Sma* I, two for the endonuclease *Eco* RI, three for *Pst* I and three for the endonuclease *Hpa* I. No sites for the restrictions enzymes *Hind* III and *Bam* HI were detected in pRSP1.

The third plasmid, pRSP20 with a molecular weight of 12 x 10^6 daltons, has one substrate site for the restriction enzymes *Sal* I and *Bgl* II. It carried two substrate sites for the endonucleases *Eco* RI and *Sma* I, three for the endonuclease *Pst* I and four, possibly five, *Hind* III sites. Three of these *Hind* III sites have been mapped in the molecule; a fourth was localized in either one of the two *Hind* III positions shown as dashed lines in the map. A fifth site could be localized very close to one of these four sites. No *Bam* HI site was present in the hybrid molecule.

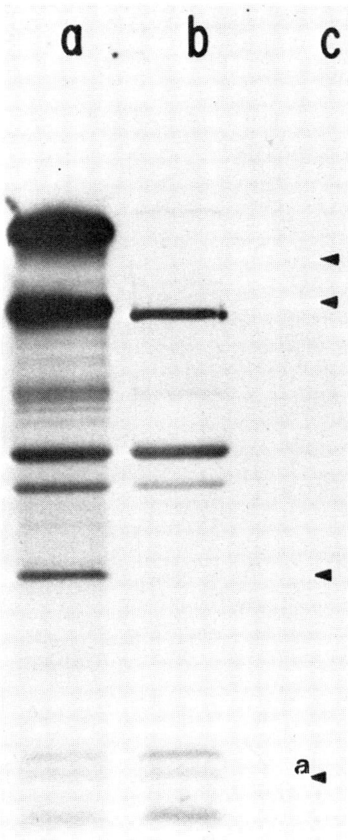

Fig. 1A. Autorradiography of a polyacrilamide SDS gel showing the [^{35}S] methionine labeled proteins produced by pACR1 (lane a) and ColE1 (lane b), in minicells.
Molecular weight markers (lane c) from the top to the bottom: 60,000 d, 50,000 (*GS, E. coli*), 25,000 and 17,800.

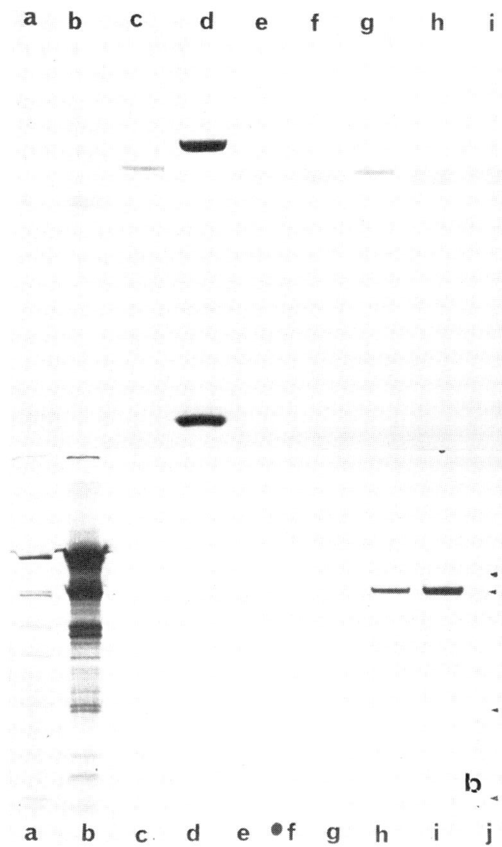

Fig. 1B. Immunoprecipitation of *de novo* synthesized GS. The immunoprecipitation was carried out as described in MATERIALS AND METHODS. The immunoprecipitates were processed for slab gel electrophoresis in the presence of sodium dodecyl sulfate. The top part of the figure shows the stained gel and the bottom the autorradiography of the [^{35}S] labeled proteins. Lanes a and b show proteins coded by ColEl and pACRI, respectively. Lanes c and g show purified (not labeled) GS monomer. Lanes d and j show the same molecular weight markers as in fig. 1A. Lanes e and f show the immunoprecipitation with anti *E. coli* GS of the ColEl minicells extract, negative control). Lanes h and i show the immunoprecipitation with anti *E. coli* GS of the pACRI minicells extract.

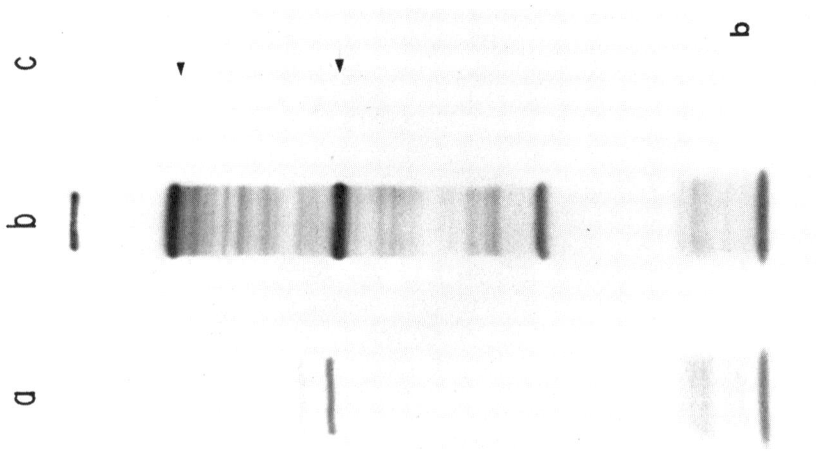

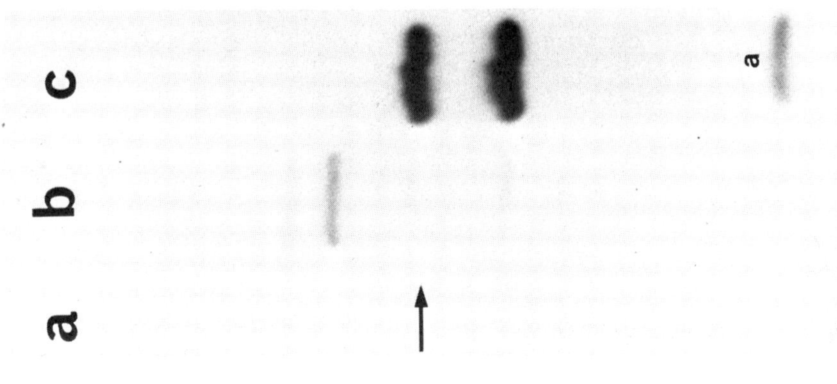

Fig. 2A Autorradiography of a polyacrylamide-SDS gel showing the [35S] methionine labeled proteins coded by ColE1 (lane b) and pRSP1 (lane c), in minicells. The arrow in lane a shows the position of purified GDH monomer. Fig. 2B *De novo* synthesized labeled proteins codified by ColE1 (lane a) and pRSP20 (lane b) in minicells. The position of purified GOGAT subunits are shown on lane c.

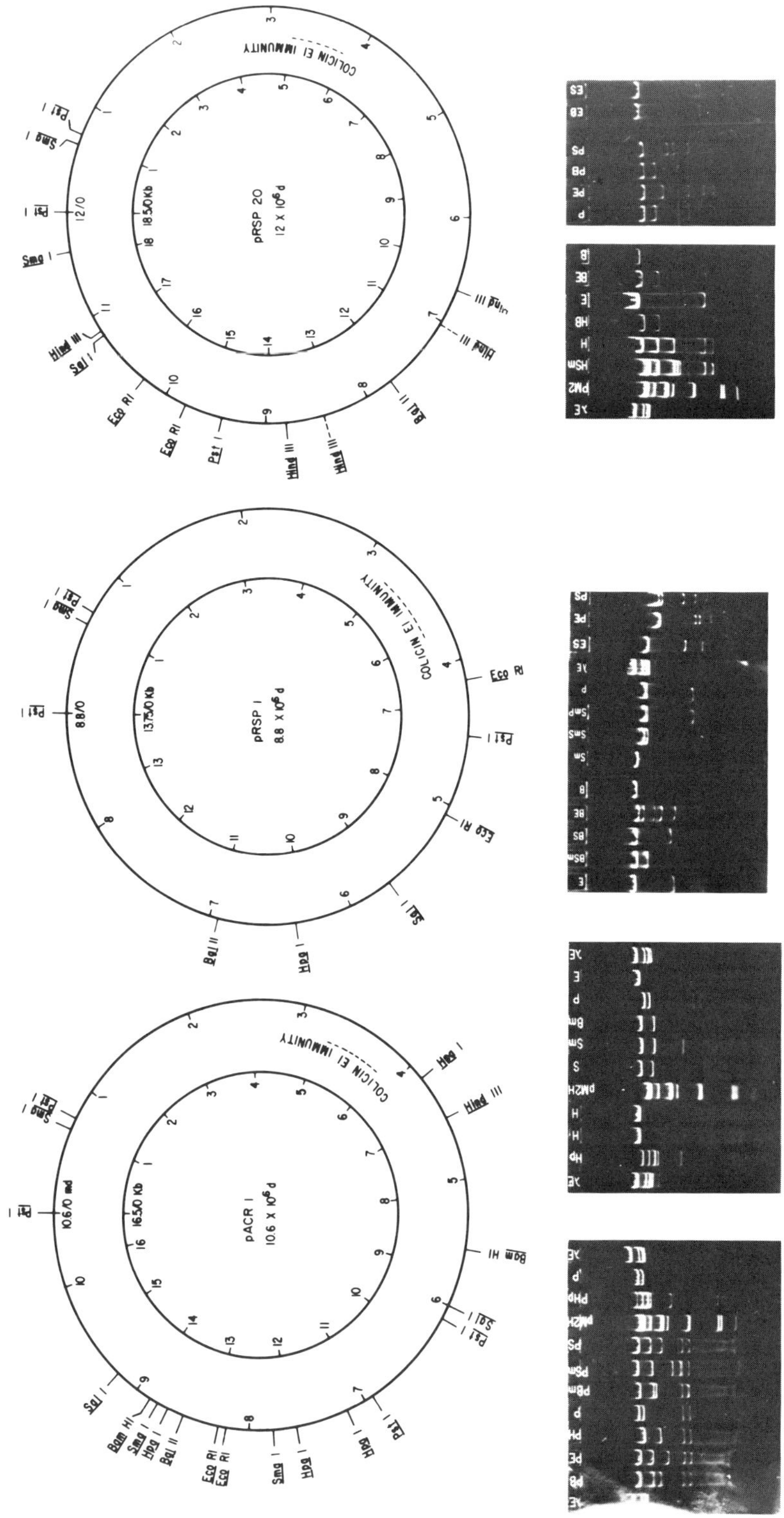
pRSP 20
12 X 10⁶ d
COLICIN E1 IMMUNITY
pRSP 1
8.8 X 10⁶ d
COLICIN E1 IMMUNITY
pACR 1
10.6 X 10⁶ d
COLICIN E1 IMMUNITY

Fig. 3. Circular restriction maps of pRSP1 (A), pRSP20 (B) and pACR1 (C). The relative positions of restriction sites are drawn to scale on circular maps divided into units of 1×10^6d (outer circle) and 1 kilobase (inner circle).

One percent agarose electrophoresis slab gels showing single and double restriction endonuclease digestions for each of the plasmid DNAs, are presented at the bottom of each map.

For the plasmid pRSP1: E: *EcoRI;* Bsm:Bgl*II* and *Sma*I; BS: *Bgl*II and *Sal*I; BE:*Bgl*II and *Eco*RI(partial); B:*Bgl*II; Sm: *Sma*I; *SmS:Sma*I and *Sal*I *SmP:Sma*I and *Pst*I; ES: *Eco RI* and *Sal*I; PE: *Pst*I and EcoRI; PS:*Pst*I and *Sal*I. Molecular weight estimates are based on the six λ DNA I agments generated by *Eco*RI (λ E) 13.7, 4.68,3.7, 3.56, 3.03 and 2.09 md., (13).

For the plasmid pRSP20: HSm: *Hind*III and *Sma*I; H: *Hind*III, HB: *Hind*III *and Bgl*II; E. *Eco*RI; BE: *Bgl*II and *Eco*RI; B: *Bgl*II, P:*Pst*I; PB:*Pst*I and *Bgl*II: PS:*Pst*I and *Sal*I: EB: *EcoRI* and *Bgl*II; *ES:* Eco*RI* and *Sal*I. Molecular weight markers as in the previous section. In addition, PM2 DNA digested with *Hind*III (PM2) was used. The molecular weights of the resulting fragments are: 3.14, 1.34, 0.6, 0.31, 0.29, 0.14 and 0.06 md; the last one not seen in the gel (13).

For the plasmid pACR1; PB: *Pst*I and *Bgl*II; PE:*Pst*I and *Eco*RI PH:*Pst*I and *Hind*III; P:*Pst*I; PBm:*Pst*L and *Bam*HI; PSm: *Pst*I and *Sma*I; PS:*Pst*I and *Sal*I; PHp:*Pst*I *and Hpa*I; H: *Hind*III; S:*Sal*I *Sm:Sma*I; Bm; *Bam*HI: P:*Pst*I; E:*Eco*RI Molecular weight markers are the same ones used for pRSP1 and pRSP20.

DISCUSSION

In this communication we reported the isolation and initial restriction enzyme characterization of three ColE1-hybrid plasmids that carry the GS, GDH and GOGAT structural genes.

The conclusive evidence that plasmid pACR1 carries the *glnA* gene was demonstrated via the specific immunoprecipitation with anti GS antibodies of a plasmid coded-protein of approximately 53,000d. Evidence presented here strongly suggests that pRSP1 carries the structural gene that codes for the GDH subunit and that plasmid pRSP20 carries the GOGAT structural genes. Double dimension electrophoresis, in which plasmid coded proteins comigrate with purified GDH and GOGAT monomers, has recently confirm this hypothesis (Sánchez-Pescador *et al.*, unpublished observations).

At the present time we are interested in isolating those minimal chromosomal DNA regions, from each of the three hybrid plasmids, that still code for these proteins. Accordingly, we have been able to subclone the *glnA* gene in different restriction enzyme fragments (Covarrubias *et al.*, unpublished results). The same is true for the *gdh* gene. The isolation of these fragments will allow us to establish a physical map of these genes, a step that we believe is essential for understanding the physical organization of the regulatory and structural regions.

Acknowledgements. We wish to thank Mario Rocha and Aurora Osorio for their technical help. This research was supported by Grants Nº 1649 and 1650 from Proyecto Nacional de Ciencias Básicas (PNCB) of the Consejo Nacional de Ciencia y Tecnología, México.

REFERENCES

1. Brenchley, J. E., and Magasanik, B. (1974) *J. Bacteriol. 117,* 544-550.
2. Berberich, M. A. (1972) *Biochem. Biophys. Res. Commun. 47,* 1498-1503.
3. Brenchley, J. E., Prival, M. J., Magasanik, B. (1973) *J. Biol. Chem. 248,* 6122-6128.
4. Pahel, G., Zelenetz, A. D., and Tyler, B. M. (1978) *J. Bacteriol. 133,* 139-148.
5. Tyler, B. (1978) *Ann. Rev. Biochem. 47,* 1127-1162.
6. Pahel, G., and Tyler, B. (1979) *Proc. Natl. Acad. Sci. U.S.A. 76,* 4544-4548.
7. Kustu, S., Burton, D., Garcia, E., McCarter, L., and McFarland, N. (1979) *Proc. Natl. Acad. Sci. U.S.A. 76,* 4576-4580.
8. Bachmann, B. J., Low, K. B., and Taylor, A L. (1976) *Bacteriol. Rev. 40,* 116-167.
9. Clarke, L., and Carbon, J. (1975) *Proc. Natl. Acad. Sci. U.S.A. 72,* 4361-4365.

10. Covarrubias, A. A., Sánchez-Pescador, R., Osorio, A., Bolívar, F., and Bastarrachea, F. (1980) *Plasmid 3,* 150-164.
11. Betlach, M., Hershfield, V., Chow, L., Brown, W., Goodman, H. M., and Boyer, H. W. (1976) *Fed. Proc. 35,* 2037-2043.
12. Greene, P. J., H. L., Bolívar, F., Rodríguez, R. L., Betlach, M. C., Covarrubias, A. A., Backman, K., Russel, D. J., Tait, R., and Boyer, H. W. (1978) *Nucleic Acids Res. 5,* 2373-2380.
13. Bolivar, F., Rodriguez, R. L., Betlach, M. C., and Boyer, H. W. (1977) *Gene 2,* 75-93.
14. Bender, R. A., Janssen, K. A., Resnick, A. D., Blumenberg, M., Foor, F., and Magasanik, B. (1977) *J. Bacteriol 129,* 1001-1009.
15. Shapiro, B. M., Stadtman, E. R. (1970). In: Methods in enzymology (H. Tabor and C. W. Tabor eds.) vol. 17A, pp. 910-922 Academic Press, New York.
16. Bender, R. A., and Magasanik, B. (1977) *J. Bacteriol. 132,* 100-105.
17. Bloom, F. R., Levin, M. S., Foor, F. and Tyler, B. (1978) *J. Bacteriol. 132,* 569-577.
18. Stadtman, E. R., Ginsburg, A., Ciardi, J. E., Yeh, J., Henning, S. B., and Shapiro, B. M. (1970) *Adv. Enz. Regul. 8,* 99-118.
19. Meers, J. L., Tempest, D. W., and Brown, C. M. (1970) *J. Gen. Microbiol. 64,* 187-192.
20. Palacios, R. (1976) *J. Biol. Chem. 251,* 4787-4791.
21. Sakamoto, N., Kotre, A. M., and Savagean, M. A. (1975) *J. Bacteriol. 124,* 775-783.
22. Palacios, R., Campomanes, M., and Quinto, C. (1977) *J. Biol. Chem. 252,* 3028-3034.
23. Meagher, R. B., Tait, R. C., Betlach, M., and Boyer, H. W. (1977) *Cell 10,* 521-536.
24. Schwartz, S. A., and Helinski, D. R. (1971) *J. Biol. Chem. 246,* 6318-6327.
25. Lowry, O. H., Rosebrough, N. J., Farr, A. L., and Randall, R. J. (1951) *J. Biol. Chem. 193,* 265-275.
26. Laemmli, U. K. (1970) *Nature (London) 227,* 680-685.
27. Hershfield, V., Boyer, H. W., Yanofsky, C., Lovett, M. A. and Helinski, D. R. (1974) *Proc. Natl. Acad. Sci. U.S.A. 76,* 4576-4580.
28. Bell, G. I., DeGennaro, L. J., Gelfand, D. H., Bishop, R. J., Valenzuela, P., and Rutter, W. J. (1977) *J. Biol. Chem. 252,* 8118-8125.

9. Role of Glutamine Synthetase in Nitrogen Fixation

ROBERT A. DARROW

Charles F. Kettering Research Laboratory
150 East South College Street
Yellow Springs, Ohio 45387
Contribution Number 682

SUMMARY

Glutamine synthetase, together with glutamate synthase, provides the main pathway for ammonia assimilation in free-living nitrogen-fixing bacteria. Catalytically active glutamine synthetase also appears to be needed for the repression of nitrogenase by ammonia in these bacteria. In addition, glutamine synthetase (or the product of a regulatory gene closely linked genetically to the structural gene for glutamine synthetase) has been implicated as a possible regulator of transcription of nitrogenase in *K. pneumoniae,* in parallel with studies on the nitrogen regulation of several enzymes of *K. aerogenes.* Rhizobia cannot grow on N_2, but fix nitrogen in symbiotic association with the roots of legumes, in nodules, which are differentiated organs containing both plant and bacterial tissue. Here also, glutamine synthetase, together with glutamate synthase, provides the pathway for initial ammonia assimilation, but in this case the enzymes appear to be of plant origin. Rhizobial nitrogenase is relatively insensitive to repression by ammonia in comparison to the nitrogenase of free-living bacteria. Some role for glutamine synthetase in the overall process of symbiotic nitrogen fixation by rhizobia is indicated by the observation that mutants of rhizobia, auxotrophic for glutamine and lacking glutamine synthetase, lose the capacity both for effective nodulation and for free-living nitrogen fixation. Some of the prototrophic revertants of these mutants regain these capacities. Rhizobia possess two distinct glutamine synthetase proteins,

ISBN: 0-12-506040-8

having different physical and catalytic properties, and also differing in their response to ammonia in the medium. The existence of these two proteins must be taken into account in evaluating a role for glutamine synthetase in symbiotic nitrogen fixation.

INTRODUCTION

Biologists with a primary interest in biological nitrogen fixation have, from time to time, focused their attention on glutamine synthetase (Reaction 2), both as a catalytic entity for the assimilation of the ammonia formed during nitrogen fixation, and as a potential regulatory protein, needed for the regulation of the expression of the genes involved in the functioning of the nitrogenase reaction (Reaction 1).

In the following discussion I shall include both aspects of the role of glutamine synthetase in nitrogen fixation, with emphasis on the two most intensively studied systems 1) bacteria that fix nitrogen in the free-living state, especially *Klebsiella pneumoniae,* and 2) bacteria of the genus *Rhizobium,* that only fix nitrogen in nature in association with leguminous plants. My brief summary of the work on free-living nitrogen-fixing bacteria will be that of an interested outsider. Most weight will be given to the rhizobia, the system where my own experience lies. More exhaustive reviews of various aspects of this subject have been published recently (44, 57, 72).

REACTIONS OF N_2 ASSIMILATION

1) *Nitrogenase*
$$1/2N_2 + 6ATP + 3e^- = NH_3 + 6ADP + 6P_i$$

2) *Glutamine Synthetase* (GS)
$$ATP + \textit{Glutamate} + NH_3 = ADP + P_i + \textit{Glutamine}$$

3) *Glutamate Synthase* (GOGAT)
$$\begin{array}{l} \textit{Glutamine} + \alpha\textit{-Ketoglutarate} = \textit{2Glutamate} + NAD^+ \ (NADP^+) \\ \quad + NADH\ (NADPH) \end{array}$$

$$\begin{array}{l} \text{SUM: } 1/2N_2 + 7ATP + 3e^- + NADH\ (NADPH) + \alpha\textit{-ketoglutarate} \\ \quad = 7ADP + P_i + NAD^+\ (NADP^+) + \textit{Glutamate} \end{array}$$

FREE – LIVING NITROGEN FIXING BACTERIA
Catalytic Aspects

The discovery by Tempest *et al.* (67) of the glutamate synthase reaction (Reaction 3, EC 2.6.1.53) revealed a means by which the high affinity of the glutamine synthetase reaction for ammonia could be put to use, *via* the coupled formation of glutamate, in the general assimilation of ammonia by the cell. Previous attention had focused on the specific role of the amide group of glutamine as the nitrogen donor for the biosynthesis of particular cellular metabolites. A survey of the various classes of nitrogen-fixing bacteria by Nagatani *et al.* (45) showed them all to posses the glutamate synthase reaction, and thus to be capable of assimilating low levels of ammonia through glutamine synthetase. Although exceptions may exist, it seems clear that this is the major pathway operating during growth on N_2. The level of activity of these enzymes that catalyse ammonia assimilation (as well as the availability of the carbon skeleton acceptors of ammonia) may, under some conditions, limit the rate of fixation of nitrogen. For example, in *Clostridium pasteurianum* the extractable nitrogenase is present at two-fold excess beyond what might be expected from the rate of growth on N_2 (12).

A convincing demonstration of the inhibition of the catalytic activity of glutamine synthetase in whole cells by the glutamate analogues, methionine sulfoximine and methionine sulfone, was provided by Brenchley (9), working with mutants of *Klebsiella aerogenes.* Gordon and Brill (26) took advantage of this technique to show that catalytically active glutamine synthetase is necessary for the repression of nitrogenase by ammonia. They found that the presence of either of the two inhibitors caused both *K. pneumoniae* and *Azotobacter vinelandii* to continue to produce nitrogenase, even in the presence of excess ammonia. Their conclusion was that glutamine itself, or other metabolic products of glutamine, in addition to ammonia (or perhaps rather than ammonia), functions as the effector for nitrogenase repression. Consistent with this idea is a report by Bender and Magasanik (3) that, in *K. aerogenes* growing with a poor carbon source, glutamine was a more potent repressor than ammonia, both of glutamine synthetase itself and of other enzymes that appear to be regulated by glutamine synthetase in the same way as nitrogenase (see below). Along the same lines, Shanmugan and Morandi (58) showed that certain amino acids, especially glutamine in combination with aspartate or glutamate, were able to repress nitrogenase synthesis completely in various mutant strains of *K. pneumoniae* which resist repression by ammonia. With a high concentration of glutamine (1 mg/ml) in the medium, sensitivity to ammonia was restored in that ammonia could partially substitute for aspartate to bring about the repression of nitrogenase. The conclusion drawn from these studies was that the repression of nitrogenase by ammonia occurs in an undefined way *via* some amino acid product, and that the negative regulatory function of glutamine synthetase is simply to provide that product.

Positive Regulation

Paradoxically, there is a considerable body of evidence in *K. pneumoniae* involving glutamine synthetase in the derepression of nitrogenase, as well as in its repression by amonia. This evidence was originally sought because of the similarity of the function of nitrogenase to the function of a number of other ammonia or glutamate-providing enzymes in *K. aerogenes.* The regulation of these other enzymes has been linked by a series of studies in Magasanik's laboratory to the glutamine synthetase protein (40). The diagnostic test for positive regulation by glutamine synthetase of these enzymes required that their activity parallel that of glutamine synthetase in the following way: 1) It should be high in wild-type cells under nitrogen limitation, where glutamine synthetase is high; 2) It should be low, even under nitrogen limitation, in mutants having a defect in the structural gene for glutamine synthetase (*gln* A) and thus lacking the glutamine synthetase protein; 3) It should be high in mutants that produce high levels of glutamine synthetase constitutively, even when grown with excess ammonia and a good carbon source.

Two sorts of evidence have been important to the conclusion that the glutamine synthetase protein itself is functional in regulation, rather than some product of the glutamine synthetase reaction: 1) One strain of *K. aerogenes* having a mutation in the structural gene for glutamine synthetase (*gln* A51) synthesizes large amounts of catalytically inactive but antigenically active glutamine synthetase, constitutively (64). The altered protein appears to retain its regulatory properties, in spite of its inability to form glutamine (4, 64). 2) Purified glutamine synthetase was found to stimulate the transcription of the *hut* operon in an *in vitro* system where the catalytic activity of the enzyme should have been irrelevant (71).

Analogous applications of the diagnostic test for regulation by glutamine synthetase in *K. aerogenes* were made for the nitrogenase of *K. pneumoniae* by Streicher *et al* (65), who introduced three independently-isolated mutations causing constitutive synthesis of glutamine synthetase from *K. aerogenes* into *K. pneumoniae* by transduction. The resulting strains also had constitutive levels of glutamine synthetase, and two of the three continued to synthesize nitrogenase to some extent in the presence of concentrations of ammonia that completely repress nitrogenase in the wild-type. In other experiments, the same authors showed that glutamine auxotrophs of *K. pneumoniae* which did not produce glutamine synthetase activity were also unable to produce nitrogenase. Complementation of the glutamine synthetase genome of these strains by an *E. coli* plasmid restored both glutamine synthetase activity and the ability to synthesize nitrogenase. Tubb (70) also showed that constitutively-produced glutamine synthetase allows nitrogenase to be synthesized in the presence of ammonia. In this case the nitrogen fixation genes of *K. pneumoniae* were introduced into a strain of *K. aerogenes* which was already constitutive for glutamine synthetase.

Shanmugam *et al* (56) also obtained mutants of *K. pneumoniae* that produced nitrogenase in the presence of ammonia, by nitrosoguanidine mutagenesis of glutamate auxotrophs. Some of the mutants gave results consistent with those of the other workers, in that nitrogenase derepression was associated with constitutive formation of glutamine synthetase, which was sometimes catalytically active and sometimes only detectable immunologically. However, two nitrogenase-derepressed mutants, with mutations genetically associated with the structural gene for glutamine synthetase, had neither catalytically active glutamine synthetase enzyme nor immunologically cross-reactive glutamine synthetase protein. This finding led the authors to speculate that there might be an additional factor, previously unrecognized, which might itself be the primary regulator of nitrogenase, or which might substitute for glutamine synthetase in its absence. It is also possible, of course, that the mutants had lost the relevant antigenic determinants in the glutamine synthetase protein while retaining a regulatory capacity, as had been suggested for a similar mutant of *K. aerogenes* (17).

Very recently, reports have appeared of new types of regulatory mutants involved with nitrogen control in *K. aerogenes* (24), *E. coli* (48), and *Salmonella typhimurium* (33, 34). In each case the mutations involved are tightly-linked to the structural gene for glutamine synthetase (*gln* A), but, to varying degrees, independent of it. Detailed discussions of these findings, and of other aspects of nitrogen control in enteric bacteria will be found in chapters 3-8. The parallelism between the regulation of nitrogenase in *K. pneumoniae* and nitrogen control in the other enteric bacteria leads one to expect that the resolution of the unanswered questions concerning the role of glutamine synthetase in the regulation of nitrogen fixation will come, in part, from these rapidly developing studies in other systems. The relative complexity of the process of nitrogen fixation, with as many as 14 genes necessary for its expression (18, 39), may be an intrinsic disadvantage in the study of its regulation. Further, nitrogenase activity in *K. pneumoniae* is not only nitrogen-regulated, but also requires limitation of oxygen, raising the possibility of additional regulatory mechanisms which are still unknown (63). In recent studies by Ausubel *et al* (1), however, a new class of regulatory mutants of *K. pneumoniae* are described, all very closely linked to the structural gene for glutamine synthetase, which are abnormal with respect to their regulation of nitrogenase (and of histidase), but, unlike similar mutants reported previously, retain glutamine synthetase biosynthetic activity. Thus, the physiological data in this case also are consistent with the mutations being on a regulatory gene, separate from the structural gene for glutamine synthetase, although the mapping data did not permit this conclusion.

Adenylylation

I have not mentioned the significance of the state of adenylylation of glutamine synthetase (see the contribution of Stadtman to this volume, chapter 2) in the foregoing discussion. Studies in *K. aerogenes* have implicated both the intracellular levels and the adenylylation state of glutamine synthetase in its regulatory activity (21, 71). In a recent study from Ausubel's laboratory on regulatory glutamine synthetase mutations affecting nitrogenase expression in *K. pneumoniae* (1), there was no correlation between the measured degree of adenylylation of glutamine synthetase and the extent of nitrogenase derepression. Goldberg and Hanau (25) came to a similar conclusion concerning the derepression of other nitrogen-assimilating enzymes in *K. pneumoniae*. It seems wisest, at the moment, to reserve judgment on the importance of the adenylylation state of glutamine synthetase to its possible regulatory function in *K. pneumoniae*, except, of course with respect to the inactivation of the catalytic function of the enzyme by adenylylation (61).

Binding Studies

A different line of evidence linking glutamine synthetase to the regulation of transcription concerns its physical interaction with

DNA, first observed by Streicher and Tyler (66) in crude extracts of the enteric bacteria. Janssen *et al* (28) have used the protection of DNA, by DNA-binding proteins, from cleavage by specific restriction endonucleases (29) as a means of further studying the interaction of glutamine synthetase and RNA polymerase with the specific DNA of the *nif* (nitrogen fixation) genome. They used DNA from amplifiable plasmids, one containing genes from the *nif* region of *K. pneumoniae* as well as the nearby *his* region, and the other containing the *his* sequence only. Glutamine synthetase was found to protect most of the same sites on the DNA as RNA polymerase, as might be expected if glutamine synthetase affects transcription by binding at or near the binding sites for RNA polymerase. A problem with this sort of experiment, however, is that the binding is sensitive to salt concentration. RNA polymerase binds to and therefore protects more sites on the plasmid DNA as the salt concentration is decreased (200 mM NaCl to 100 mM NaCl), making the identification of binding sites with promoter sites questionable. Nevertheless, the binding of glutamine synthetase in these experiments was specific for *nif* genes, since there was no protection by glutamine synthetase of sites on the DNA containing only the *his* sequence, even at low salt concentrations.

RHIZOBIA

Ammonia Assimilation

In contrast to *K. pneumoniae* and other bacteria that fix N_2 while in the free-living state and can subsist on N_2, members of the genus *Rhizobium* fix N_2 in association with the roots of legumes, as part of a highly differentiated organ (the nodule) whose exclusive function is to supply the plant with usable nitrogen (74). Rhizobia cannot grow on N_2, although they can produce nitrogenase in the free-living state, with the proper manipulation of the environment (30, 32, 42, 47, 52, 60, 68). The bacteroid, which is the differentiated, nitrogen-fixing form of the bacterium within the nodule, exports the fixed N_2 as NH_3 into the plant tissue, where it is converted to an array of nitrogen-containing substances for translocation within the plant (49). Both isolated bacteroids suspensions (5) and free-living N_2-fixing rhizobia (46) convert $^{15}N_2$ into $^{15}NH_3$, which is largely excreted into the culture medium. Rhizobia, in general, differ from free-living bacteria in that their nitrogenase is relatively insensitive to repression by ammonia (57).

Strong evidence that the assimilation of the fixed nitrogen in lupin nodules occurs in the plant tissue was furnished by Robertson *et al.* (54), who monitored the activities of both glutamine synthetase and glutamate synthase during nodule development. Fig. 1 shows the parallel increase in those two activities in the plant protein fraction between day 10 and day 20 after inoculation of the plant roots with bacteria. The same authors previously had shown (53) that the increase in glutamine synthetase between day 10 and day 20 was closely matched by a similar increase in both nitrogenase and leghemoglobin,

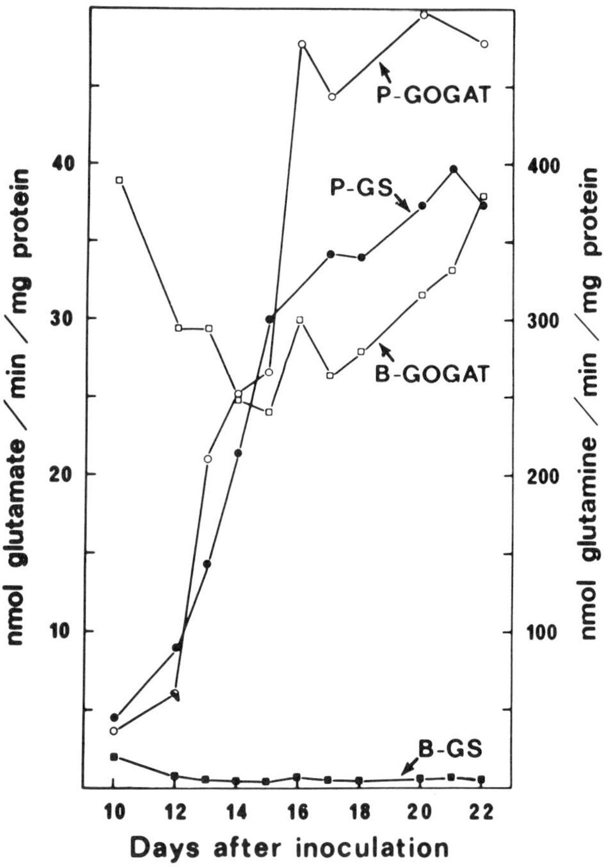

Fig. 1. Induction of glutamine synthetase and glutamate synthase during nodule development in lupin. (■) glutamine synthetase activity in the bacteroid soluble protein fraction (B-GS). (□) glutamate synthase activity in the bacteroid soluble protein fraction (B-GOGAT). (●) glutamine synthetase activity in the plant soluble protein fraction (P-GS). (○) glutamate synthase activity in the plant soluble protein fraction (P-GOGAT). See original publication for experimental details. From Robertson, Warburton and Farnden (54).

an oxygen-binding hemeprotein always present in legume nodules that are actively fixing N_2. It should be noted (Fig.1) that glutamine synthetase in the bacteroids remained low throughout the period of measurement (although, unaccountably, glutamate synthase was high). Several other groups of investigators have also found that up to 90% and more of the total glutamine synthetase is in the plant fraction, and not in the bacteroid (10, 19, 43). McParland *et al* (43) have purified glutamine synthetase from the plant cytosol of soybean nodules, and found that it had physical properties like glutamine synthetase isolated from other eukaryotic tissue. All these lines of evidence indicate that N_2 fixed in the bacteroid is first converted to organic nitrogen in the plant cytosol by a glutamine synthetase of plant origin.

A Role for Rhizobial Glutamine Synthetase

Even though the glutamine synthetase of bacteroids or N_2-fixing rhizobia does not appear to function in ammonia assimilation during fixation, other roles are certainly not excluded. Prompted by the possibility of a regulatory fuction for rhizobial glutamine synthetase analogous to that postulated in *K. pneumoniae,* a number of groups of workers have addressed this problem. Some have concluded that such a function was unlikely (8, 46, 55), and others have found positive evidence for such a role (6, 7). The interpretation of these results is rendered difficult, however, by the potential presence of two glutamine synthetases in rhizobia ((13) and see below). The best evidence for a role, however undefined, for rhizobial glutamine synthetase in nitrogen fixation comes from Ludwig and Signer (37) and Kondorosi, *et al.* (31), both of whom isolated mutants of *Rhizobium* auxotrophic for glutamine (and lacking glutamine synthetase) which formed ineffective nodules on their respective host plants. Ludwig and Signer (37), working with the cowpea organism, *Rhizobium* sp 32H1, found that their glutamine auxotroph also failed to produce nitrogenase in liquid culture, compared to the wild-type parent strain (Fig. 2). Revertants were isolated from the auxotroph; two out of six of these revertants tested for nodulation gave effective nodules. These findings call for further work with similar mutants, with special attention paid to the GSI/GSII phenomena described below, since glutamine auxotrophs of *Rhizobium* could be double mutants, with the possibility that prototrophic revertants might have only one or the other enzyme.

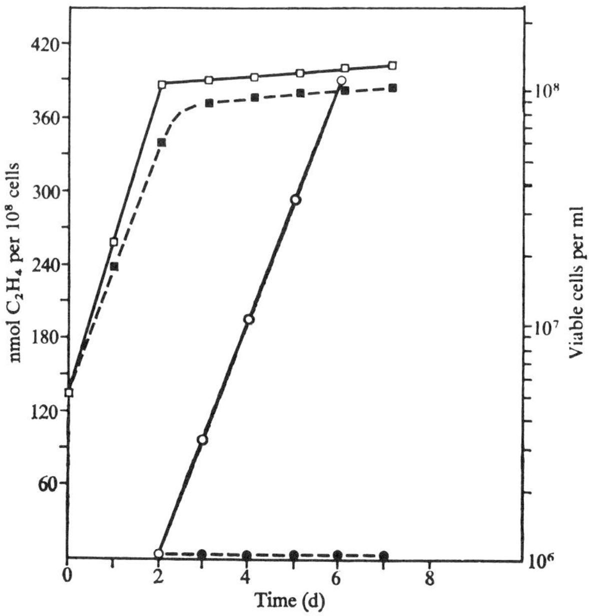

Fig. 2. Nitrogenase induction in liquid culture of *Rhizobium sp.* 32H1 compared with 32H1 *gln* 5 (glutamine auxotroph). Nitrogenase was assayed by acetylene reduction in iiquid culture, and aliquots were withdrawn and plated for cell counts: (□), 32H1+; (■), 32H1 *gln* 5; or assayed for ethylene production: (○), 32H1+; (●), 32H1 *gln* 5. See original publication for experimental details. From Ludwig and Signer (37).

Two Forms of Glutamine Synthetase: Early Observations.

Our experiments on glutamine synthetase from *Rhizobium* began with the notion that, given the role of the nodule as a whole in ammonia assimilation, together with its biological complexity (part-plant, part-bacterium), there might be unique forms of glutamine synthetase, specifically adapted for functioning in the nodule. Early experiments using column isoelectric focusing of crude extracts of various tissues, including whole nodules, gave patterns containing such a profusion of transferase-active bands as to be uninterpretable (16). (The γ-glutamyl transferase activity of glutamine synthetase (59) provides a sensitive assay method for the enzyme, especially useful in crude extracts.) Using the same procedure on crude extracts of free-living *R. japonicum* 61A76, we obtained two major peaks of transferase activity with distinctly different properties, as shown in Fig. 3. Both peaks had a requirement for ADP in the transferase reaction, and, in the experiment shown, both were inhibited by $MgCl_2$. Material from each peak re-focused at the same position in the pH gradient, showing that there was no interconversion between the two forms. One of the peaks, which we refer to as GSI, has properties analogous to those of the glutamine synthetase of *E. coli* (61) and other enteric bacteria. The other peak, GSII,

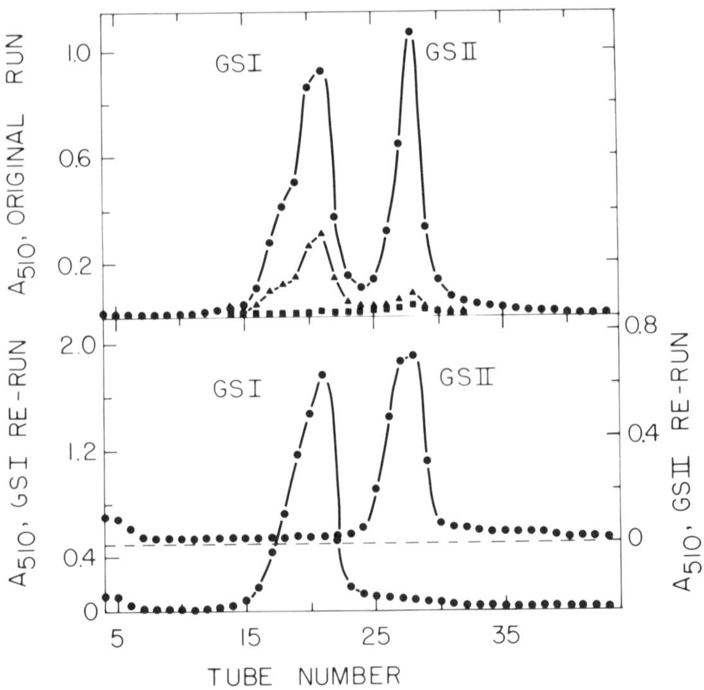

Fig. 3. Resolution of GSI and GSII by column isoelectric focusing, and refocusing of the separated peaks. A crude extract of *R. japonicum* 61A76 (5 ml), grown to stationary phase on 5mM NH_4Cl, prepared essentially as described by Tronick *et al.* (69) but without streptomycin treatment and containing 44 units/ml of transferase and 10 mg/ml of protein, was focused in a 100 ml colum (LKB) for 22 h. in a sucrose gradient between pH4 and pH7. Samples (10μl) were analyzed for γ-glutamyl transferase activity in the complete assay (59) (●) (30 min), in the absence of ADP (■) (120 min), and with 80 mM $MgCl_2$ added to the complete assay system (△) (120 min). Fractions 17-22 (GSI) and 26-29 (GSII) were combined and processed on a mixed-bed ion-exchanger column as described by Bauman and Chrambach (2) for the separation of carrier ampholytes from protein. Active fractions were combined and refocused individually. Similar results were obtained when treatment on the mixed-bed ion-exchanger column was omitted and the peak fractions were refocused directly. See original publication for further experimental details. From Darrow and Knotts (13).

is different in a number of respects. One of the most useful differences between GSI and GSII is the relative heat-lability of GSII, illustrated in Fig. 4. When employed with care and at appropriate

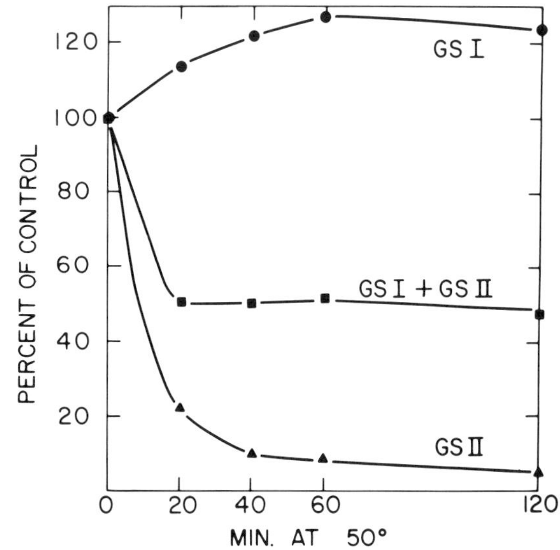

Fig. 4. Relative heat stability of GSI. and GSII. GSI and GSII were obtained from an extract of glutamate-grown *R. japonicum* 61A76 SM-5 (ineffective mutant (41)) by isoelectric focusing, by a procedure similar to that shown in Fig. 3. Samples of the two peaks were diluted 10-fold in 10 mM Imidazole-HCl, 1 mM $MnCl_2$, pH 7.0, and heated at 50° for the stated times, individually, and in a 1 to 1 (v/v) mixture. They were rapidly cooled in ice, and 100μl of each was assayed by observing NADH oxidation in the coupled assay system described by Shapiro and Stadtman (59). See original publication for experimental details. From Darrow and Knotts (13).

protein concentrations, transferase activity before and after heating at around 50° can be used as a measure of the relative concentrations of GSI and GSII in crude extracts, in place of more laborious procedures such as isoelectric focusing. The assay employed in Fig. 4, however measures the formation of ADP in the biosynthetic reaction, showing that both GSI and GSII catalyse the formation of glutamine. The enzyme samples tested in the experiment shown in Fig. 4 were obtained from a crude extract of *R. japonicum* 61A76 SM-5, a mutant isolated by Maier and Brill (41) which is unable to produce effective nodules on soybeans. We tested this mutant and other bacteria for the presence of both GSI and GSII upon isoelectric focusing of crude extracts, in order to rule out the possibility of a contaminant in our culture being responsible for GSII. We found both GSI and GSII present in five strains of *R. japonicum*, as well as the cowpea organism, *Rhizobium sp* 32H1 (a result confirmed by Ludwig (38)). No GSII was found in *A. vinelandii* OP, *K. pneumoniae* M5A1 or *E. coli* W, when grown and extracted in the same way as the rhizobia. Recently, Fuchs and Keister (23) have extended the survey, using both differential heat stability and sedimentation behavior (see Fig. 6) as diagnostic tests. They have found that GSII (together with GSI) is present in all members of the *Rhizobiaceae* (including the crown gall bacterium, *Agrobacterium tumefaciens*), and absent from all other bacteria tested, including a number of other plant pathogens.

The relative concentration of GSI and GSII was linked physiologically to free-living N_2 fixation, in experiments using *R. japonicum* 61A76, reported by Rao *et al.* (51). The essential finding was that GSII activity disappears and GSI becomes adenylylated (and thereby inactivated) at approximately the same partial pressure of O_2 which leads to the appearance of nitrogenase activity. These experiments are being refined and extended by Fuchs and Keister, using *R. japonicum* 311b 83. One of their experiments is shown in Fig. 5 where it can be seen that the transition is exceedingly sharp, suggesting the involvement of an all-or-none switching mechanism. Although these experiments do not in themselves establish a causal relationship between the changes in GSI and GSII and the appearance of nitrogenase activity, they do suggest the potential for some sort of interdependence.

Physical Properties

GSI and GSII can be separated in crude extracts by sedimentation in a sucrose density gradient between 5 and 20% sucrose (Fig. 6), affording a means for analysing the GSI/GSII ratio of crude extracts which is somewhat more rapid and convenient than isoelectric focusing. The sucrose gradient technique has also found use as part of a method for the purification of GSI from *R. japonicum* 311b 83 (22). In Fig. 6, β-galactosidase from *E. coli* was included as a marker protein in order to obtain a numerical estimate of the sedimentation coefficients of GSI and GSII. The value for GSI is 20.1S, in good agreement with the value of 20.3S reported for glutamine synthetase from *E. coli* W (61). The value for GSII is 11.4S, suggesting a much lower molecular weight for this protein than for

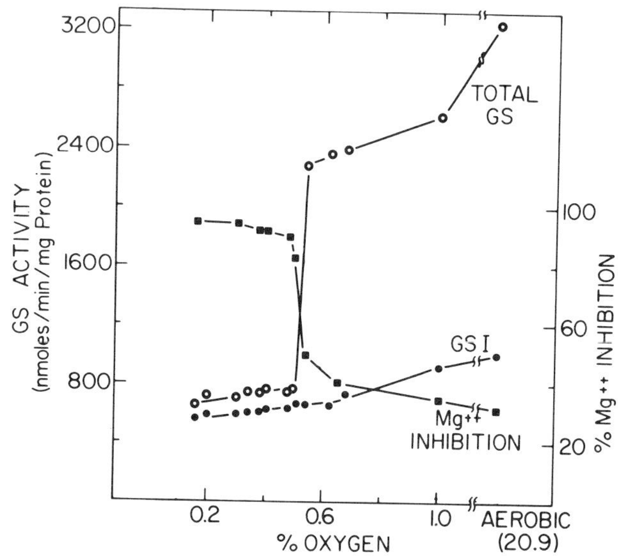

Fig. 5. Effect of oxygen tension on GSI and GSII in *R. japonicum* 3Ilb 83. Shake cultures of *R. japonicum* were grown with continuous gassing at the constant 0_2 levels indicated. Only minor changes in GSI and GSII activities occurred in cells grown at 0_2 concentrations from aerobic to 0.54%. The decrease from 0.54 to 0.5% caused a disappearance of GSII (heat labile) activity and a marked increase in the adenylylation of GSI as measured by the increase in Mg++ inhibition. Unpublished observations of Fuchs, R. L. and Keister, D. L.; experimental details similar to Rao *et al.* (51).

GSI. Other experiments with purified GSII, especially electron microscopy, are in conflict with this conclusion, however, and therefore the molecular weight of native GSII remains unclear.

We have developed a purification scheme, independent of isoelectric focusing, for obtaining both GSI and GSII from the same extract of *R. japonicum* 61A76, grown with glutamate as a nitrogen source. Under our conditions of growth, GSI is unadenylylated, and GSII constitutes up to 80% of the total transferase activity of the crude extract. The details of the procedure will be published elsewhere, but the essential features are as follows: 1) The supernatant fraction of a French pressure-cell extract of washed cells is diluted to 2 mg protein per ml and treated with 0.5% streptomycin sulfate. Under these conditions, all the GSI is precipitated and all the GSII remains in the supernatant solution. (The precipitation of GSI by streptomycin sulfate is DNA-dependent, because it can be abolished by treatment

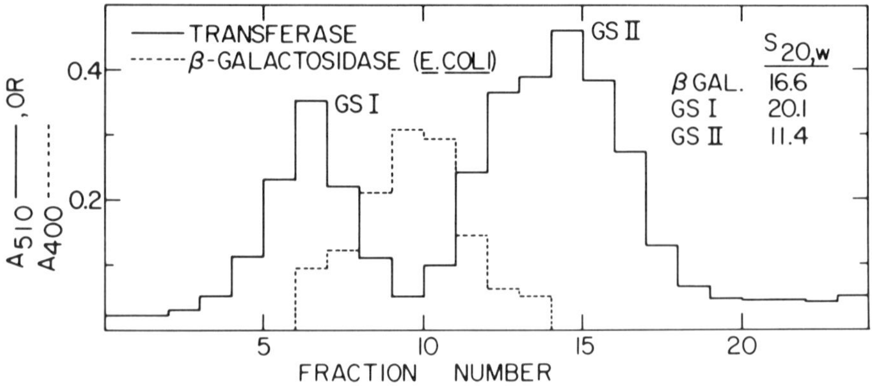

Fig. 6. Separation of GSI and GSII by sedimentation on a sucrose density gradient. 12.7 units (transferase (59)) of a crude extract of glutamate-grown *R. japonicum* 61A76 prepared as in Fig. 3 were sedimented for 3 h. at 23° at 40,000 rpm in the spinco SW 50.1 rotor, using a gradient of sucrose between 5 and 20%. Crystalline β-galactosidase from *E. coli* (2.5 units (75)) was included as a sedimentation marker. At the end of the run the contents of the tube were divided into 24 equal fractions and the fractions assayed for γ-glutamyl transferase (59) and β-galactosidase (75) activity. Values of S_{20}, are assumed to be proportional to distance of sedimentation. GSI and GSII were identified by heating at 50° (not shown).

of the extract with pancreatic DNAase.) 2) The supernatant, containing the GSII, can be purified to near-homogeneity by chromatography on Cibacron Blue-agarose, with elution by 10mM ATP. 3) The precipitate, containing the GSI, can also be purified by chromatography on Cibacron Blue-agarose, after first solubilizing the enzyme by treatment with NaCl at pH 7.0. Using this procedure, material of sufficiently high specific activity, and sufficient homogeneity on SDS gel electrophoresis was obtained to warrant the determination of the subunit molecular weight electrophoretically (76). The results obtained are shown in Fig. 7, which also gives the calibration curve from which the numbers were derived. For an idea of the relative banding patterns of GSI and GSII, refer to lanes A and B of Fig. 8, which is a slab gel, run with the same buffer system as the disc gels used in Fig. 7, but with 15% instead of 10% acrylamide. It is clear that the subunit of GSII (36,000) is significantly smaller than that of GSI (59,000), and also smaller than the value usually obtained for glutamine synthetases isolated from other prokaryotes.

The smaller size of the GSII peptide suggests that it could be derived from GSI by proteolytic cleavage, resulting in an enzyme that retains some catalytic activity but is structurally destabilized, perhaps as a first step in the intracellular degradation of GSI. In

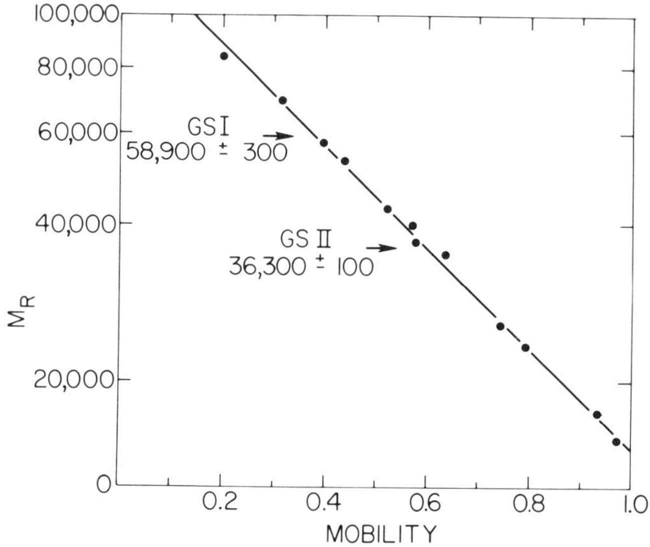

Fig. 7. Estimation of the subunit molecular weight of GSI and GSII by SDS-gel electrophoresis. Sample preparation, electrophoretic procedures (10% acrylamide), staining and de-staining were done according to Laemmli (35), except that the stacking gel was omitted. GSI and GSII were prepared from glutamate-grown *R. japonicum* 61A76 according to an unpublished procedure of Darrow and Knotts, as outlined in the text. Average mobilities were used to calculate molecular weight according to Weber and Osborn (76).

none of our experiments have we seen any evidence for such a conversion of GSI to GSII, either in aged crude extracts or in purified preparations. We have added the proteolytic inhibitor, phenylmethylsulfonyl fluoride (20), to cell suspensions prior to disruption of the cells, with no effect on the ratio of GSI to GSII. Finally, we have compared the peptides obtained by treatment of purified GSI and GSII with various endopeptidases in the presence of sodium dodecylsulfate, according to the method of Cleveland *et al.* (11).

Fig. 8 shows that digestion by chymotrypsin produced essentially independent banding patterns, with 8 major bands from GSI and 6 from GSII, and one peptide with the same mobility from both preparations. Chymotrypsin by itself produced 2 bands, at the concentrations used in the experiment. GSII was more resistant to digestion by chymotrypsin than GSI, with some material of the original mobility remaining after 60 minutes of digestion, both during incubation alone (lane D) and in the presence of GSI (lane E). In other experiments with GSI (not shown here), neither decreasing the time of incubation nor lowering the concentration of

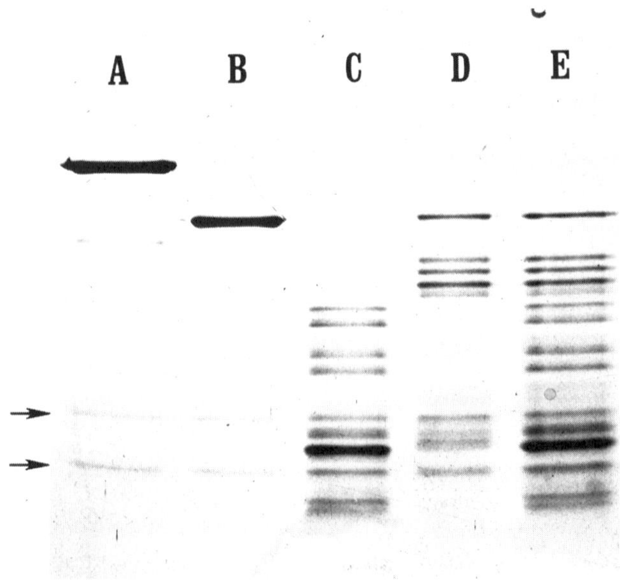

Fig. 8. Separation by SDS-gel electrophoresis of peptides produced by chymotryptic digestion of GSI and GSII. The procedure was essentially that of Cleveland *et al.* (11), using GSI and GSII (ca 10μg each per slot) prepared as in Fig. 7. Digestion proceeded for 60 min at 37° in the presence of chymotrypsin at 25 μg/ml, and 15% acrylamide was used for the separation. Arrows indicate the banding positions of chymotrypsin. A) GSI, t_0; B) GSII, t_0; C) GSI, t_{60}; D) GSII, t_{60}; E) GSI plus GSII, t_{60}.

chymotrypsin produced the four slowly-migrating (large) peptides derived from GSII in Fig. 8. Similar experiments to that shown in Fig. 8 have been done using either papain or staphylococcal protease (27), instead of chymotrypsin. In both cases the results were the same: few, if any, common bands from GSI and GSII. We have concluded the GSII is not derived from GSI and probably represents an independent gene product.

We have also differentiated GSI from GSII by electron microscopy. Early efforts (15) by Dr. Maurice Lalonde, using phosphotungstic acid for contrast, gave images of GSI analogous to those originally reported by Valentine *et al.* (73) for glutamine synthetase isolated from *E. coli*. The characteristic pattern of 12 subunits arranged in 2 hexagonal layers was seen, as well as the tendency to undergo face-to-face stacking, as noted for unadenylylated glutamine synthetase from *E. coli* (73). When GSII preparations were examined by Dr. Lalonde, using the same methods, no regular structure could be distinguished. More recently, Dr. Harry Calvert has tried additional procedures in an attempt to obtain images of GSII, and some of his results are shown in Fig. 9, where GSI and GSII have been fixed with glutaraldehyde and contrasted with uranyl acetate. In Fig. 9A, GSI can be seen in its characteristic hexameric face

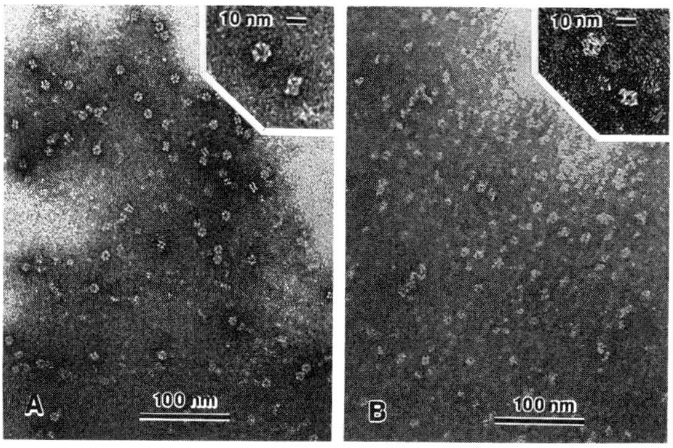

Fig. 9. Transmission electron microscopy of GSI (A) and GSII (B) from *R. japonicum* 61A76. Enzymes were prepared as in Fig. 7, diluted to 20μg/ml in 10 mM Imidazole·HCl, pH 7.0, and glutaraldehyde was added to a final concentration of 0.5%. Five to 10 minutes after glutaraldehyde addition, the enzymes were mounted on a copper grid with a carbonized formvar film. The molecules were contrasted with 0.5% uranyl acetate, pH 4.5, and examined in a Philips 200 electron microscope at 80kv with a 25μm objective aperture. Fig. 9A shows GSI X 208,000; Insert X 450,000. Fig. 9B shows a field of GSII at the same magnifications as GSI in Fig. 9A.

views, along with two types of edge views. Fields of GSII (9B) contain aggregates with dimensions like those of GSI, along with what appear to be fragments of the molecule. Among the aggregates, in this field and in other fields, one finds pentagonal figures (see insert Fig. 9B), along with what look like edge views, some of which appear tetrameric. We do not yet have good data for the molecular weight of catalytically active GSII, and therefore we cannot say whether or not the aggregates seen in Fig. 9B represent the native form of the protein. Our results do identify GSI with the classical molecular structure of glutamine synthetase from *E. coli* (73), and further emphasize the distinction between GSI and GSII.

Catalytic Properties

Both GSI and GSII have the catalytic properties of true glutamine synthetases (61), with a requirement for ADP, divalent cation, and arsenate in the γ-glutamyl transferase reaction, and a requirement for ATP, glutamate, and ammonia for the biosynthetic reaction, as determined by the appearance of either inorganic phosphate or ADP (59). In testing GSII for possible additional catalytic activities, we have found, assaying the appearance of inorganic phosphate, in the presence and in the absence of inorganic pyrophosphatase, that L-aspartate cannot function as amide acceptor, with either ammonia, L-glutamine, or L-asparagine as amide donor nor can L-asparagine function as amide donor with L-glutamate as amide acceptor. No detectable hydroxamate is formed when L-asparagine is substituted for glutamine in the transferase reaction. Extensive kinetic analyses of GSI and GSII have not yet been done, but, measuring the formation of inorganic phosphate at the pH optimum for each enzyme (see Table 1) with all other substrates in excess, we have found near-hyperbolic saturation for L-glutamate, with half-maximal velocity at about 2 mM for both GSI and GSII. With varying $ATP \cdot 3MgCl_2$, we have found sigmoidal saturation, with half-maximal velocity at about 2 mM for GSI and 1 mM for GSII. The assay for inorganic phosphate in its standard form is not sensitive enough to determine the saturation curve for ammonia using initial velocities, but we can estimate from the course of the reaction with limiting ammonia that the point of half-maximal velocity is approximately 0.2 mM for both GSI and GSII.

Table 1 shows the maximum velocities observed with GSI and GSII, together with the pH optima, for the biosynthetic reaction with either ammonia or hydroxylamine as amide donor and for the transferase reaction. GSI catalyses the biosynthetic reaction with the natural substrate, ammonia, at twice the speed of GSII, in the presence of Mg^{++}. On the other hand, GSII catalyses the transferase reaction at about twice the speed of GSI, in the presence of Mn^{++}. As is true for the unadenylylated enzyme from *E. coli* (61), both GSI and GSII show a marked preference for Mg^{++} in the biosynthetic reaction and for Mn^{++} in the transferase reaction. GSI is virtually inactive in the presence of Mn^{++} in the biosynthetic reaction with ammonia as the amide donor, but with hydroxylamine as the amide donor the distinction between Mn^{++} and Mg^{++} becomes much less. Especially striking are the extreme differences in pH optima for both GSI and GSII caused by substituting Mn^{++} for Mg^{++} in the biosynthetic reaction with hydroxylamine.

TABLE I

Relative catalytic activities of purified GSI and GSII

Glutamine synthetases (GSI (unadenylylated) and GSII) were purified as outlined in the text by an unpublished procedure (Darrow and Knotts) from *R. japonicum* 61A76 grown with glutamate as nitrogen source. The SDS gel patterns in Fig. 8 and the electron micrographs in Fig. 9 are of the same preparations. Protein was determined by Peterson's adaptation (50) of the method of Lowry *et al* (36). Transferase assays were performed as specified by Shapiro and Stadtman (59) with equimolar $MgCl_2$ substituted for $MnCl_2$ where appropriate, except that the buffer was 50 mM final concentration each of imidazole, bicine, and MES, added as a 5-fold concentrated mixture, with 50 mM acetate in addition at pH values below 6.0. Biosynthetic reactions with ammonia as amide donor (measuring inorganic phosphate) were also performed as described by Shapiro and Stadtman (59), with the same buffer modifications as for transferase, but with 40 mM glutamate, 10 mM ATP and NH_4Cl, and either 30 mM $MgCl_2$ or 10 mM $MnCl_2$, as appropriate. Biosynthetic reactions with hydroxylamine as amide donor had the same reaction mixture, except that neutralized NH_2OH: HCl was substituted for NH_4Cl and hydroxamate formation was determined as in the transferase reaction.

		micromoles min^{-1} mg $protein^{-1}$, 30°		
		Biosynthetic (NH_3)	*Biosynthetic* (NH_2OH)	*Transferase*
GSI	Mg++	18 (pH8.4)	8.1 (pH8.0)	32 (pH7.9)
	Mn++	1.1 (pH6.7)	4.3 (pH5.6)	180 (pH7.6)
GSII	Mg++	9.1 (pH7.9)	5.3 (pH8.6)	93 (pH7.6)
	Mn++	3.0 (pH6.1)	2.0 (pH5.6)	380 (pH6.9)

Fig. 10 illustrates the sort of data which first suggested to us that GSI could undergo adenylylation reactions and that GSII could not. The experiments reported in this paper using purified GSI have been done with unadenylylated enzyme, isolated from cells grown with glutamate as nitrogen source. However, the samples of GSI and GSII used in the experiment shown in Fig. 10 were obtained by isoelectric focusing (Fig. 3) from an extract of cells grown with 5 mM ammonia as a nitrogen source. When added to a culture grown on glutamate (see below), ammonia causes disappearance of GSII and the adenylylation of GSI. Here, however, the cells were harvested at stationary phase, at a time when ammonia had probably become limiting, and contained both GSI and GSII.

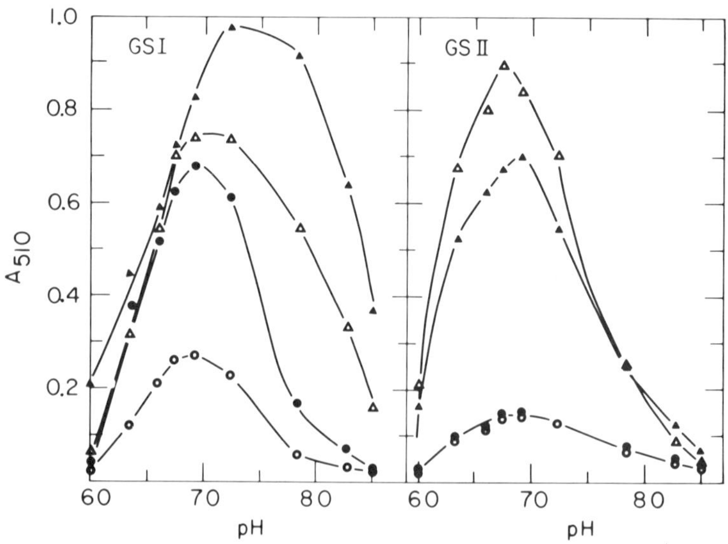

Fig. 10. The effect of digestion with snake venom phosphodiesterase on the pH/activity profile of γ-glutamyl transferase activity of GSI and GSII. GSI and GSII were obtained from the preparative isoelectric focusing run shown in Fig. 3, and 2.4 units of each (59) were incubated for 60 minutes at 37° at pH 9.0 in the presence of 0.1M Tris HCl, 1 mM $MgCl_2$, with and without 0.05 units of snake venom phosphodiesterase (SVD) (ICN). The samples were chilled in ice and 20μl assayed at 37° for 60 minutes at varying pH values, with and without 80 mM $MgCl_2$. The reaction mixture was that described by Stadtman *et al.* (62) for similar measurements with enzyme from *E. coli*, except that the buffer was 50 mM each in imidazole, bicine, and MES, added as 5-fold concentrated mixture, adjusted to give the indicated pH in the final reaction mixture. (△), no addition, (▲), +SVD; (○), +80mM $MgCl_2$; (●), +80mM $MgCl_2$ +SVD.

The experiment is based on the work of Stadtman and his colleagues (62), who have shown that, for the glutamine synthetase of *E. coli*, the degree of inhibition by high concentrations of Mg^{++} of the transferase reaction (in the presence of the usual cation, Mn^{++}) is proportional to the degree of adenylylation of the enzyme, under strictly defined conditions of substrate saturation, pH, and buffer ions. Under the specified conditions there is an isoactivity pH where the transferase activity is identical for highly adenylylated and unadenylylated enzyme in the normal assay with Mn^{++} and also for unadenylylated enzyme in the presence of high Mg^{++}, but essentially zero for highly adenylylated enzyme in the presence of high Mg^{++}. Treatment with snake venom phosphodiesterase, which causes de-adenylylation and concomitant release of the inhibition by Mg^{++}, has been used by Tronick *et al.* (69) as an additional tool to estimate the degree of adenylylation of glutamine synthetase in crude extracts of various bacteria.

In Fig. 10, it can be seen that treatment of GSI with snake venom phosphodiesterase caused the activity to increase, with an alkaline shift in the pH optimum. This behavior would be predicted for partially adenylylated GSI, from the set of curves published by Stadtman *et al.* (62) for glutamine synthetase from *E. coli* at varying degrees of adenylylation. Similarly, as seen in Fig. 10, the partial inhibition of GSI by high Mg^{++}, which is relieved by phosphodiesterase, also suggests that the sample under question is partially adenylylated. There is an approximate isoactivity point at pH 6.7. Fuchs and Keister (22, unpublished data) have made a detailed study of the adenylylation of GSI in *R. japonicum* 3I1b 83 and have found that the principles used for the assay of adenylylation of *E. coli* apply also to GSI of rhizobia, although the experimental conditions must be modified for quantitation of results.

There is as yet no indication that GSII is ever adenylylated. In Fig. 10 it can be seen that phosphodiesterase treatment has no effect on the inhibited activity in high Mg^{++}, and actually decreases the activity in the absence of Mg^{++}. In other experiments, the addition of bovine serum albumin was shown to prevent the decrease, suggesting that it could have been caused by a protease contaminant in the phosphodiesterase preparation. The absorption spectrum of purified GSII does not suggest the presence of any non-protein material absorbing at 260 nm.

GSI and GSII also differ with respect to their susceptibility to inhibition by various nitrogen-containing metabolites. Fig. 11 illustrates the inhibition of GSI and GSII by alanine and glycine, measuring ADP formation in the coupled biosynthetic reaction. In this experiment, 50% inhibition of GSI was obtained with alanine and glycine at 0.5 and 0.9mM, respectively, and 50% inhibition of GSII was obtained for the same two amino acids at 5 and 1.8 mM. However, it should be stressed that these results simply illustrate another sort of difference in the catalytic properties of GSI and GSII. One should not conclude that GSI is more sensitive to feedback inhibitors than GSII since, for example, when assaying the biosynthetic reaction by appearance of inorganic phosphate with substrates below saturation, GSII is generally more sensitive to inhibitors than GSI.

Physiological Regulation of GSI and GSII

We have done two types of experiment to show the regulation of GSI and GSII by the presence or absence of ammonia. In the first type of experiment we added 10 mM ammonia to a culture which

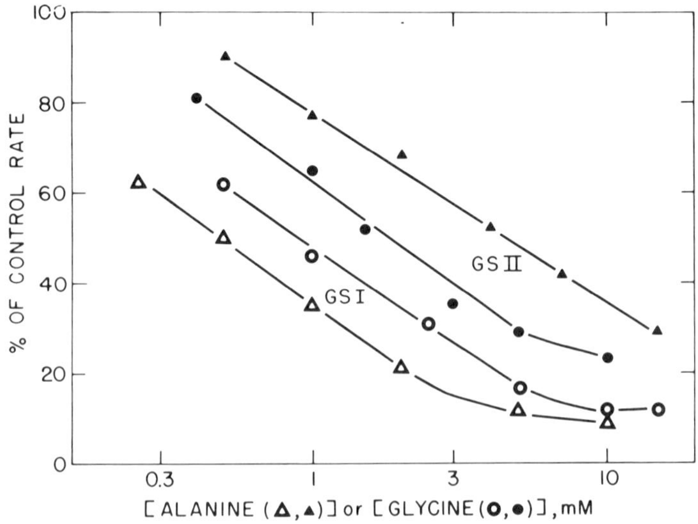

Fig. 11. Differential inhibition of GSI and GSII biosynthetic activities by alanine and glycine. GSI and GSII were purified from *R. japonicum* 61A76, grown on glutamate, by an unpublished procedure similar to the one outlined in the text. Glutamine synthetase activity was determined by the rate of disappearance of NADH, coupled to the appearance of ADP with excess phosphoenol-pyruvate, pyruvate kinase and lactate dehydrogenase (59). Excess glycine or alanine had no effect on the rate or stoichiometry of the oxidation of NADH, with either limiting or excess ADP, under the conditions used. (○), GSI +glycine; (△) GSI +alanine; (●), GSII +glycine, (△) GSII +alanine.

had been growing on glutamate and contained 14 units per mg protein of total transferase activity, with 30% heat-stable activity (GSI). Two hours after the addition of ammonia, total transferase had dropped to 5.5 units per mg protein, with 55% heat-stable activity (GSI). At four hours, the total activity had dropped to 3 units per mg protein, where it remained, with 90-100% heat stable activity (GSI). The estimation of the relative amounts of GSI and GSII at the various time points obtained by heating was confirmed by analytical isoelectric focusing. The relative degree of adenylylation of GSI (obtained in the isoelectric focusing run) was determined by the effect of Mg^{++} on the transferase reaction at the approximate isoactivity point (as in Fig. 10). There was relatively little inhibition (10%) at time zero, indicating a low degree of adenylylation, and maximum inhibition (80%) at 2 h. Thus the addition of 10 mM ammonia caused a rapid adenylylation of GSI, but little change (36% decrease) in its level as measured by transferase activity, and the rapid disappearance of GSII transferase activity (75% decrease in 2 h; essentially absent in 4 h).

In the second type of experiment on the regulation of GSI and GSII by ammonia, a culture was grown under conditions where the ammonia added (5 mM) was known to limit the extent of growth. The ammonia concentration was monitored and samples were taken when it had dropped to 0.5 mM, and at intervals afterward, during which time it remained essentially undetectable in the medium. The total transferase increased from 1.7 units per mg protein at the first sample (0.5 mM ammonia) to a maximum of 5.4 units per mg protein 32 h later. Over the same time interval the percentage of GSI, as determined by heat stability, went from about 80-90% to about 45%. Put another way, during this period of nitrogen starvation the specific activity of GSI increased 1.7-fold, while the specific activity of GSII increased 12-fold. Again, the relative percentages of GSI and GSII were checked by isoelectric focusing, and found to agree with the values obtained by heat treatment. The relative adenylylation of GSI, which was still high at 0.5 mM ammonia, had reached a minimum value 8 h later. Thus, just as the addition of ammonia causes the disappearance of GSII and the adenylylation of GSI, the depletion of ammonia causes the appearance of GSII and the de-adenylylation of GSI.

Speculation Concerning the Significance of GSI and GSII

Table II summarizes some of the properties which distinguish GSI and GSII. We do not yet know why all the rhizobia, but no other bacteria yet tested, have GSII as well as GSI. Since the rhizobia all have in common both a capacity for growth in the free-living state, and a capacity for intimate interaction with plant tissue, it is natural to suppose that the two forms of glutamine synthetase are somehow related to the two life-styles of the organisms.

The most obvious hypothesis is that the catalytic and regulatory functions which coexist in a single molecule in the enteric bacteria, reside separately in GSI and GSII in the rhizobia. Thus GSI is adenylylated and catalytically inactive under nitrogen-fixing conditions, but retains the capacity for whatever role glutamine synthetase may have in maintaining nitrogenase in the derepressed state. GSII is inactive during nitrogen fixation, thus assuring that the ammonia produced by nitrogenase is exported to the plant tissue as such. In the free-living state, GSII is the primary assimilating activity at critically low external concentrations of ammonia. It is responsive to changes in environmental ammonia by means of a regulatory mechanism which is independent of the one governing nitrogenase,

TABLE II

GSI and GSII from R. japonicum *61A76*

Experimental evidence for the items listed has been presented elsewhere in preliminary form (13, 14, 15, 51), or else is cited in the present text.

GSI	*GSII*
Like GS of enteric bacteria	Unlike GS of enteric bacteria
12-subunit structure	MW unknown
SDS MW 59,000	SDS MW 36,000
adenylylation	no adenylylation
DNA binding	no DNA binding
stable (heat, urea)	labile (heat, urea)
Limited response to NH_3 (transferase activity)	Strong response to NH_3 (transferase activity)
Adenylylated under N_2-fixing conditions	Activity disappears under N_2-fixing conditions

and it does not itself participate in nitrogenase regulation. Since rhizobia lack the capacity to grow on N_2 in the free-living state, it would be useless and energetically wasteful to synthesize nitrogenase outside the nodule.

A second possibility is that GSI and GSII are both primarily catalytic entities, but serve different biochemical pathways, as has been suggested for two isozymes of glutamine synthetase isolated from *Bacillus caldolyticus* (77) on the basis of their differential response to feed-back inhibitors. GSI and GSII, which also show a differential response to feed-back inhibitors, might be physically separated within the cell, as components of different multienzyme systems.

Finally, although GSII *is* a glutamine synthetase, it is possible that GSII has some unknown additional function. This could be catalytic, or regulatory, or both, and might be expressed only in conjunction with the plant tissue.

Acknowledgments: Special thanks are due Mr. Richard Knotts for excellent technical assistance; Dr. Harry Calvert for performing the electron microscopy; Drs. Fred Ausubel, John Robertson, Roy Fuchs, and Donald Keister for providing manuscripts and other experimental data prior to publication.

REFERENCES

1. Ausubel, F. M., Bird, S. C., Durbin, K. J., Janssen, K. A., Margolskee, R. F., and Peskin, A. P. (1979) *J. Bacteriol. 140*, 597-606.
2. Bauman, G., and Chrambach, A. (1975) *Anal. Biochem. 69*, 649-651.
3. Bender, R. A. and Magasanik, B. (1977) *J. Bacteriol. 132*, 100-105.
4. Bender, R. A., and Magasanik, B. (1977) *J. Bacteriol. 132*, 106-112.
5. Bergersen, F. J., and Turner, G. L. (1967) *Biochim. Biophys.* Acta *141*, 507-515.
6. Bergersen, F. L., and Turner, G. L. (1976) *Biochem. Biophys. Res. Commun. 73*, 524-531.
7. Bergersen, F. L., and Turner, G. L. (1978) *Biochim. Biophys. Acta 538*, 406-416.
8. Bishop, P. E., Guevara, J. G., Engelke, J. A., and Evans, H. J. (1975) *Plant Physiol. 57*, 542-546.
9. Brenchley, J. E. (1973) *J. Bacteriol. 114*, 666-673.
10. Brown, C. M., and Dilworth, M. J. (1975) *J. Gen. Microbiol. 86*, 39-48.
11. Cleveland, D. W., Fischer, S. G., Kirschner, M. W. and Laemmli, U. K. (1977) *J. Biol. Chem. 252*, 1102-1106.
12. Daesch, G., and Mortenson, L. E. (1972) *J. Bacteriol. 110*, 103-109.
13. Darrow, R. A., and Knotts, R. R. (1977) *Biochem. Biophys. Res. Commun. 78*, 554-559.
14. Darrow, R. A., and Knotts, R. R. (1977) *Plant Physiol. 59S*, 95 (abstract).
15. Darrow, R. A., and Knotts, R. R. (1978) *Proceedings of the Steenbock-Kettering International Symposium on Nitrogen Fixation*, University of Wisconsin, Madison, Wisconsin, *B95* (abstract).
16. Darrow, R. A., Knotts, R. R. and Jarrell, S. D. (1976) *Plant Physiol. 57S*, 70 (abstract).
17. DeLeo, A. B., and Magasanik, B. (1975) *J. Bacteriol. 121*, 313-319.
18. Dixon, R., Merrick, M., Filser, M., Kennedy, C. and Postgate, J. R. (1980) in *Nitrogen Fixation* (Newton, W. E., and Orme-Johnson, W. H., eds) Vol. 1, University Park Press, Baltimore.
19. Dunn, S. D., and Klucas, R. V. (1973) *Can. J. Microbiol. 19*, 1493-1499.
20. Fahrney, P. E., and Gold, A. M. (1963) *J. Am. Chem. Soc. 85*, 997-1000.
21. Friederich, B., and Magasanik, B. (1977) *J. Bacteriol. 131*, 446-452.
22. Fuchs, R. L., and Keister, D. L. (1979) *Plant Physiol. 63S*, 113 (abstract).
23. Fuchs, R. L. and Keister, D. L. (1980) *J. Bacteriol. 141*, 996-998.
24. Gaillardin, C. M., and Magasanik, B. (1978) *J. Bacteriol. 133*, 1329-1338.
25. Goldberg, R. B., and Hanau, R. (1979) *J. Bacteriol. 137*, 1282-1289.
26. Gordon, J. K., and Brill, W. J. (1974) *Biochem. Biophys. Res. Commun. 59*, 967-971.
27. Houmard, J., and Drapeau, G. R. (1972) *Proc. Natl. Acad. Sci. U.S.A., 69*, 3506-3509.
28. Janssen, K. A., Riedel, G. E., Ausubel, F. M., and Cannon, F. C. (1980) in *Nitrogen Fixation* (Newton, W. E., and Orme-Johnson, W. H., eds.) Vol. 1, University Park Press, Baltimore.

29. Jones, B. B., Chan, H., Rothstein, S. Wells, R. and Reznikoff, W. (1977) *Proc. Natl. Acad. Sci. U.S.A., 74*, 4914-4918.

30. Keister, D. L. (1975) *J. Bacteriol. 123*, 1265-1268.

31. Kondorosi, A. Sváb, Z., Kiss, G. B., and Dixon, R. A. (1977) *Molec. Gen. Genet. 151*, 221-226.

32. Kurz, W. G. W., and LaRue, T. A. (1975) *Nature 256*, 407-409.

33. Kustu, S., Burton, D., Garcia, E., McCarter, L., and McFarland, N. (1979) *Proc. Natl. Acad. Sci. U.S.A., 76*, 4576-4580.

34. Kustu, S. G., McFarland, N. C., Hui, S. P., Esmon, B. and Ferro-Luzzi Ames, G. (1979) *J. Bacteriol. 138*, 218-234.

35. Laemmli, U. K. (1970) *Nature 227*, 680-685.

36. Lowry, O. H., Rosebrough, N. J., Farr, A. L., and Randall, R. J. (1951) *J. Biol. Chem. 193*, 265-275.

37. Ludwig, R. A. and Signer, E. R. (1977) *Nature 267*, 245-248.

38. Ludwig, R. A. (1978) *Proceedings of the Steenbock-Kettering International Symposium on Nitrogen Fixation*, University of Wisconsin, Madison, Wisconsin, *A87* (abstract).

39. MacNeil, T., MacNeil, D., Roberts, G. P., Supiano, M. A. and Brill, W. J. (1978) *J. Bacteriol. 136*, 253-266.

40. Magasanik, B., Prival, M. J., Brenchley, J. E., Tyler, B. M., DeLeo, A. B., Streicher, S. L., Bender, R. A., and Paris, C. G. (1974) *Curr. Top. Cell. Regul. 8*, 119-138.

41. Maier, R. J., and Brill, W. J. (1976) *J. Bacteriol. 127*, 763-769.

42. McComb, J. A., Elliot, J., and Dilworth, M. J. (1975) *Nature 256*, 409-410.

43. McParland, R. H., Guevara, J. G., Becker, R. R., and Evans, H. J. (1976) *Biochem. J. 153*, 597-606.

44. Mortenson, L. E. (1978) *Curr. Top. Cell. Regul. 13*, 179-232.

45. Nagatani, H., Shimizu, M., and Valentine, R. C. (1971) *Arch. Microbiol. 79*, 164-175.

46. O'Gara, F. and Shanmugam, K. T. (1976) *Biochim Biophys. Acta 437*, 313-321.

47. Pagan, J. P., Child, J. J., Scowcroft, W. R., and Gibson, A. H. (1975) *Nature 256*, 406-407.

48. Pahel, G., and Tyler, B. (1979) *Proc. Natl. Acad. Sci. U.S.A. 76*, 4544-4548.

49. Pate, J. S. (1977) in *A Treatise on Dinitrogen Fixation* (Hardy, R. W. F., and Silver, W. S., eds.) Section III pp 473-515. John Wiley and Sons, N. Y.

50. Peterson, G. L. (1977) *Anal. Biochem. 83*, 346-356.

51. Rao, V. R., Darrow, R. A., and Keister, D. L. (1978) *Biochem. Biophys. Res. Commun. 81*, 224-231.

52. Reporter, M., and Hermina, N. (1975) *Biochem. Biophys. Res. Commun. 64*, 1126-1133.

53. Robertson, J. G., Farnden, K. J. F., Warburton, M. P. and Banks, J. M. (1975) *Aust. J. Plant Physiol. 2*, 265-272.

54. Robertson, J. G., Warburton, M. P., and Farnden, K. J. F. (1975) *FEBS letts. 55*, 33-37.

55. Scowcroft, W. R., Gibson, A. H. and Pagan, J. D. (1976) *Biochem. Biophys. Res. Commun. 73,* 516-523.
56. Shanmugam, K. T., Chan, I., and Morandi, C. (1975) *Biochim. Biophys. Acta 408,* 101-111.
57. Shanmugam, K. T., O'Gara, F., Andersen, K., and Valentine, R. C. (1978) *Ann. Rev. Plant Physiol. 29,* 263-276.
58. Shanmugam, K. T., and Morandi, C. (1976) *Biochim. Biophys. Acta 437,* 322-332.
59. Shapiro, B. M., and Stadtman, E. R. (1970) *in Methods in Enzymology* (Tabor, H., and Tabor, C. W., eds.) Vol. XVIIA, pp. 910-922 Academic Press, New York.
60. Skotnicki, M. L., Rolfc, B. G., and Reporter, M. (1979) *Biochem. Biophys. Res. Commun. 86,* 968-975.
61. Stadtman, E. R., and Ginsburg, A. (1974) in *The Enzymes* (Boyer, P. D., ed) Vol. X, 3rd ed. pp. 755-807. Academic Press, New York.
62. Stadtman, E. R., Ginsburg, A., Ciardi, J. E., Yeh, J., Hennig, S. B., and Shapiro, B. M. (1970) *Advan. Enzyme Regul. 8,* 99-118.
63. St. John, R. T., Shah, R. T., and Brill, W. J. (1974) *J. Bacteriol. 119,* 266-269.
64. Streicher, S. L. DeLeo, A. B., and Magasanik, B. (1976) *J. Bacteriol. 127,* 184-192.
65. Streicher, S. L., Shanmugam, K. T., Ausubel, F., Morandi, C., and Goldberg, R. B. (1974) *J. Bacteriol. 120,* 815-821.
66. Streicher, S. L., and Tyler, B. (1976) *Fed. Proc. 35,* 1471.
67. Tempest, D. W., Meers, J. L., and Brown, C. M. (1970) *Biochem. J. 117,* 405-407.
68. Tjipkema, J. D., and Evans, H. J. (1975) *Biochem Biophys Res. Commun. 65,* 625-628.
69. Tronick, S. R., Ciardi, J. E., and Stadtman, E. R. (1973) *J. Bacteriol. 115,* 858-868.
70. Tubb, R. S. (1974) *Nature 251,* 481-485.
71. Tyler, B. A., DeLeo, B., and Magasanik, B. (1974) *Proc. Natl. Acad. Sci. U.S.A. 71,* 225-229.
72. Tyler, B. (1978) *Ann. Rev. Biochem. 47,* 1127-1162.
73. Valentine, R. C., Shapiro, B. M., and Stadtman, E. R. (1968) *Biochemistry 7,* 2143-2152.
74. Vincent, J. M. (1977) in *A Treatise on Dinitrogen Fixation* (Hardy, R. W. F., and Silver, W. S., eds.), Section III, pp. 277-366 John Wiley and Sons, New York.
75. Wallenfels, K. (1962) in *Methods in Enzymology* (Colowick, S. P. and Kaplan, N. O. eds.) Vol. 5, Academic Press, N. Y.
76. Weber, K., and Osborn, M. (1969) *J. Biol. Chem. 244,* 4406-4412.
77. Wedler, F. C., Kenney, R. M., Ashour, A. E., and Carfi, J. (1978) *Biochem. Biophys. Res. Commun. 81,* 122-126.

10. *Neurospora crassa* Glutamine Synthetase: a model to study Gene Expression in Eukaryotes

RAFAEL PALACIOS, LOURDES BLANCO, EDMUNDO CALVA, MIGUEL CAMPOMANES, OFELIA CHÁVEZ, JULIO COLLADO, SUSANA LÓPEZ, JOSÉ MA. DE LA PEÑA, CARMEN QUINTO, AND FEDERICO SÁNCHEZ

Departamento de Biología Molecular, Instituto de Investigaciones Biomédicas, Universidad Nacional Autónoma de México Apartado Postal 70228, México, D. F.

SUMMARY

Neurospora crassa glutamine synthetase has been selected as a model system to study gene organization and expression in eukaryotes. The role of this enzyme in various metabolic pathways is beginning to be understood, and mutants in glutamine synthetase have been characterized (Mora *et al.*, Chapter 11). These features allow the study of the molecular mechanisms that govern the expression of specific eukaryotic genes.

The enzyme protein has been purified and partially characterized; antibodies have been obtained and specific mRNA has been translated in a cell-free system derived from rabbit reticulocytes. We have found that the nitrogen source regulates the levels of glutamine synthetase mRNA. Cultures grown on glutamate as sole nitrogen source have a 10-fold higher concentration of translatable glutamine synthetase mRNA than those grown on glutamine. There is a corresponding increase in the rate of synthesis and specific activity of the enzyme.

Previous studies on *N. crassa* glutamine synthetase indicated that the enzyme is made up of subunits of identical molecular weight.

ISBN: 0-12-506040-8

Recent experiments show that these subunits are heterogeneous. Such evidence is presented and the general scope of our research project is discussed.

INTRODUCTION

Recently, there have been important conceptual advances with regard to the organization and expression of eukaryotic genomes (1-5). Most studies have been performed in higher organisms, particularly in terminally differentiated tissues that synthesize large amounts of specific gene products. A limitation of these investigations could possibly arise from the difficulty at present of obtaining and selecting specific mutations. In this regard, eukaryotic microorganisms have a relatively small genome and offer the possibility of performing a highly sophisticated genetic analysis. Among these microorganisms, those that have morphological changes during their life cycle could be used as simple models to study cell differentiation at the molecular level. For these reasons we have selected *Neurospora crassa* as our model biological system.

Our research has focused on glutamine synthetase, a key enzyme in cellular metabolism. In addition to its central metabolic role, this enzyme serves various specialized functions in different cells. Thus, the mechanisms that control its activity and concentration are rather complex and depend on the type of organism or tissue under study (6-11).

In *Neurospora crassa*, the studies of Mora and coworkers (12-16, Chapter 11) have shown that glutamine synthetase is regulated by both the nitrogen and the carbon source; that the regulatory mechanisms operate not only in the growing organism but also in nongrowing conidia. Moreover, this laboratory has defined two different metabolic functions by which glutamine synthetase assimilates inorganic nitrogen either at limiting or at excess substrate concentrations. These functions have been correlated with different quaternary structures of the enzyme (13, 14).

It is our aim to define the different levels at which the glutamine synthetase concentration can be controled in *N. crassa*. These include the sequence organization of structural and regulatory genes; the expression, processing, and translation of this information; and the turnover of the specific macromolecules involved. In its initial steps, the general strategy that we are following comprises the purification and characterization of the enzyme protein, the identification and isolation of specific mRNA, and the mobilization and amplification of specific gene sequences into molecular cloning vehicles.

While pursuing these goals, we have found that the effect of the nitrogen source on synthesis of glutamine synthetase during exponential growth of *Neurospora*, is the result of changes in the levels of specific mRNA (17). We have confirmed the results of Mora, Chávez, and Mora (16) who first indicated the participation of the carbon source on the degradation of *N. crassa* glutamine synthetase.

Most of the conclusions derived from these studies have been based on the assumption that *Neurospora crassa* glutamine synthetase subunits are identical. However, recent evidence from our laboratory (18) indicates that in *Neurospora crassa* different polypeptides participate in glutamine synthetase activity. Furthermore, Dávila *et al.* (19) have correlated the presence of each polypeptide with different quaternary structures of the enzyme and hence with its capacity to fix ammonia at low or high concentrations (12-14).

In this paper we review some of our previous work on *N. crassa* glutamine synthetase and present evidence for the heterogeneity of its subunits.

RESULTS

Purification of Glutamine Synthetase

Purification of *Neurospora crassa* glutamine synthetase has been based on affinity chromatography (20). Different metabolites that interact with the enzyme were coupled to CH-Sepharose 4B and tested for their capacity to bind active enzyme. Anthranilate-bound Sepharose completely retains *Neurospora crassa* glutamine synthetase activity from either crude extracts or partially purified fractions. Enzyme activity can be eluted from the affinity chromatography resin by addition of AMP (20) or other compounds including: UMP, ADP, GTP, and adenine (unpublished observation). It is interesting that this purification method is not restricted to the enzyme from *N. crassa*, since the *E. coli* enzyme can be purified to homogeneity by the same procedure (Fig. 1). Moreover, the fact that glutamine synthetases from other sources such as rat liver (not shown) and mouse brain (Fig. 1) bind quantitatively to anthranilate-Sepharose, suggests that this procedure could be used as a general method for the isolation of glutamine synthetase.

The quaternary structure of glutamine synthetase from different sources has been thoroughly studied. In prokaryotes, the enzyme is

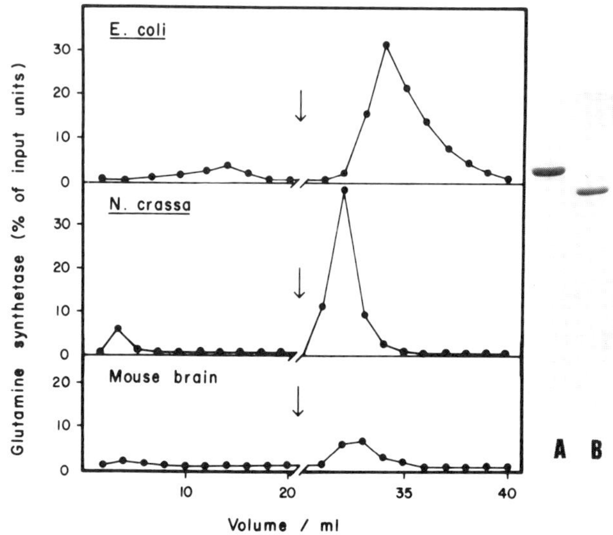

Fig. 1. Purification of glutamine synthetase from different sources by affinity chromatography. Left, crude extracts from *E. coli, N. crassa,* or mouse brain, were passed through a column containing a Sepharose-anthranilate resin. The resin was washed extensively, and then glutamine synthetase was eluted by adding 40 mM AMP (arrow). Glutamine synthetase was measured in the fractions by the transferase assay. Right, polyacrylamide gel electrophoresis in the presence of sodium dodecyl sulphate of *E. coli* glutamine synthetase purified by two passages through Sepharose-anthranilate (A), and of *N. crassa* glutamine synthetase purified by DEAE-Cellulose chromatography followed by Sepharose-anthranilate chromatography (B). Details of the procedure are given in reference (20).

an oligomer made up of 12 subunits; in some microorganisms these subunits can be modified by adenylylation (21). On the other hand, the eukaryotic enzyme is usually present in the form of an octamer (22) and no heterogeneity or modifications in the subunits have been reported. Different criteria, including sucrose gradient sedimentation, electrophoresis in acrylamide gradients, cross-linkage of subunits with bivalent reagents, and electron microscopy, indicated that purified *Neurospora crassa* glutamine synthetase is present as an octamer made up of subunits of similar molecular weight (20). Kapoor and Bray (23) first reported some properties of *N. crassa* glutamine synthetase; recently Lin and Kapoor (24) have reached conclusions similar to ours with respect to the octameric structure of the enzyme.

Under the electron microscope (Fig. 2), *Neurospora crassa* glutamine synthetase appears similar to those reported from higher organisms (25-27). In some projections (inset Fig. 2. A), four round structures are arranged in a square of about 100 A per side, while in others (inset Fig. 2, B) two parallel rod-shaped electron-dense structures are evident. This indicates an octameric structure and is in accordance with the model proposed by Haschemeyer (25). Mora and coworkers have found that the quaternary structure of the enzyme depends on the culture conditions of the organism. They have correlated the presence of an octameric form of the enzyme

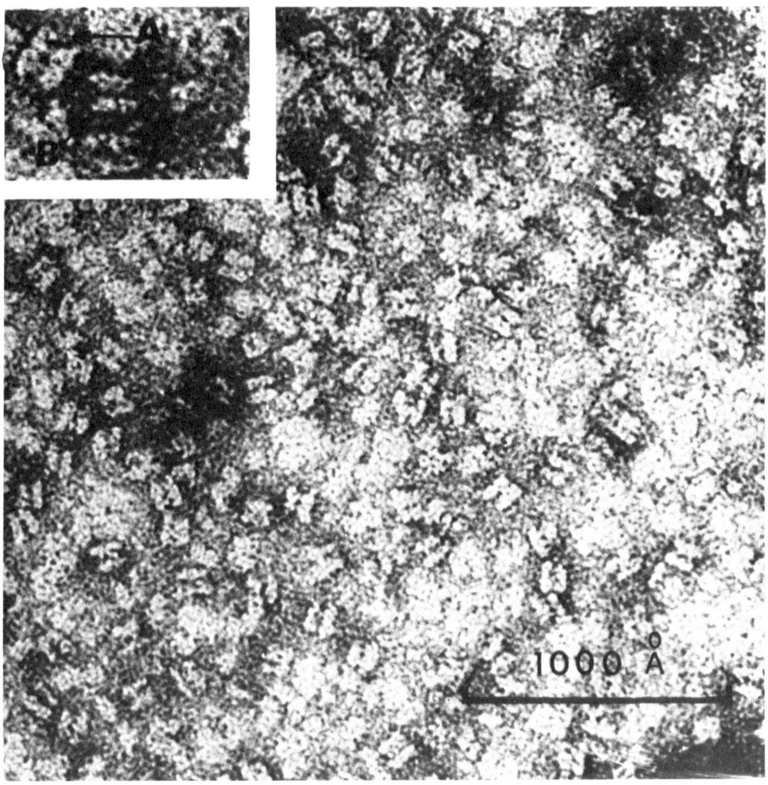

Fig. 2. Electron microscopy of purified *Neurospora crassa* glutamine synthetase. One drop of a purified preparation of *N. crassa* glutamine synthetase, at a concentration of 100μg per ml, was applied to a 400 mesh grid coated with a thin carbon film. The sample was then negatively stained with 2% uranyl acetate for 2 min and the grid was observed in a Joel 100-B electron microscope. Magnification: 270,000 in the main micrograph and 495,000 in the inset. The arrows (A, B) indicate different projections of the enzyme (see Results). Taken from reference (20).

with the synthesis of glutamine when nitrogen is in excess, and a tetrameric form with the fixation of ammonia at low concentrations (13, 14).

The purified enzyme from *N. crassa* was used as antigen to obtain antibodies. These antibodies completely precipitate glutamine synthetase activity from either crude extracts or purified fractions and have been found to be monospecific for *N. crassa* glutamine synthetase (28, 29). These antibodies were used as a tool to quantify the synthesis and degradation of the enzyme.

Glutamine Synthetase Synthesis and Degradation

The relative rates of glutamine synthetase synthesis and degradation were studied in *Neurospora crassa* by double isotope incorporation studies (30). The advantages and limitations of this experimental approach have been discussed by Schimke (31). To estimate the relative rate of enzyme synthesis, parallel *N. crassa* cultures grown under different conditions received short pulses with either [^{14}C] leucine or [^{3}H] leucine. After the extracts were combined, total protein and glutamine synthetase were acid-precipitated and immunoprecipitated, respectively. Samples were subjected to polyacrylamide gel electrophoresis, in the presence of sodium dodecyl sulphate, and the amount of radioactivity (^{3}H and ^{14}C) present in gel fractions was determined. The ratio of ^{3}H to ^{14}C in the glutamine synthetase band, divided by this ratio in the total protein, was used to calculate the relative rates of enzyme synthesis in both cultures. As an example of this experimental approach, Fig. 3 shows the data obtained when the synthesis of glutamine synthetase was compared in *N. crassa* cultures grown exponentially in the presence of either glutamate or glutamine as the sole nitrogen source. The ten-fold difference obtained correlates well with the difference in enzyme activity that was found under both experimental conditions (12, 30). Espín and Mora (32) have demonstrated that conidia of *N. crassa*, incubated in non-growing conditions, accumulate glutamine and arginine and induce glutamine synthetase activity. Similar methodology as that presented in Fig. 3 has been used to demonstrate that the increase in glutamine synthetase activity is the result of *de novo* synthesis of the enzyme protein (15).

To measure the rate of glutamine synthetase degradation relative to that of total protein, parallel *N. crassa* cultures received a pulse of either [^{14}C] leucine or [^{3}H] leucine. The ^{14}C-labeled mycelium was filtered and stored; the ^{3}H-labeled mycelium was washed exten-

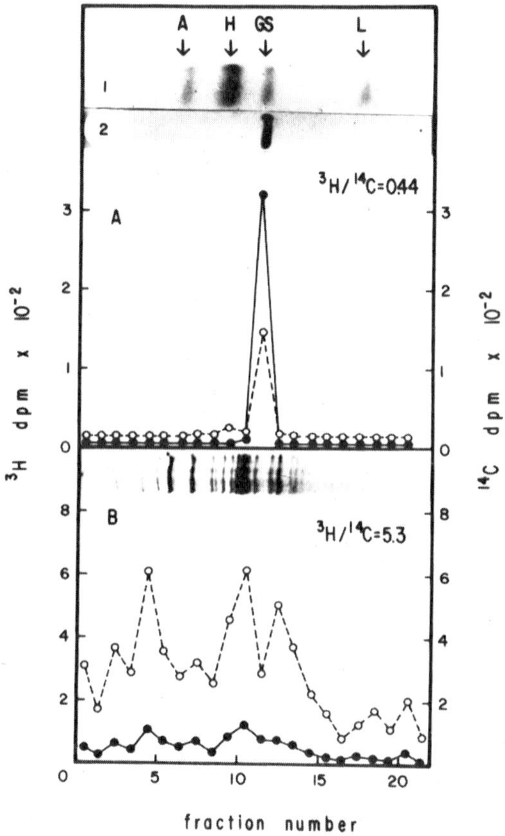

Fig. 3. Relative rate of glutamine synthetase synthesis in *Neurospora crassa* cultures grown on glutamate or glutamine as the sole nitrogen source. Parallel cultures of *Neurospora crassa* were grown on either 5 mM glutamate or 5 mM glutamine as the sole nitrogen source. After 7½ h, the culture growing on glutamate received a 30-min pulse with [^{14}C] leucine (0.5 μCi/ml) while that growing on glutamine received a 30-min pulse with [^3H] leucine (10 μCi/ml). Extracts obtained from both cultures were combined, total protein and glutamine synthetase were isolated by trichloroacetic acid precipitation and specific immunoprecipitation, respectively. Aliquots from the trichloroacetic acid precipitate and the immunoprecipitate were then subjected to polyacrylamide slab gel electrophoresis in the presence of sodium dodecyl suphate, followed by fluorography. After fluorography, the gels were sliced and ^{14}C and ^3H radioactivities were measured in each slice. Panel A, electrophoresis of specific immunoprecipitate after addition of 5 μg of non-labeled purified *N. crassa* glutamine synthetase: O—O, ^3H radioactivity; ●—●, ^{14}C radioactivity; 1, gel stained with Coomassie blue; arrows indicate the position of albumin (A), the heavy chain of gamma globulin (H), glutamine synthetase (GS), and the light chain of gamma globulin (L); 2, gel after fluorography. The ratio $^3H/^{14}C$ was obtained from the peak fraction. Panel B, electrophoresis of total protein precipitate: O—O, ^3H radioactivity; ●—●, ^{14}C radioactivity; a fluorogram of the gel is shown. To obtain the $^3H/^{14}C$ ratio, the ^3H radioactivity in each slice was added and divided by the sum of the ^{14}C radioactivity. The isotope ratio in glutamine synthetase divided by that of total protein, quantifies the relative rate of synthesis of the enzyme between the culture grown on glutamine and that grown on glutamate. This ratio is 0.08. Figure taken from reference (30).

sively and then cultured for several hours more in the presence of an excess on non-labeled leucine. Samples before and after the chase were combined, and the ratio of $^{3}H/^{14}C$ was measured in both glutamine synthetase and total protein. In exponentially growing cultures of *N. crassa,* no differences in the relative rate of enzyme degradation were found when different nitrogen sources were used (30). On the other hand, Mora *et al* found an increase in glutamine synthetase degradation when exponentially growing *Neurospora* was deprived of a carbon source (16). Under this condition, glutamine synthetase was rapidly degraded with a half-life of about 30 minutes, compared with a half-life of about ten hours found in the presence of a good carbon source (Quinto, Chávez and Palacios, to be published).

Glutamine Synthetase Specific mRNA

In order to identify and characterize glutamine synthetase m RNA, *Neurospora crassa* RNA was translated *in vitro* in a cell-free protein synthesizing system derived from rabbit reticulocytes (28, 17). Glutamine synthetase was identified among the *in vitro* synthesized products by specific direct immunoprecipitation using as carrier the purified *in vivo* synthesized enzyme. As shown in Fig. 4, the protein product that immunoprecipitates with anti-*N. crassa* glutamine synthetase comigrates, when subjected to polyacrylamide gel electrophoresis in the presence of sodium dodecyl sulphate, with the *in vivo* synthesized glutamine synthetase. Moreover, this product binds to anthranilate-Sepharose and is eluted with AMP (28), a chromatographic procedure highly specific for glutamine synthetase (20).

To determine the effect of the nitrogen source on the regulation of glutamine synthetase mRNA levels, the reticulocyte lysate system was incubated in the presence of different concentrations of polysomal RNA isolated from parallel cultures grown in the presence of either glutamate or glutamine as the sole nitrogen source (17). The amount of radioactivity incorporated into the specific immunoprecipitated product relative to that incorporated into total protein was estimated. When this amount is plotted as a function of RNA concentration, the slope of the curve obtained indicates the relative amount of translatable specific mRNA. A ten-fold difference between the culture grown on glutamate and that grown on glutamine was found (Fig. 5), in accordance with the data obtained when relative rates of glutamine synthetase synthesis were measured.

As stated before, Espin *et al.* (15) demonstrated the *de novo* synthesis of glutamine synthetase in non-growing conidia. Recently,

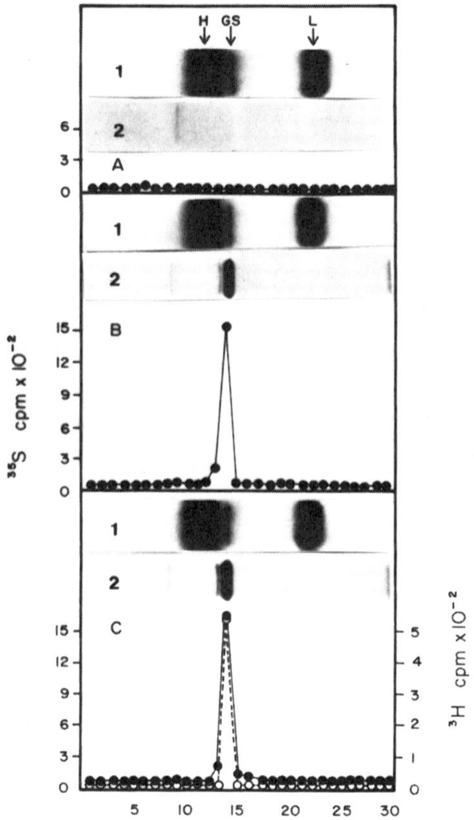

Fig. 4. Translation of *Neurospora crassa* glutamine synthetase specific mRNA in a cell-free protein synthesizing system derived from rabbit reticulocytes. To detect *in vitro* synthesized products [^{35}S] methionine was used in the reticulocyte lysate system. The reticulocyte lysate system was incubated in the absence (A) or in the presence (B, C) of polysomal RNA (320 μg/ml) extracted from a culture grown on glutamate as the sole nitrogen source. Direct immunoprecipitation of the *in vitro*-synthesized protein was carried out after addition of 5 μg of purified *N. crassa* glutamine synthetase as carrier, in the absence (A, B) or in the presence (C) of a trace amont (700 cpm) of glutamine synthetase labeled *in vivo* with [^{3}H] methionine. Immunoprecipitates were subjected to electrophoresis on polyacrylamide slab gels, in the presence of sodium dodecyl sulphate, and stained with Coomassie blue. The gels were cut in 3 mm slices and counted for ^{3}H/^{35}S. Panels show the stained gel, 1; the gel after fluorography, 2; ^{35}S counts per min, ● — ●; and ^{3}H counts per min, O — O. The arrows in (A) indicate the position of the heavy chain of gamma globulin (H), glutamine synthetase (GS), and the light chain of gamma globulin (L). Figure adapted from reference (17).

Hansberg *et al.* (33) have reported that the RNA isolated from conidia of *N. crassa*, although very active as a template for *in vitro* protein synthesis, does not prime the synthesis of glutamine synthetase. Shortly after conidia are incubated, even in the presence of

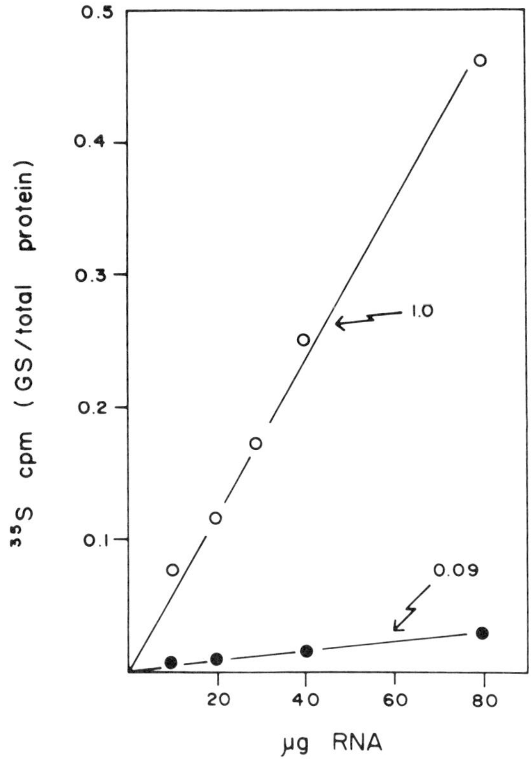

Fig. 5. Quantification of glutamine synthetase mRNA activity of *N. crassa* polysomal RNA. The reticulocyte lysate system was incubated in the presence of different concentrations of polysomal RNA extracted from parallel cultures of *N. crassa* grown on glutamate (O) or glutamine (●) as the sole nitrogen source. The *in vitro* synthesized product was immunoprecipitated and the immunoprecipitates were subjected to polyacrylamide gel electrophoresis in the presence of sodium dodecyl sulphate, stained, and treated for fluorography. After fluorography, the stained glutamine synthetase protein band was cut, dissolved and counted for ^{35}S. Immunoprecipitated radioactivity is expressed as % of that incorporated into total protein. The indicated slopes of the lines were normalized to that of the culture grown on glutamate, set arbitrarily at 1.0. Figure adapted from reference (17).

media that impair germination, glutamine synthetase mRNA activity appears. Under this condition, the levels of glutamine synthetase mRNA, measured by its translation *in vitro,* depend on the nature of the nitrogen compounds present in the incubation medium. The absence of nitrogen or the presence of glutamate results in an increase of specific mRNA, while the presence of glutamine prevents such an increase (33).

Further Studies on the Structure of Glutamine Synthetase Subunits

From our initial investigations on the structure of *Neurospora crassa* glutamine synthetase (20) we concluded, as have most studies on eukaryotic glutamine synthetase, that its subunits were identical. However, the following data suggested the possibility of heterogeneous subunits: 1) the very low yield of glutamine auxotrophs that can be obtained by mutagenesis, as compared with auxotrophs of other amino acids (14); 2) the present impossibility to obtain auxotrophs completely devoid of glutamine synthetase activity by mutagenizing the wild type strain (J. Mora, personal communication); and 3) the existence of two different oligomeric forms of glutamine synthetase that could be correlated with the capacity of the enzyme to fix ammonia at low or high substrate concentration (13, 14).

As previously described (20), purified glutamine synthetase from wild type *N. crassa* moves as a single band when subjected to acrylamide gel electrophoresis in the presence of sodium dodecyl sulphate (Fig. 6 A). On the other hand, two protein bands are observed when electrophoresis is perfomed in the presence of 7 M urea in addition to sodium dodecyl sulphate (Fig. 6 B). Both bands, which we have named α and β, are also evident when *N. crassa* RNA is translated *in vitro* and the products identified by specific immunoprecipitation (Fig. 6 C). These two polypeptides are separated in a two-dimensional electrophoretic analysis, provided that the second dimension is carried out in the presence of sodium dodecyl sulphate and 7 M urea (Fig. 7). This analysis shows that both polypeptides have a similar charge.

The ratio between the slower moving polypeptide, α, and the faster moving one, β, does not vary with the method used for RNA extraction, nor with the concentration of RNA and the time of incubation in the cell-free protein-synthesizing system (18). These data suggest that there is not a precursor-product relationship between both polypeptides; moreover, the ratio between α and β varies according to the culture conditions of the organism. Dávila *et al.* (19) have correlated the presence of the α polypeptide with the tetrameric form of glutamine synthetase and that of the β polypeptide with the octameric form of the enzyme.

Neurospora crassa glutamine auxotrophs have been obtained and partially characterized (14, 29, and J. Mora, personal communication). Purified glutamine synthetase preparations from some of these auxotrophs present the α but not the β band (18, 19). These mutants are

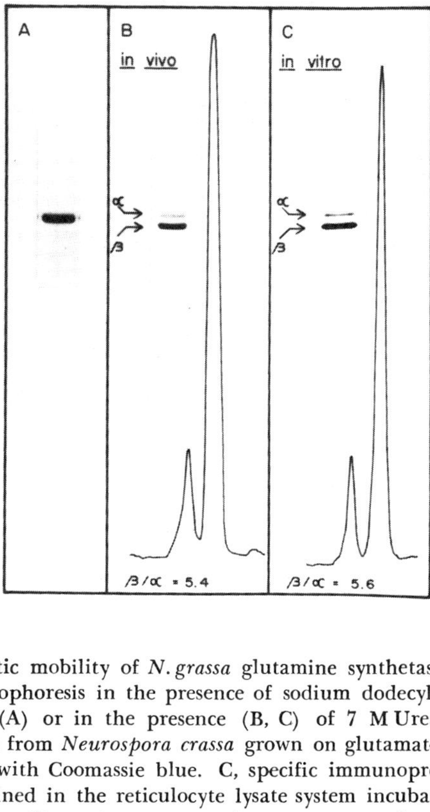

Fig. 6. Electrophoretic mobility of *N. grassa* glutamine synthetase polypeptides. Polyacrylamide gel electrophoresis in the presence of sodium dodecyl sulphate was carried out in the absence (A) or in the presence (B, C) of 7 M Urea. A and B, purified glutamine synthetase from *Neurospora crassa* grown on glutamate as the sole nitrogen source; gels stained with Coomassie blue. C, specific immunoprecipitate of the translation products obtained in the reticulocyte lysate system incubated in the presence of RNA extracted from a *Neurospora crassa* culture grown on glutamate as sole nitrogen source; fluorogram of the gel. The arrows indicate the position of α and β polypeptides of glutamine synthetase. Densitometer scans of the stained gel in B, and of the fluorogram in C, are shown; ratios of α/β are indicated.

unable to grow in excess, but do grow on limiting amounts of ammonia (14).

Recent experiments from our laboratory (18) show the reconstitution of glutamine synthetase activity from either the α or the β polypeptide eluted from polyacrylamide gels in the presence of sodium dodecyl sulphate and 7 M urea. From these experiments we concluded that both polypeptides contain an active site for glutamine synthetase activity.

We are currently determining the degree of similarity between

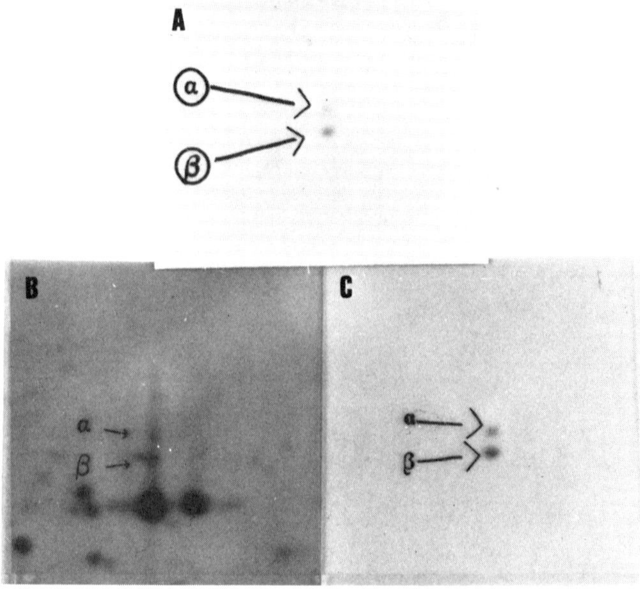

Fig. 7. Double-dimension electrophoretic analysis of *N. crassa* glutamine synthetase polypeptides. *N. crassa* was grown on glutamate as the sole nitrogen source; poly-(A) enriched RNA was isolated as described (17). The RNA was further purified by sedimentation through a 5-20% sucrose gradient. One fraction out of twenty, containing the highest glutamine synthetase mRNA activity, was translated in a reticulocyte-lysate system in the presence of $[^{35}S]$ methionine. Translation products were immunoprecipitated with anti-glutamine synthetase antibody using purified *N. crassa* glutamine synthetase as carrier. Proteins were analyzed by isoelectric focusing in the first dimesion and by electrophoresis in the presence of SDS and 7 M urea in the second dimension. Areas of the gels centered around the glutamine synthetase polypeptides are shown. Arrows indicate the position of α and β. A and C, specific immunoprecipitate; A, stained gel; C, fluorogram; B, fluorogram of total protein.

both α and β polypeptides. Analysis of tryptic products, amino acid composition, and immunochemical specificity, should define this situation. Since the subunit heterogeneity of *N. crassa* glutamine synthetase was unknown at the time of their work, it is not clear whether the amino acid composition obtained by Lin and Kapoor (24) was on the α, β, or α and β polypeptides.

DISCUSSION

Our data indicate that the nitrogen source regulates the levels of mRNA specific for glutamine synthetase. This in turn results in different rates of enzyme synthesis and different glutamine synthetase activities (Table I). Although direct information on transcrip-

TABLE I

Specific activity, relative rates of synthesis and degradation, and specific mRNA levels of Neurospora crassa glutamine synthetase

The data on specific activity and relative rates of synthesis and degradation were taken from reference (30); the data on specific mRNA levels were taken from reference (17). Data were normalized to those of the culture grown on glutamate, set arbitrarily at 1.0.

Nitrogen Source	Specific Activity	Synthesis	Degradation	mRNA
Glutamate	1	1	1	1
Glutamine	0.08	0.08	1.1	0.09

tion is not yet available, the fact that *Neurospora crassa* conidia induce glutamine synthetase mRNA upon incubation (33) suggests that this system can be regulated at the level of specific gene expression.

It has been recently found (Mora *et. al.*, chapter 11) that when a *N. crassa* culture grown on glutamine is transferred to a medium containing glutamate as sole nitrogen source, glutamine synthetase activity increases rapidly. Polysomal RNA isolated before and after this shift has been translated *in vitro* in a cell-free protein synthesizing system. It was found that the capacity of the RNA to prime the synthesis of α and β polypeptides of glutamine synthetase increased after the shift, indicating that both polypeptides are under control of the nitrogen source (unpublished experiments). As pointed out by Mora *et al.* (chapter 11), the ratio of α to β polypeptides depends on the availability of nitrogen in the medium. A high nitrogen concentration can be correlated with the preferential presence of the β polypeptide, while limiting nitrogen correlates with the presence of the α polypeptide (19).

Most studies on the structure of eukaryotic glutamine synthetase indicate that the enzyme is made up of subunits of equal molecular weight (20, 22, 24, 34). However, the identity of the monomers has not been conclusively demonstrated. In the light of the finding of two different polypeptides, each with glutamine synthetase activity in *N. crassa,* the subunit structure of other glutamine synthetases should be reevaluated. In this regard, it is important to point out that two different glutamine synthetases have been reported in *Bacillus caldolyticus* (35), *Rhizobium japonicum* (36, chapter 9), and *Hordeum vulgare* (37).

Our data suggest that there is not a precursor-product relationship between α and β polypeptides of *N. crassa* glutamine synthetase. Experiments are now being undertaken to identify the number of mRNAs and specific genes; a deep genetic analysis as well as the isolation and amplification of specific nucleotide sequences are of prime importance.

We are following different strategies in order to clone the gene or genes that code for *N. crassa* glutamine synthetase. These include the construction of *Neurospora crassa* gene banks, in both plasmid and phage cloning vehicles, as well as the construction and amplification of recombinant plasmids containing DNA complementary to *N. crassa* mRNA. The cloning of cDNA derived from mRNA preparations highly enriched for glutamine synthetase sequences will greatly facilitate the isolation of specific genes. Although partial purification of this mRNA has been obtained by sucrose gradient sedimentation, glutamine synthetase polypeptides represent only a minor fraction of the *in vitro* translated products (Fig. 7) . In this respect, obtaining conditions of maximal glutamine synthetase induction for each of the two polypeptides (Mora *et al,* chapter 11) represents a primary step for purification of the specific mRNA.

Different experimental approaches are being followed to identify specific sequences out of heterogeneous populations of bacterial clones or phage plaques. The possibility of translating glutamine synthetase mRNA in a cell-free system (Fig. 4) allows one to search for its DNA coding sequences. For instance, hybridization of cloned DNA with *N. crassa* RNA followed by translation of the hybrids, *i. e.*, arrested translation (38); or translation of the RNA eluted from such hybrids *i. e.*, positive selection (39). The fact that the concentration of *N. crassa* glutamine synthetase is subject to a very fine regulation will allow the isolation of RNA fractions with different proportions of specific mRNA; these in turn will be useful in identifying specific clones by differential hybridization (40).

Covarrubias *et. al.* (41; see also chapter 8) have recently isolated the gene for *E. coli* glutamine synthetase. Since this enzyme can be purified by the same affinity chromatography procedure used for the *Neurospora* enzyme (Fig. 1), it is possible that both enzymes have some structural similarities. Experiments are now in progress to test the possibility of using *E. coli* DNA sequences as a hybridization probe for *N. crassa* glutamine synthetase genes.

From the data presented and discussed in this paper and by Mora *et al.* (chapter 11), it is clear that *Neurospora crassa* glutamine synthetase constitutes an atractive system to study gene organization and expression. Most important is the fact that metabolic programs have been defined (Mora *et al.*, chapter 11) in which glutamine synthetase participates as part of three different pathways. One is the assimilation of inorganic nitrogen at high substrate concentrations; glutamate dehydrogenase being the first enzyme to fix ammonia and the octameric glutamine synthetase the one responsible for the synthesis of glutamine. The second one is the assimilation of inorganic nitrogen at low substrate concentrations, where the tetrameric glutamine synthetase is responsible for the fixation of ammonia, and glutamate synthase (42) is involved in the synthesis of glutamate. Finally, the operation of a glutamine cycle in growth-restricted conditions, with the participation of a glutamine transaminase, an ω-amidase, glutamate dehydrogenase, and glutamine synthetase. It is our goal to determine the molecular events involved in the genetic expression of these metabolic programs.

Acknowledgments. We acknowledge the inspiration, encouragement and criticism that we have received through the collaboration with the research group of Dr. Jaime Mora. We are grateful to Patricia Ferrera and Elena Palacios for excellent technical assistance. This research was supported in part by the Consejo Nacional de Ciencia y Tecnología, México.

REFERENCES

1. Jeffreys, A. J., and Flavell, R. A. (1977). *Cell 12,* 1097-1108.
2. Breatnach, R., Mandel, J. L., and Chambon, P. (1977). *Nature 270,* 314-318.
3. Brack, C., and Tonegawa, S. (1977). *Proc. Natl. Acad. Sci. U. S. A. 74,* 5652-5656.
4. Tilghman, S. M., Tiemeier, D. C., Seidman, J. G., Peterlin, B. M., Sullivan, M., Maijel, J. V., and Leder, P. (1978). *Proc. Natl. Acad. Sci. U. S. A. 75,* 725-729.
5. Valenzuela, P., Venegas, A., Weinberg, F., Bishop, R., and Rutter, W. J. (1978). *Proc. Natl. Acad. Sci. U. S. A. 75,* 190-194.
6. Woolfolk, C. A. and Stadtman, E. R. (1964). *Biochem. Biophys. Res. Commun. 17,* 313-319.
7. Tate, S. S., Leu, F. Y., and Meister, A. (1972). *J. Biol. Chem. 247,* 5312-5321.
8. Moscona, M., Frenkel, N., and Moscona, A. A. (1972). *Dev. Biol. 28,* 229-241.
9. Sims, A. P., Toone, J., and Box, V. (1974). *J. Gen. Microbiol. 80,* 465-499.
10. Streicher, S. L., Bender, R. A., and Magasanik, B. (1975). *J. Bacteriol. 121,* 320-331.
11. Milman, G., Portnoff, L. S., and Tiemeier, D. C. (1975). *J. Biol. Chem. 250,* 1393-1399.
12. Vichido, I., Mora, Y., Quinto, C., Palacios, R., and Mora, J. (1978). *J. Gen. Microbiol. 106,* 251-259.
13. Limón-Lason, J., Lara, M., Resendiz, B., and Mora, J. (1977). *Biochem. Biophys. Res. Commun. 78,* 1234-1240.

14. Dávila, G., Sánchez, F., Palacios, R., and Mora, J. (1978). *J. Bacteriol. 134,* 693-698.

15. Espín, G., Palacios, R., and Mora, J. (1979). *J. Gen. Microbiol. 115,* 59-68.

16. Mora, Y., Chávez, O., and Mora, J. (1980) *J. Gen. Microbiol.,* in press.

17. Sánchez, F., Campomanes, M., Quinto, C., Hansberg, W., Mora, J., and Palacios, R. (1978). *J. Bacteriol. 136,* 880-885.

18. Sánchez, F., Calva, E., Campomanes, M., Blanco, L., Guzmán, J., Saborío, J. and Palacios, R. (1980). *J. Biol. Chem., 255,* 2231-2234.

19. Dávila, G., Lara, M., Guzmán, J. and Mora, J. (1980). *Biochem. Biophys. Res. Commun., 92* (1), 134-140.

20. Palacios, R. (1976). *J. Biol. Chem. 251,* 4787-4791.

21. Stadtman, E. R., and Ginsburg, A. (1974). *The Enzymes 10,* 755-807.

22. Meister, A. (1974). *The Enzymes 10,* 699-754.

23. Kapoor, M., and Bray, D. (1968). *Biochemistry 7,* 3583-3589.

24. Lin, W. S., and Kapoor, M. (1978). *Can. J. Biochem. 56,* 927-933.

25. Haschemeyer, R. H. (1970). *Adv. Enzymol. Relat. Areas Mol. Biol. 33,* 71-118.

26. Tiemeier, D. C., and Milman, G. (1972). *J. Biol. Chem. 247,* 2272-2277.

27. Stahl, J., and Jaenicke, L. (1972). *Eur. J. Biochem. 29,* 401-407.

28. Palacios, R., Campomanes, M., and Quinto, C. (1977). *J. Biol. Chem. 252,* 3028-3034.

29. Sánchez, F., Dávila, G., Mora, J., and Palacios, R. (1979). *J. Bacteriol. 139,* 537-543.

30. Quinto, C., Mora, J., and Palacios, R. (1977). *J. Biol. Chem. 252,* 8724-8727.

31. Schimke, R. T. (1975). *Methods Enzymol. 40,* 241-266.

32. Espín, G., and Mora, J. (1978). *J. Gen. Microbiol. 104,* 233-240.

33. Hansberg, W., Espín, G., Palacios, R., and Sánchez, F. (1979). *Develop. Biol. 73,* 68-75.

34. Mc Parland, R. H., Guevara, J. G., Becker, R. R., and Evans, H. J. (1976). *Biochem. J. 153,* 597-606.

35. Wedler, F. C., Kenney, R. M., Ashour, A. E., and Carfi, J. (1978). *Biochem. Biophys. Res. Commun. 81,* 122-126.

36. Darrow, R. A. and Knotts, R. R. (1977). *Biochem. Biophys. Res. Commun. 78,* 554-559.

37. Mann, A. F., Fentem, P. A., and Stewart, G. R. (1979). *Biochem. Biophys. Res. Commun. 88,* 515-521.

38. Paterson, B. M., Roberts, B. E., and Kuff, E. L. (1977). *Proc. Natl. Acad. Sci. U.S.A. 74,* 4370-4374.

39. Ricciardi, R. P., Miller, J. S., and Roberts, B. E. (1979). *Proc. Natl. Acad. Sci. U.S.A. 76,* 4927-4931.

40. St. John, T. P., and Davis, R. W. (1979). *Cell 16,* 443-452.

41. Covarrubias, A., Sánchez-Pescador, R., Osorio, A., Bolivar, F., and Bastarrachea, F. (1980). *Plasmid, 3,* 150-164.

42. Hummelt, G., and Mora, J. (1980). *Biochem. Biophys. Res. Commun., 92,* (1), 127-133.

11. Glutamine Metabolism in *Neurospora crassa*

JAIME MORA, GUILLERMO DÁVILA, GUADALUPE ESPÍN, ALICIA GONZÁLEZ, JOSEFINA GUZMÁN, GEORGINA HERNÁNDEZ, GISELA HUMMELT, MIGUEL LARA, ESPERANZA MARTÍNEZ, YOLANDA MORA AND DAVID ROMERO.

Departamento de Biología Molecular, Instituto de Investigaciones Biomédicas, Universidad Nacional Autónoma de México
México 20, D. F., Apto. Postal 70228
MEXICO

SUMMARY

Glutamine and the carbon and the nitrogen source, regulate *Neurospora crassa* glutamine synthetase (GS) at different levels. In nitrogen excess, glutamine regulates its own synthesis by decreasing the specific mRNA for an octameric GS. Restricting the carbon source increases the degradation of the octametric GS. On the other hand, nitrogen-restriction induces the disappearance of the octametric GS and the synthesis of a tetrametric GS. It has been found that the tetrametric GS is composed mainly of a polypeptide different than that present in the octameric form.

When nitrogen is in excess, ammonium is fixed by glutamate dehydrogenase (GDH-NADPH) and by the octameric GS. However, ammonium in low amounts is assimilated by the tetrameric GS and a glutamate synthase which we now report to be present in *Neurospora*.

Neurospora adapts to grow in the presence of limiting ammonium, and in this condition synthesizes preferentially cell wall over protein. This occurs through the mobilization of nitrogen sources that are converted to glutamine. The activity of GS is determinant for the mobilization and turnover of nitrogen compounds.

A different situation holds when *Neurospora* is growth-restricted in

ISBN: 0-12-506040-8

excess of nitrogen. In this condition, glutamine, arginine, and other amino acids are accumulated as nitrogen reservoirs, as a result of the operation of a glutamine cycle. It is proposed that through the participation of a glutamine transaminase and an ω-amidase, glutamine is converted irreversibly into amino acids and ammonium. Next GDH-NADPH fixes the ammonium into glutamate, thus providing the other substrate for GS to resynthesize glutamine. The proposed cycle therefore provides a mechanism for non-growing cells to utilize and distribute the carbon and nitrogen compounds from the medium towards the synthesis and accumulation of amino acids.

INTRODUCTION

The metabolism of glutamine has been throughly studied in prokaryotes; glutamine is the final product in the catabolism of nitrogen compounds and in the incorporation of inorganic nitrogen. In addition, this amino acid has a very important role in the distribution of nitrogen for biosynthetic purposes (1). The regulation of the activity (2, 3) and synthesis of glutamine synthetase (4), the enzyme that catalyzes the synthesis of glutamine from glutamic acid and ammonium, has received considerable attention. Furthermore, a regulatory model of nitrogen catabolism has been proposed in which glutamine synthetase plays a central role (5, 6, 7).

The metabolism of glutamine has been also explored in eukaryotic microorganisms. Folkes & Sims (8), using ^{15}N-labeled ammonium in turbidostat cultures of *C. utilis* grown in excess nitrogen, calculated that about 75% of the cellular nitrogen was assimilated directly through glutamate and 12-15% through the amide group of glutamine. The enzymes involved are glutamate dehydrogenase (GDH-NADPH), that fixes the NH_4^+ in glutamate (9) (the GDH-NAD being recognized as a catabolic enzyme), and glutamine synthetase (10). In *Neurospora*, the regulation in the activity of this enzyme by its final products has been described (11).

Studies in *Schizosaccharomyces* have shown that in some species of this genus, glutamine synthetase is the first enzyme to fix ammonium in low concentrations; the glutamine formed is converted into 2 molecules of glutamate by the action of a glutamate synthase in the presence of 2-oxoglutarate and NADH (12). Glutamate synthase has also been found in *Saccharomyces cerevisiae* but in very low amounts, and its significance is unclear (13). In contrast, the function in this organism of GDH-NADPH in the synthesis of glutamate from NH_4^+ is very well established, and even a regulatory role in nitrogen repression has been proposed for this enzyme and for ammonium. Thus, it seems that GDH mediates nitrogen catabolite repression or arginase through an NH_4^+ signal (14, 15). Other factors involved in the regulation of arginase have also been described (16).

Some reports have appeared indicating that in *Neurospora* glutamine plays a role in nitrogen repression. Thus, the induction of arginase (17) and of nitrate reductase (18) is repressed by glutamine. An effect of ammonium on the repression of the induction of arginase was shown to be the result of its conversion to glutamine (17). Similar results were reported by Van de Poll (19) in *S. carlsbergensis* and by Hynes in *A. nidulans* (20). The presence of high concentrations of arginine in *Neurospora* prevents the ammonium-mediated repression of the induction of arginase (21). A nitrogen regulatory mutant (*nit-2*) has been found which is unable to derepress amino acid transport; it has been suggested that this system is regulated by ammonium (21).

This paper presents several aspects of the metabolism of glutamine in *Neurospora crassa* and deals with the following subjects: a) the regulation of glutamine synthetase by the nitrogen and the carbon source; b) the participation in the synthesis of glutamate and glutamine of glutamine synthetase (GS), glutamate dehydrogenase (GDH-NADPH), and glutamate synthase (GOGAT) in excess and in ammonium limited cultures; c) the structure-function relations that exist between different oligomeric states of GS; d) the role of glutamine synthesis in the turnover that occurs in the absence of added nitrogen; and e) the distribution of nitrogen throughout the operation of a glutamine cycle in growth-restricted cultures.

RESULTS AND DISCUSSION

Synthesis of Glutamine

The synthesis of glutamine was explored by studing the activity, concentration, and synthesis of glutamine synthetase in the presence of its substrates and products. When *Neurospora crassa* was grown on NH_4^+ as the nitrogen source, GS was partially repressed. When grown in glutamate, this compound was in excess but NH_4^+ must also be provided. Since growth on glutamate is suboptimal, we concluded that the high activity of GS found under this condition only partially compensated for the NH_4^+ deficiency. When glutamine was present as the sole nitrogen source, it was in excess inside the cell, the organism grew optimally, and GS activity was very low. Under all these conditions GS was present as an octamer and changes in activity corresponded to changes in concentration of the enzyme (22).

Since there was a correspondence in our studies between the glutamine content of the cell and the concentration of GS (22), glutamine appears as a good candidate to modulate its own synthesis. This proposal is also supported by the repression of GS when glutamine is added to cultures grown on glutamate and by the lack of induction when glutamate is added to cultures grown on glutamine (Fig. 1A). Only when glutamine was removed from the culture medium prior to the addition of glutamate was a rise in GS activity observed (Fig. 1B). The decrease in units of enzyme during the first 2 h after the addition of glutamine indicates that GS activity is lost, an observation that will be discussed later. Since it has been found that the nitrogen source regulates the synthesis of glutamine synthetase (23) by affecting the level of specific mRNA (24), it is possible that glutamine regulates transcription of GS. Ferguson and Sims (9, 25) , have also concluded from their studies with *C. utilis* that glutamine regulates its own synthesis and they have provided evidence that this occurs at the level of GS synthesis.

Glutamine has a key role in the metabolism of carbon and nitrogen. These elements participate in its synthesis, and the distribution of carbon metabolites for biosynthetic purposes depends on the concentration of glutamine. Therefore, it was decided to study the effect of the carbon and/or the nitrogen source on the concentration of GS. It was found that the octameric form of GS was degraded in the absence of a carbon source; this degradation was enhanced if glutamine instead of glutamate was present during the deprivation (Fig. 2). If the mycelium was deprived of a carbon and nitrogen source, GS and protein degradation ocurred (Fig. 3a, c), accompanied by an excretion of ammonium into the medium. Ammonium was incorporated again into the protein when the carbon source (sucrose) was added back (Fig. 3b) (26). We can speculate that the specific degradation of GS in the absence of sucrose is a regulatory mechanism that prevents the synthesis of glutamine and spares carbon metabolites to provide energy for cell maintenance. Also, glutamine has an additional regulatory effect in the degradation of GS. Similar results have been reported in *C. utilis* where GS (9, 25) and the biosynthetic glutamate dehydrogenase (GDH-NADPH) (27) are degraded in the absence of a carbon source. In addition, GS has been found to be degraded when an excess of nitrogen is added to ammonium limited cultures; this was associated with a change in the oligomeric structure of the enzyme (9, 28).

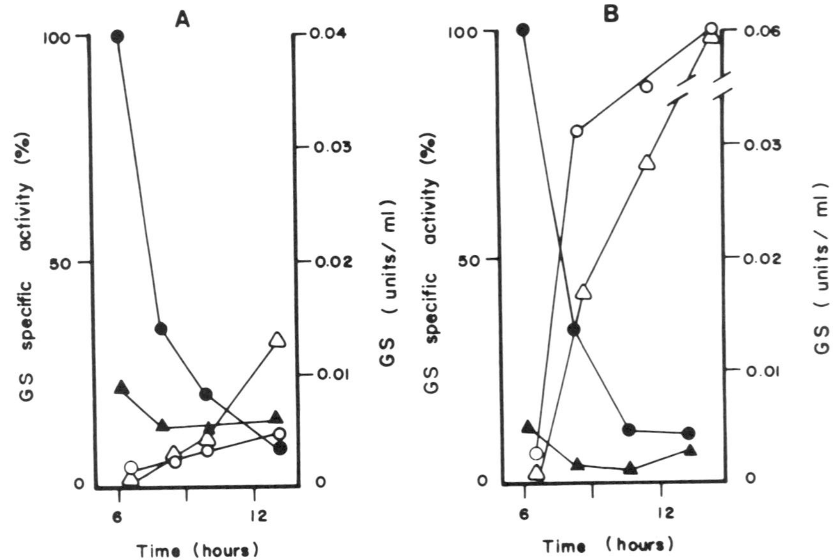

Fig. 1. Effect of the nitrogen source on the activity of glutamine synthetase. A, to conidia germinated 6 h in 10 mM glutamate as nitrogen source 5 mM glutamine was added (closed symbols), and to conidia germinated 6 h in 5 mM glutamine as nitrogen source 10 mM glutamate was added (open symbols). Specific activity of transferase (●, ○) and units of activity per ml (▲, △). B, the cultures were shifted from a medium with glutamate to a medium with glutamine, (closed symbols); and *vice versa* (open symbols); other conditions as above).

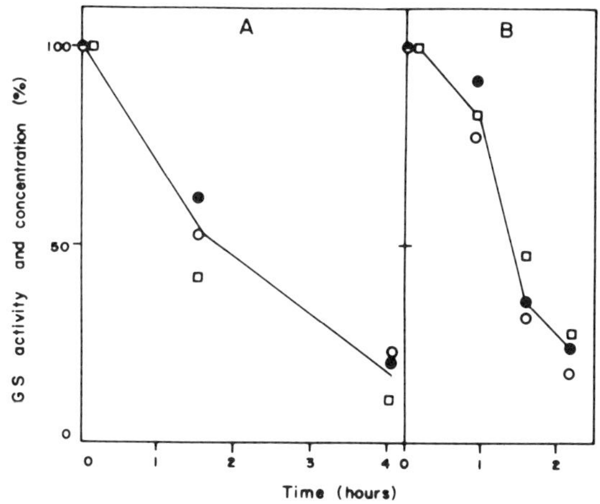

Fig. 2. Effect of the carbon source on the activity and antigen concentration of glutamine synthetase. A culture grown 10 h on sucrose-glutamate was shifted to 5 mM glutamate (A) or to 5 mM glutamine (B) as the sole carbon and nitrogen source. Activity as synthetase (O), or transferase (●) and antigen concentration (□) of GS (From: Mora, Y., Chávez, O., end Mora, J. (1980). J. Gen. Microbiol. *118*: 455-463).

A very different situation holds when the mycelium of *Neurospora* is deprived of the nitrogen source. Under this condition, the activity of GS rises and the octameric form is replaced by a highly active tetrameric GS (26). This effect of the nitrogen and carbon source was studied in more detail in fed-batch cultures limited for ammonium or carbon. It was found that the limitation of ammonium in an excess of the carbon source, led to an accumulation of 2-oxoglutarate with a concomitant elevation of GS in the tetrameric form. On the other hand, limitation of sucrose in the presence of an excess of ammonium, resulted in a decrease in both GS and glutamine (29). Conidia of *Neurospora* germinate and grow exponentially in low ammonium (Fig. 4a). Only when very small amounts of ammonium were supplied continuously, did the fungi grow linearly, and did the growth rate become dependent on the concentration of the administered ammonium (Fig. 4b). Early, during the adaptation to grow in very low levels of ammonium, the tetrameric form of GS appeared, increasing subsequently to become the predominant oligomeric form (Fig. 5). The pools of glutamate and glutamine under this condition were smaller than those found under conditions of nitrogen excess (Fig. 6). Conidia germinated in an excess of nitrogen were also able to adapt to grow in low ammonium. The dry weight of the adapted culture increased at a rate similar to that of cultures

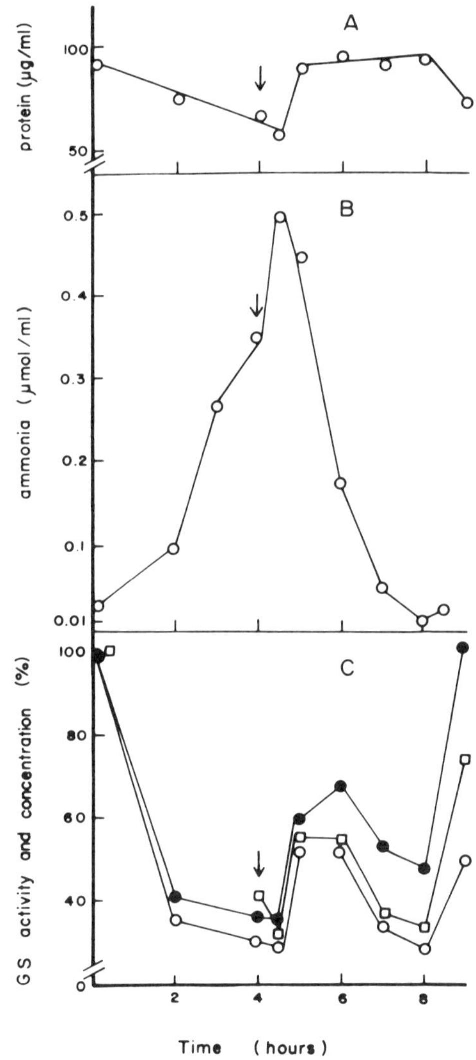

Fig. 3. Effect of carbon and nitrogen deprivation on: A, growth; B, ammonium excreted to the medium; and C, activity of synthetase (O) or transferase (●), and antigen concentration (□) of GS. A 10 h culture grown on sucrose and 5 mM glutamate was shifted to a medium lacking the nitrogen and carbon sources: subsequent addition of sucrose is indicated (↓) (From: Mora, Y., Chávez, O., and Mora, J. (1980). J. Gen. Microbiol. *118*: 455-463).

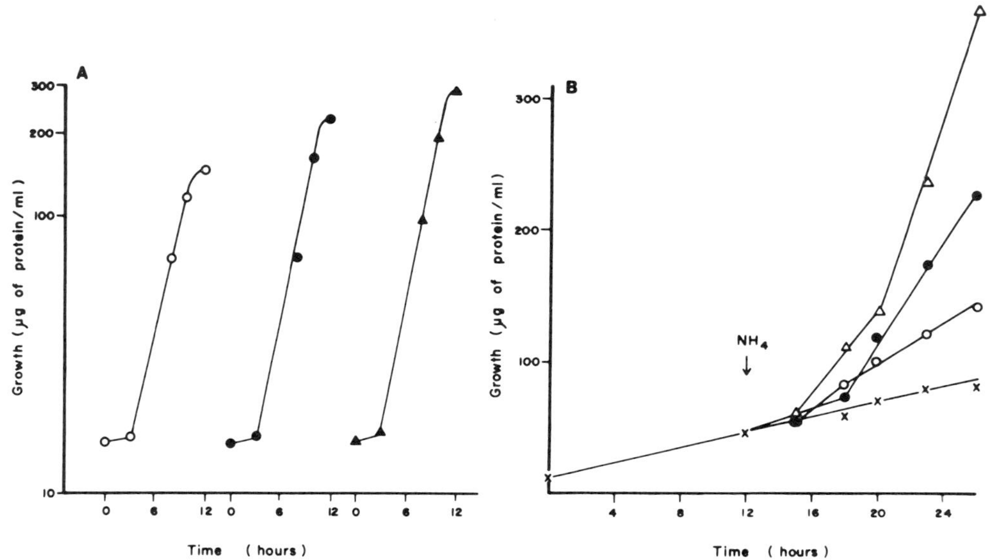

Fig. 4. Adaptation of conidia to growth on limited ammonium. A, wild-type conidia incubated in: fed-batch with a NH_4Cl dilution rate of 0.09 μmol ml−1 h−1 (○) and of 0.18 μmol ml−1 h−1 (●); batch culture with 25 mM NH_4Cl (▲). B. growth in fed-batch with a NH_4Cl dilution rate of 0.03 μmol ml−1 h−1 (x); 12 h later the rate was changed to 0.09 μmol ml−1 h−1 (○) and to 0.18 μmol ml−1 h−1 (●) or 25 mM NH_4Cl was added (△). All incubations were done at 25°C.

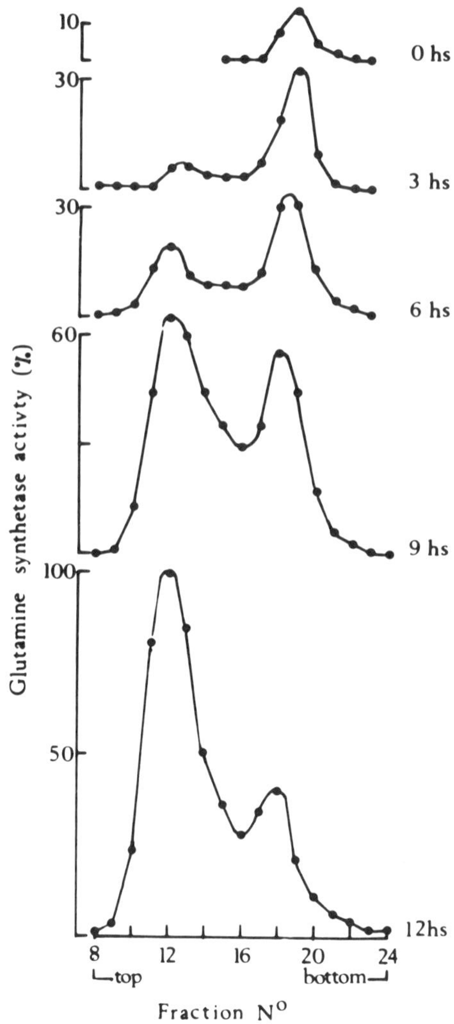

Fig. 5. Sucrose gradient sedimentation of glutamine synthetase from ammonium-limited cultures (rate 0.03 μmol ml^{-1} h^{-1}). Crude extracts of the wild-type strain were collected at various times, centrifuged through a 5-20% sucrose gradient, and transferase activity assayed in the collected fractions. Data are presented as percentage of the peak fraction in the 12 h experiment.

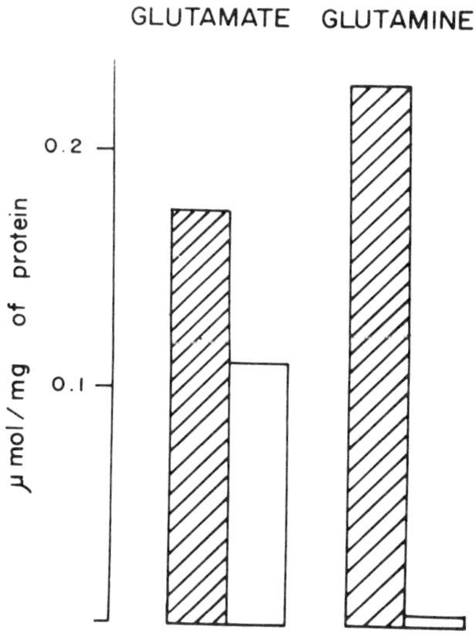

Fig. 6. Accumulation of glutamate and glutamine. The wild-type strain was grown at 25°C for 12 h in 25 mM NH_4Cl (▨) and in ammonium-limited cultures (rate 0.03 μmol ml−1 h−1) (□).

growing in nitrogen excess; however, the rate of protein synthesis was very low initially, increased some hours later, and rates similar to those observed in ammonium limited cultures were reached only 12 h after the shift (Fig. 7). The appearance of tetrameric GS corresponded to this adaptation. These data indicate that tetrameric GS has a role in the fixation of ammonium when ammonium is present in very low concentrations, a condition in which ammonium is also utilized for the synthesis of cell wall. The possibility that ammonium is accumulated only in some part of the hyphae was ruled out by autoradiographies (30) which showed that a pulse of [^{14}C] -methyl ammonium or [^{3}H] -leucine was uniformly distributed along the tubes (Fig. 8).

There is some evidence suggesting that the adaptation to grow in the presence of small amounts of ammonium is not easily reversible. If excess glutamine or ammonium was added to ammonium limited cultures, the resulting increase in the growth rate seldom became optimal. Under these conditions the tetrameric GS was partially preserved, there was some decrease in the enzyme's specific activity, and the glutamine and glutamate content increased several-fold above

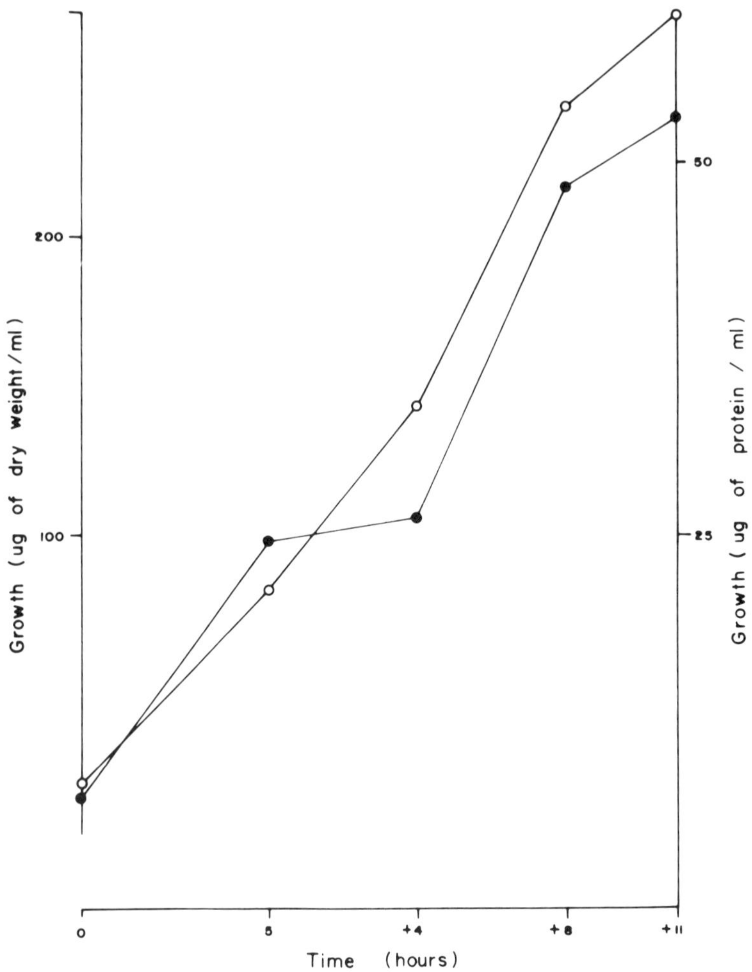

Fig. 7. Adaptation of mycelium to growth in limited-ammonium. Wild-type conidia were germinated at 25°C for 5 h on 10 mM glutamine as nitrogen source and then shifted to fed-batch ammonium-limited cultures at the dilution rate of 0.03 μmol ml^{-1} h^{-1}. Dry weight (O) and protein (●).

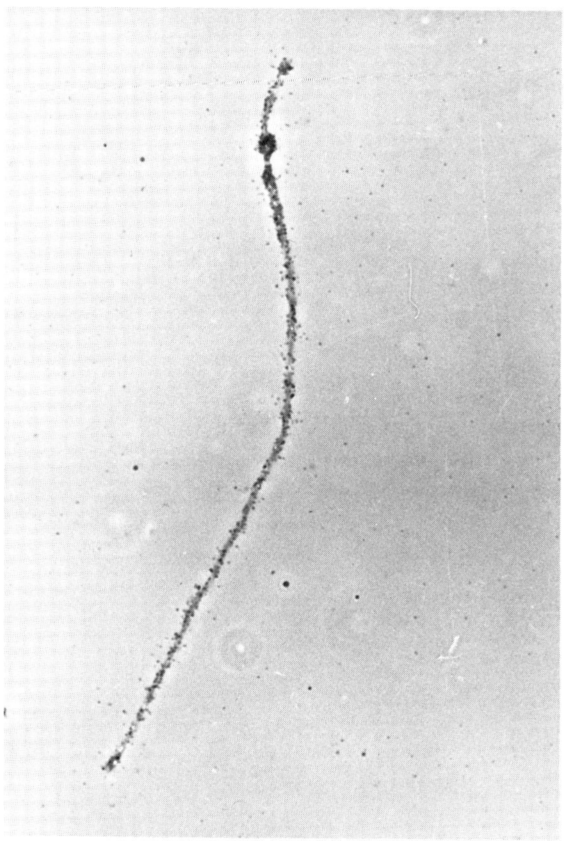

Fig. 8. Autoradiography of a hypha grown under NH_4^+ – limited condition. Granules correspond to radioactivity incorporated into protein after cultures were labelled with 5 μCi/ml of [^{3}H]-leucine for 15 min, followed by fixation with osmium tetroxide and precipitation over jelly-covered slides with TCA. The dry slides were immersed in emulsion and revealed after a 5 day exposure (30).

those found in log cultures (Lara, *et al.*, to be published). In regard to the reversibility of the metabolic state after carbon or nitrogen deprivation, the restitution of the carbon or nitrogen source to the medium was not able to bring about the oligomeric and metabolic state which was present previous to the deprivation (22, 26).

Synthesis of Glutamate

The role in *Neurospora* of NADPH-dependent glutamate dehydrogenase in the synthesis of glutamate is well established. Mutants lacking this activity do not grow well in ammonium and require glutamate for optimal growth (31). When *Neurospora* was grown in NO_3^- or NH_4^+ the activity of GDH was higher than in glutamate as the nitrogen source (Table I). Different results have been obtained by Pateman (32) in *A. nidulans* who found that the levels of GDH were similar in NH_4^+ and in glutamine. Table I also shows that the absence of a nitrogen source did not trigger an elevation of GDH and that GS activity was not coordinated with GDH, since GS was found to be higher in the presence of glutamate and lower in the presence of NH_4^+. Even in the presence of glutamine, where GS was almost totally repressed, GDH activity was only 2 to 4-fold lower than in the presence of ammonium. The elevated GDH activity observed in the presence of nitrate and ammonium could explain why in these conditions a high activity of GS is not necessary. It has been reported that in *Neurospora* the carbon source has a positive role and glutamate a negative role in the regulation of GDH-NADPH (33). It is not clear why a significant level of GDH activity is still present when growth occurs in the presence of glutamate or glutamine.

In contrast with the role of the GDH-NADPH in the fixation of excess ammonium, we have reported that a strain completely lacking in the activity of this enzyme grows as well as the wild-type on limited ammonium (29). As mentioned above (Fig. 5), a highly active tetrameric GS was found when the wild-type strain was grown on this condition. Since the catabolic GDH-NAD has a low affinity for ammonium (34), the presence of a glutamate synthase in *Neurospora* was sought. This activity was found in a strain lacking GDH-NADPH and growing on limiting ammonium. It is interesting that the assay conditions, the purification procedure, and the cofactor (NADH) requirement for the glutamate synthase of *Neurospora* (35) was similar to that present in plant nodules (36). We would like to propose that in *N. crassa* as well as in plants (37), the ammonium

TABLE I

Glutamate dehydrogenase (NADPH) and glutamine synthetase activities in wild-type strain [a]

Nitrogen Source	GDH (NADPH) activity (%)	GS (T) activity (%)
NH_4Cl	100	28
KNO_3	62	45
Glu	36	100
Gln	32	6

[a] GDH (NADPH) and GS (transferase) activities were determined after 8 h of growth at 25°C with 1.5% sucrose as carbon source and the indicated nitrogen source. NH_4Cl and KNO_3 were added at a final concentration of 25 mM; glutamate and glutamine were at a final concentration of 5 mM.

fixed in glutamine by GS is the substrate for a glutamate synthase, producing two molecules of glutamate in the presence of 2-oxoglutarate and NADH. In *Neurospora*, a tetrameric GS would fix the low ammonium, instead of the octameric GS that functions when growth occurs in the presence of excess nitrogen.

Relation Between Structure and Function of Glutamine Synthetase

The isolation and characterization of GS in glutamine auxotrophs, in which the conversion of glutamate to glutamine is impaired, has been very useful for correlating the oligomeric structure to the function of this enzyme (38). One of the mutants, *gln*-la, lacks the octameric GS and only has an active tetrameric form. The fact that this mutant is able to grow after a lag phase in limiting ammonium, supports the contention that the tetrameric GS has a physiological role in the fixation of ammonium at low concentrations. Like the wild-type, this mutant still retains the ability to grow on glutamate at 37°C. More mutants similar to this one have been obtained recently (Dávila, *et al.*, to be published). Since all these mutants only have a tetrameric GS, it is possible that the tetramer and the octamer forms of this enzyme correspond to different gene products. This is also suggested by the independent appearance of the two types of oligomers in different culture conditions (22) (Fig. 5). Recently it has been found that two different monomers contribute to the activity of GS. These monomers have been separated in acrylamide gels in the presence of SDS and urea; one monomer, β, runs ahead of the other, α (39); (see Palacios, *et al.*, Chapter 10). The possibility that the tetramer and the octamer are made up of different monomers was explored by growing the fungi in nitrogen conditions in which one of the oligomers was the predominant form. These cultures were pulsed with [^{3}H] -leucine. The cell-free extracts were sedimented in sucrose gradients and fractions corresponding to the principal peaks of activity were immunoprecipitated with antibody against GS. The proteins dissociated from these precipitates were analyzed by electrophoresis in acrylamide gels containing SDS and urea. Fig. 9 shows that the tetramer is composed mainly of α monomers and the octamer mainly of β monomers. In growth under certain nitrogen conditions, oligomeric states intermediate between the tetramer and the octamer have also been found. These oligomeric forms are composed of both α and β monomers (40) . As to the presence of some α monomers in the octamer and of some β monomers in the tetramer, the following possibilities exist: 1) some of the oligomers are hybrids of the two monomers; 2) as a result of an equilibrium between the two oligomeric states, more tetramers are formed from α monomers than from β monomers, and more octamers are composed of β than of α monomers (40) .

Role of Glutamine Synthetase in the Absence of a Nitrogen Source

In the absence of nitrogen the conidia of *Neurospora* wild-type germinate, increase their dry weight, and form tubes (unpublished observations) . Although the protein content remains unchanged, there is *de novo* protein synthesis (Fig. 10) . Furthermore, the rise of GS during nitrogen deprivation is a consequence of this synthesis (41) . Recently it has been reported that conidia of *N. crassa* synthesize mRNA specific for GS during the first h after germination in the absence of nitrogen (42) .

As stated above, protein degradation in the absence of a carbon source resulted in the excretion of ammonium to the medium. Since upon addition of sucrose, this NH_4^+ is incorporated again into protein (Fig. 3) , glutamine formation must be a necessary event for protein synthesis to occur. Evidence which suggests that glutamine is synthesized in the absence of nitrogen, has been derived from studies of a glutamine auxotroph (*gln*-1b) (38) which accumu-

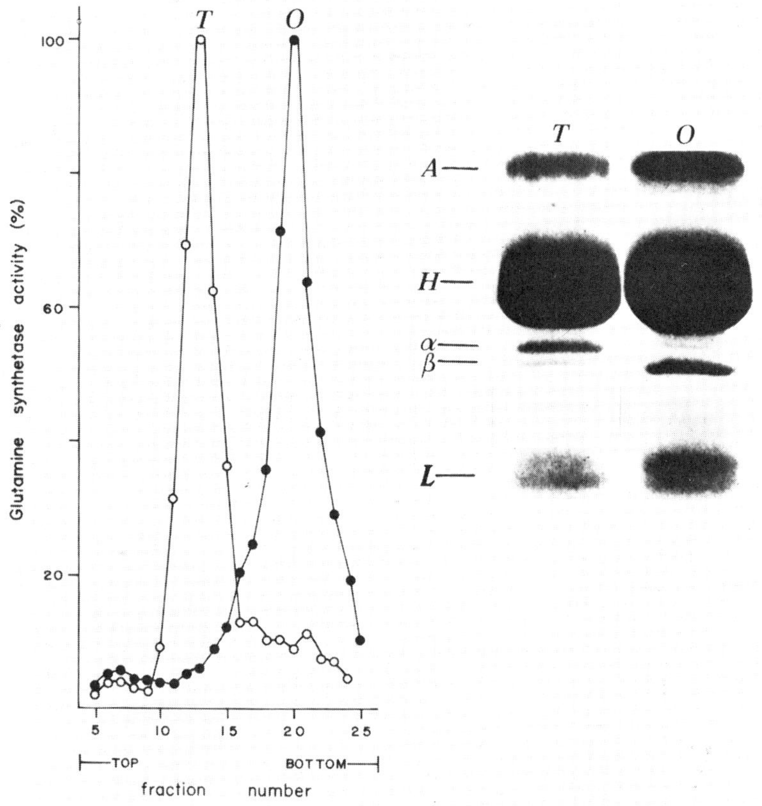

Fig. 9. Monomeric composition of different glutamine synthetase oligomers. A. GS transferase activity in the fractions of a sucrose gradient obtained upon centrifugation of crude extracts of wild-type strain grown in fed-batch NH_4^+ — limited cultures (O) and on 5 mM glutamate (●) at 25°C. B. acrylamide SDS-urea gel stained with Coomassie blue of the immunoprecipitates of fractions corresponding to peaks from the gradient. T, tetramer; O, octamer; A, albumin; H, γ-globulin heavy chain; L, γ-globulin light chain; α and β monomers of GS.

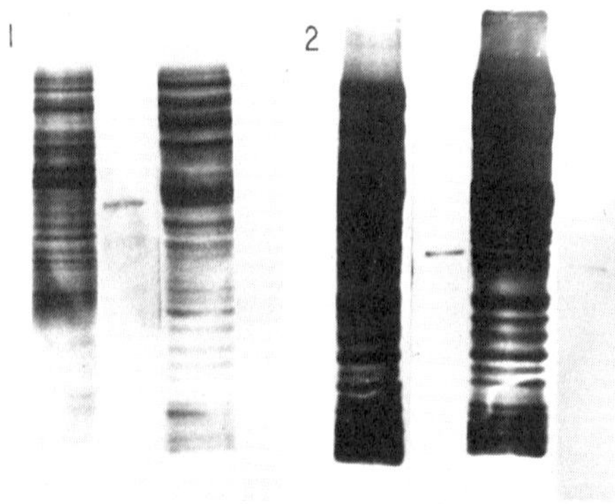

Fig. 10. Conidia of the *prol*-3 strain were incubated either in a nitrogen-free MM or in MM containing 5 mM glutamine as sole nitrogen source. After a 6 h incubation, a 1 h pulse of [^{3}H] -leucine (5 μCi per ml) was added to each culture and extracts were prepared. A to D. total protein (A, C) and glutamine synthetase (B, D) were isolated by precipitation with trichloroacetic acid or by specific immunoprecipitation (23). The samples were then subjected to acrylamide gel electrophoresis and processed for fluorography. Conidia had been incubated in nitrogen-free MM for A and B; and in MM containing glutamine as sole nitrogen source for C and D. E to H. Extracts from either culture were mixed with an extract from *74*-A strain grown for 6 h on MM containing 5 mM-glutamate as sole nitrogen source and pulsed with [^{14}C] -leucine (0.5 μCi per ml), as above. Total protein (E, G) and glutamine synthetase (F, H) from both mixtures were isolated and processed as above. Conidia had been incubated in nitrogen-free MM for E and F, and in MM containing glutamine as sole nitrogen source for G and H. (From: Espin, G., Palacios, R., and Mora, J. (1979). J. Gen. Microbiol. *115*: 59-68).

lates glutamate in this condition (Fig. 11). Since in this strain the formation of tubes is restricted, it is possible then that in the wild-type strain some of the glutamine synthesized during nitrogen deprivation is utilized for cell wall formation.

The role of GS in *N. crassa* grown in the absence of nitrogen was analyzed by measuring protein synthesis after inhibiting the activity of GS with methionine sulphoximine (MS) (43). Because it is known that MS can be degraded by *Neurospora* and thus provide nitrogen for amino acid synthesis (41), an amino acid auxotroph (*leu*-3) was used. Fig. 12 shows that 10 min after incubation with MS no activity of GS was detected, the intracellular glutamine dropped to very low levels, and a slight rise in glutamate was observed, as expected. In Fig. 13 the inhibition by MS of [^{3}H]-leucine incorporation into protein is presented. The inhibition of protein synthesis by MS in the *leu*-3 strain was even higher, since the specific radio-

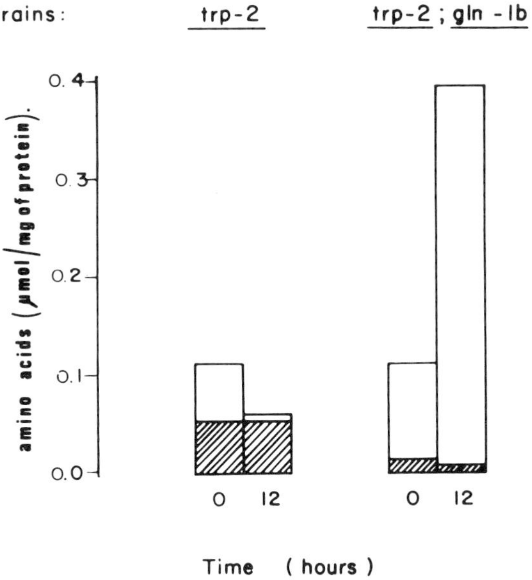

Fig. 11. Amino acid pools during nitrogen deprivation in the tryptophan auxotroph *trp*-2, and the double mutant *trp*-2; *glm*-1b. Both strains were incubated 12 h at 25°C in the absence of a nitrogen source. Glutamate (□) and glutamine (▨).

activity of this amino acid was 67% higher in MS than without (Fig. 13). Since similar results were obtained with the wild-type strain, the contribution towards the synthesis of amino acids by the nitrogen from MS may be insignificant in this short time. Also, a direct effect of MS in translation was eliminated by assaying the effect of this compound in a reticulocyte cell-free system (44).

In the absence of added nitrogen, the precursors for the synthesis of glutamine may come from trace amounts of nitrogen in the culture medium, from metabolic reserves, and/or from the catabolism of macromolecules from viable and non-viable cells. The high activity of GS compensates for the low concentration of ammonium. The synthesized glutamine may participate in at least three processes. First, in the synthesis of cell wall required for tube formation. Second, in *de novo* synthesis of some amino acids by glutamine transamination (45)-this would provide a different route of amino acid formation required for the synthesis of certain specific protein. And third, in a link between degradation and synthesis of macromolecules as a way to mobilize the biomass along the formed tube-this would explain why the mycelia from conidia germinated in the absence of added nitrogen are partially empty.

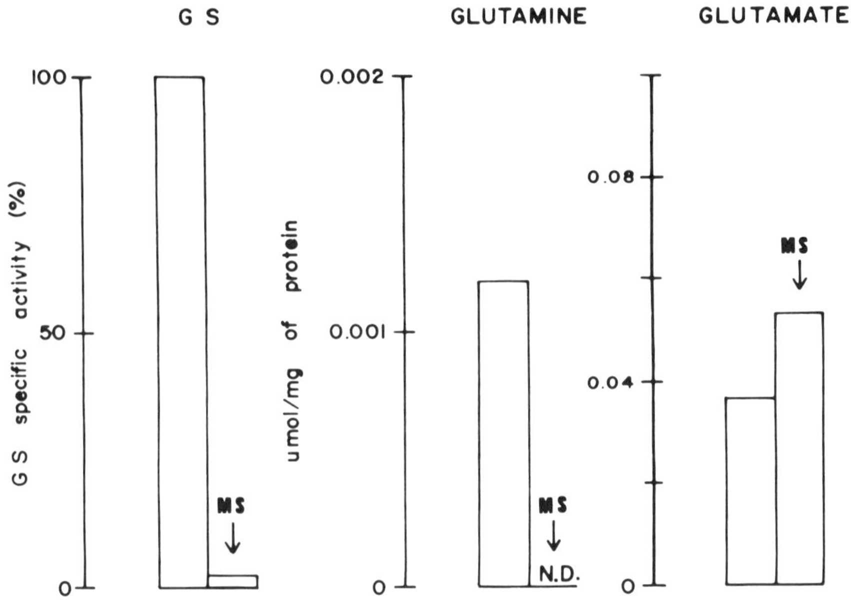

Fig. 12. Glutamine synthetase activity, glutamine and glutamate pools after addition of 0.1 mM methionine sulphoximine (MS). Conidia of the *leu*-3 strain were incubated 9 h in the absence of nitrogen; L-methionine dl-sulphoximine (MS) was then added; 10 minutes later GS activity and amino acid pools were measured (N. D. indicates not detected).

The Glutamine Cycle in the Distribution of Nitrogen During Growth Restriction

When conidia of *N. crassa* were deprived of certain amino acids in the presence of NH_4NO_3, glutamine and arginine were accumulated. Such accumulation was not found in the absence of *de novo* synthesis of pyrimidines, in the presence of cycloheximide, or in the presence of glycerol as the carbon source. Since more than half of the accumulated amino nitrogen is stored as glutamine and arginine, we have proposed that these amino acids are nitrogen and/or carbon reservoirs under non-growth conditions (46). Mycelia of *N. crassa* also accumulate arginine and glutamine when deprived of an amino acid; this accumulation also occurs in the presence of cycloheximide or at the end of exponential growth (Fig. 14). A large proportion of the accumulated arginine is sequestered in an osmotically sensitive compartment (47). These data are relevant to reports that, in *N. crassa,* arginine synthesized endogenously is not as effectively hydrolyzed as exogenous arginine (48, 49). Ornithine and

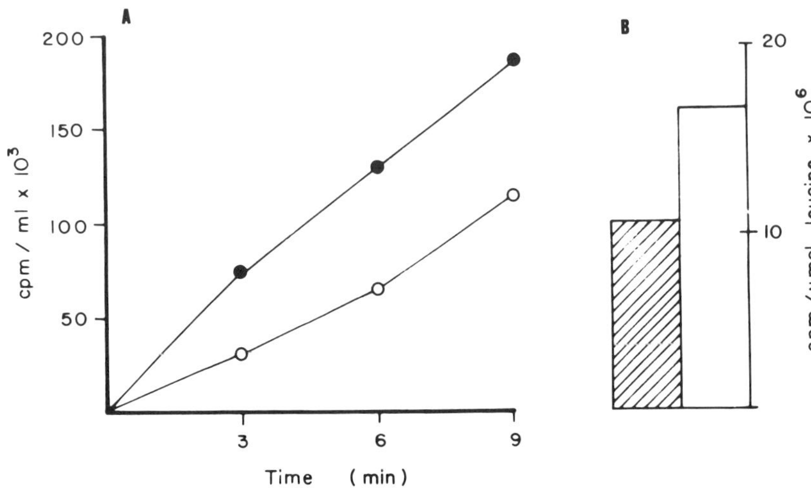

Fig. 13. Effect of MS on protein synthesis in *N. crassa*. A. [^{3}H] leucine incorporation into protein. Conidia of the *leu-3* strain were incubated 9 h in the absence of a nitrogen source; before (●) and 10 minutes after the addition of 0.1 mM L-methionine dl-sulphoximine (O). Radioactivity in total protein (cpm/ml) was measured at the indicated times after the addition of 1 (●) and 2.5 (O) μCi/ml of [^{3}H] -leucine. B. leucine specific radioactivity. After a 10 minute pulse (see fig. 13A), leucine was isolated as described (41) from cell-free extracts of cultures before (▨) and after (□) MS addition.

citrulline have been proposed as modulators of arginase activity. Thus they could contribute to the control of a plausible wasteful catabolism of endogenous arginine by arginase (50).

Davis and collaborators (51, 52) were the first to report the sequestering of most arginine and ornithine that is synthesized when *N. crassa* is growing exponentially, a subject that has been througly studied subsequently (53, 54, 55, 56). Since the biosynthesis and accumulation of glutamine is coupled with the inability to catabolize arginine, we and others have suggested that compartmentation of endogenously synthesized arginine unables it to prevent its own synthesis and induce its own catabolism (47, 57).

We have reported that glutamine prevents the catabolism of exogenous arginine by preventing the induction of arginase by arginine (17). We have also shown that arginase is not induced by exogenous arginine after the mycelia have been deprived of an amino acid. Thus we have proposed that during growth-restricted conditions, the catabolism of non-sequestered arginine is prevented by the accumulated glutamine (47).

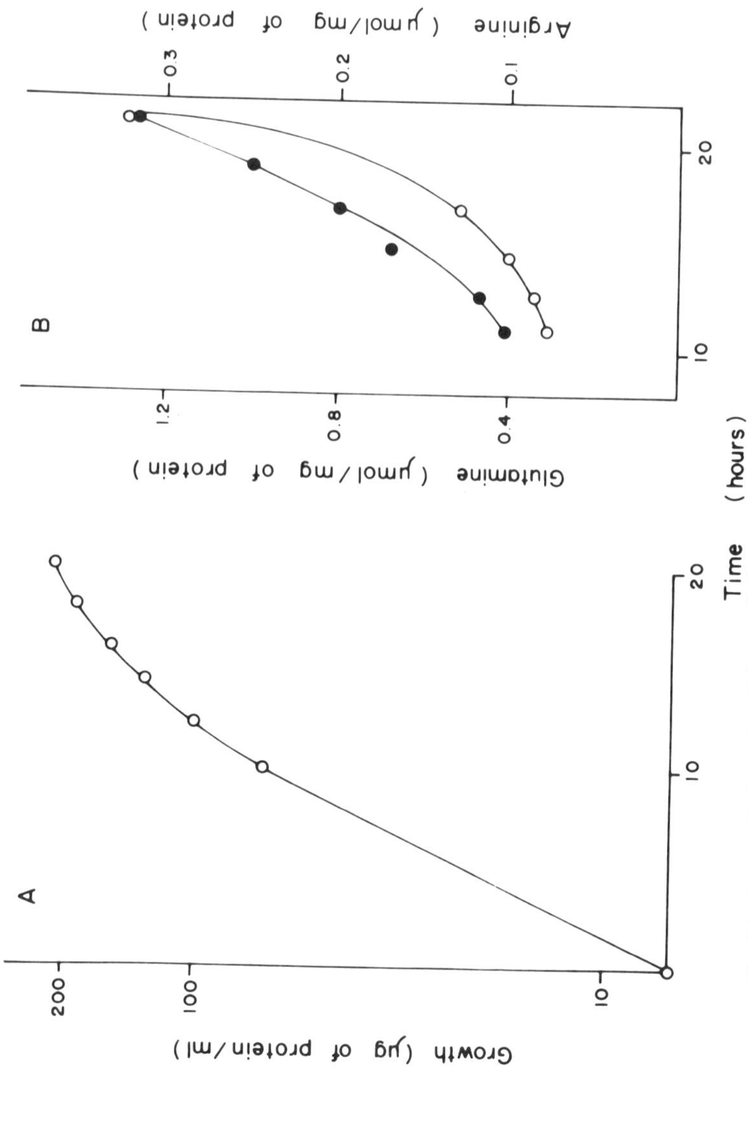

Fig. 14. Accumulation of arginine and glutamine by mycelium of strain *prol-3* in the pre-stationary phase of growth. A. growth of strain *prol-3* in MM plus proline (100 µg per ml). B. pools of glutamine (○) and arginine (●) measured at the times indicated in (A). (from: Mora, Y., Espín, G., Willms, K., and Mora, J. (1978). J. Gen. Microbiol. *104*: 241-250).

The rate of accumulation of glutamine and arginine increased when ammonium was added to conidia which had been preincubated without a required amino acid (Table II) and this accumulation was no longer prevented by cycloheximide or by glycerol as carbon source (Table III). The increased accumulation of these amino acids was paralleled by an increase in specific activity of glutamine synthetase which, in one case at least, involved *de novo* enzyme synthesis (Table III, Fig. 10) (41).

Evidence indicating that the fixation of ammonium into glutamate is a necessary step before transfer of the amide nitrogen of glutamine to arginine, is given by the fact that strains that lack the activity of the enzymes that assimilate ammonia into glutamate do excrete ammonium. The synthesis of glutamine is also required because *glm* auxotrophs are unable to accumulate arginine in the presence of glutamine but do accumulate glutamate instead (41).

These observations can be explained on the basis of a glutamine cycle (Fig. 15). The first step of this cycle is the synthesis of different amino acids through an irreversible transamination with glutamine as a donor; the second is the hydrolysis of 2-oxoglutaramate

TABLE II

Effect of preincubation in nitrogen-free medium on the accumulation of glutamine and arginine by conidia of the prol-3 *strain*

Conidia with or without 12 h preincubation in nitrogen-free MM were incubated in MM with NH_4NO_3 (25 mM) as nitrogen source

		Conc [μmol (mg protein)$^{-1}$]	
Culture conditions	Time (h)	Arginine	Glutamine
Not starved	0	0.032	0.068
	6	0.290	0.530
	12	0.520	1.035
Starved	0	0.015	0.031
	6	0.998	1.840
	12	1.120	3.190

(From: Espín, G., Palacios, R., and Mora, J. (1979). *J. Gen. Microbiol. 115*: 59-68).

TABLE III

Intracellular concentrations of glutamine and arginine, and glutamine synthetase activity in conidia of the prol-3 *strain under different conditions of incubation*

Glutamine synthetase activity is expressed as μmol of γ-glutamyl hydroxamate formed at 30°C min^{-1} (mg protein)$^{-1}$

Condition	Glutamine	Arginine	Glutamine synthetase activity
	Conc [μmol (mg protein)$^{-1}$]		
Not incubated	0.068	0.032	0.004
Incubated 12 h in MM	1.035	0.520	0.035
Incubated 12 h in MM plus cycloheximide (10 μg ml^{-1})	0.400	0.163	0.010
Preincubated 12 h in nitrogen-free MM and then NH_4NO_3 (25 mM) added for additional 12 h.	3.190	1.120	0.149 *

* Activity was measured after 12 h incubation in nitrogen-free MM, just before NH_4NO_3 was added. (From: Espín, G., Palacios, R., and Mora, J. (1979). *J. Gen. Microbiol. 115*: 59-68.

to 2-oxoglutarate and ammonium by an ω-amidase; the third being the formation of glutamate; and the fourth the resynthesis of glutamine. Evidence for the degradation of glutamine by the glutamine transaminase and ω-amidase in other cell systems, has been recently reviewed, (45). These enzymes have been detected in *N. crassa* (58). The proposed cycle would therefore provide a mechanism for non-growing cells to utilize and distribute the carbon and nitrogen from the medium towards the synthesis and accumulation of amino acids, mainly glutamine and arginine.

A nitrogen regulatory mutation (*bm*-2) that accumulates glutamine, arginine, and other amino acids while growing exponentially in the presence of glutamine or NH_4^+ as a nitrogen source and a good carbon source, has been isolated. This mutant is nitrogen limited when growing in the presence of arginine as nitrogen source, in spite of the fact that arginine from the medium is very efficiently accumulated and taken up preferentially to other amino acids. Other characteristics of the *bm*-2 mutant strain are that glutamine in excess

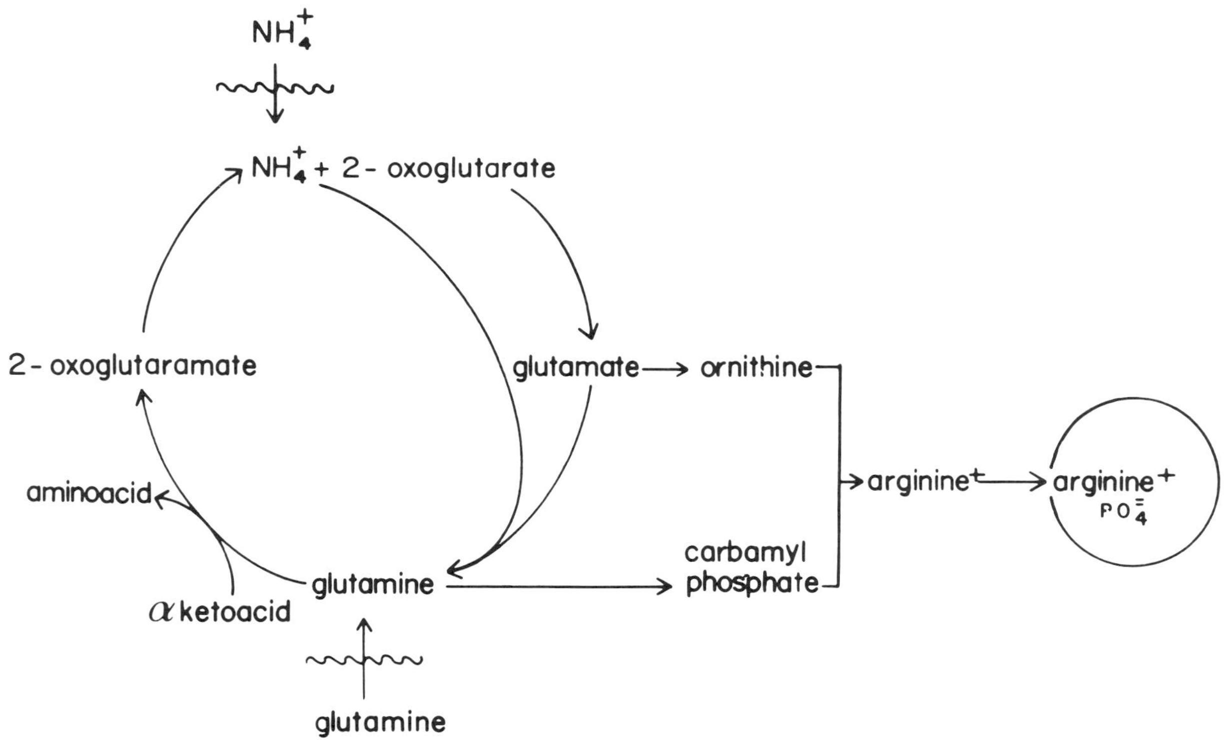

Fig. 15. Diagramatic representation of the glutamine cycle and arginine biosynthesis. Arginine accumulated in vesicles is represented inside the circle; the wavy line represents the cell membrane.

only represses partially the GS and it is resistant to MS in minimal medium. Because of the similarity of this phenotype with what occurs in non-growing conditions (41), we have proposed that in this regulatory mutant the glutamine cycle operates during exponential growth. In support of this hypothesis is the observation that double mutants that in addition to the *bm*-2 mutation have a block in the synthesis of glutamate did not accumulate arginine. Furthermore, the double mutant *bm*-2, *am*-1 is unable to synthesize glutamate, and excretes ammonium to the medium when growing in the presence of glutamine as nitrogen source. However, in distinction to the single mutant *bm*-2, the ammonium from the medium is not assimilated afterwards. These data indicate that in the *bm*-2 mutant strain the ammonium derived from glutamine is efficiently assimilated in glutamate by GDH-NADPH(González, *et al.*, to be published).

Although it is understood that the cycle provides glutamate and glutamine for the synthesis of arginine which is accumulated in vesicles, it is still not clear how the assimilation of nitrogen from the medium, the operation of the cycle, and the synthesis and accumulation of arginine during non-growing conditions are coordinated. The regulatory mutation, just described, is pleiotrophic in the sense that in addition to its ability to sequester arginine, the cycle is operating, and ammonium from the medium is readily assimilated. It is anticipated that studies of this new mutant will provide an explanation of how the metabolism of nitrogen is coordinated in *Neurospora* during growth-restricted conditions.

Finally, on the basis of the results presented in this work, we would like to propose in the similarity of *Neurospora* to plants in the following respects: A) the fixation of low amounts of ammonium by the GS-GOGAT pathway (37) ; B) the distribution of nitrogen by asparagine and glutamine (59, 60) ; and C) the role of GS in the assimilation of ammonium derived from the catabolism of proteins during senescence and seed development (60). All these considerations make *Neurospora* a suitable model for studying assimilation, distribution, and turnover of nitrogen. This biological system has also been useful for gaining a better understanding of some of the molecular mechanisms responsible for gene expression in eukaryotic microorganisms (Palacios, *et al.*, Chapter 10).

Acknowledgments: We would like to acknowledge the benefits obtained from the collaboration with Dr. R. Palacios and his group during the realization of this work. Also we wish to thank Dr. E. Calva for critically reviewing this manuscript and to Dr. K. Willms for helping us with the radioautographies. This research was supported in part by, the Consejo Nacional de Ciencia y Tecnología, México.

REFERENCES

1. Tyler, B. (1978). *Ann. Rev. Biochem. 47*: 1127-1162.
2. Ginsburg, A., and Stadtman, E. R. (1973). In *The Enzymes of Glutamine Metabolism* (Prusiner-Stadtman, eds.) pp. 9-43, Academic Press, New York.
3. Wohlhueter, R. M., Schutt, H., and Holzer, N. (1973). In *The Enzymes of Glutamine Metabolism* (Prusiner, Stadtman, (eds.) pp. 45-64, Academic Press. New York.
4. Foor, F., Jansen, K. A., Magasanik, K. B. (1975). *Proc. Natl. Acad. Sci. U.S.A. 72*: 4844-4848.
5. Magasanik, B. (1977). *Trends Biochem. Sci. 2*: 9-12.
6. Pahel, G., and Tyler, B. (1979). *Proc. Natl. Acad. Sci. U. S. A. 76*: 4544-4548.
7. Kustu, S., Burton, D., García, E., McCarter, L., and McFarland, N. (1979). *Proc. Natl. Acad. Sci. U.S.A. 76*: 4576-4580.
8. Folkes, B. F., and Sims, A. P. (1974). *J. Gen. Microbiol. 82*: 77-95.
9. Ferguson, A. R., and Sims, A. P. (1974). *J. Gen. Microbiol. 80*: 159-171.
10. Sims, A. P., Toone, J., and Box, V. (1974). *J. Gen. Microbiol. 84*: 149-162.
11. Kapoor, M., and Bray, D. (1968). *Biochemistry 7*: 3583-3589.
12. Brown, C. M., Burn, V. J., and Johnson, B. (1973). *Nature New Biol. 246*: 115-116.
13. Roon, R. J., Even, H. L., and Larimore, F. (1974). *J. Bacteriol. 118*: 89-95.
14. Dubois, E., Grenson, M., and Wiame, J. M. (1974). *Eur. J. Biochem. 48*: 603-616.
15. Dubois, E., Vissers, S., Grenson, M., and Wiame, J. M. (1977). *Biochem. Biophys. Res. Commun. 75*: 233-239.
16. Dubois, E., Hiernaux, D., Grenson, M., and Wiame, J. M. (1978). *J. Mol. Biol. 122*: 383-406.
17. Vaca, G., and Mora, J. (1977). *J. Bacteriol. 131*: 719-725.
18. Dunn-Coleman, N. S., Tomsett, A. B., and Garret, R. H. (1979). *J. Bacteriol. 139*: 697-700.
19. Van de Poll, K. W. (1973). *FEBS Lett. 32*: 265-266.
20. Hynes, M. J. (1974). *J. Bacteriol. 120*: 1116-1123.
21. Facklam, T. J., and Marzluf, G. A. (1978). *Biochem. Genet. 16*: 343-354.
22. Vichido, I., Mora, Y., Quinto, C., Palacios, R., and Mora, J. (1978). *J. Gen. Microbiol. 106*: 251-259.
23. Quinto, C., Mora, J., and Palacios, R. (1977). *J. Biol. Chem. 252*: 8724-8727.
24. Sánchez, F., Campomanes, M., Quinto, C., Hansberg, W., Mora, J., and Palacios, R. (1978). *J. Bacteriol. 136*: *880-885*.
25. Ferguson, A. R., and Sims, A. P. (1974). *J. Gen. Microbiol. 80*: 173-185.
26. Mora, Y., Chávez, O., and Mora, J. (1980). *J. Gen. Microbiol., 118*: 455-463.
27. Hemmings, B. A. (1978). *J. Bacteriol. 133*: 867-877.
28. Sims, A. P., Toone, J., and Box, V. (1974). *J. Gen. Microbiol. 80*: 485-499.
29. Limón-Lason, J., Lara, M., Reséndiz, B., and Mora, J. (1977). *Biochem. Biophys. Res. Commun. 78*: 1234-1240.
30. Orlowski, M., and Sypherd, P. S. (1978). *J. Bacteriol. 133*: 399-400.

31. Fincham, J. R. S. (1950). *J. Biol. Chem. 182*: 61-73.
32. Pateman, J. A. (1969). *Biochem. J. 115*: 769-775.
33. Kappor, M., and Grover, A. K. (1970). *Can. J. Microbiol. 16*: 33-40.
34. Wootton, J. C., Kavanagh, J. P., Baron, A. J., and Lovett, M. G. (1976). *Biochem. J. 159*: 803-806.
35. Hummelt, G., and Mora, J. (1980). *Biochem. Biophys. Res. Commun., 92*: 127-133.
36. Boland, M. J., and Benny, A. G. (1977). *Eur. J. Biochem. 79*: 355-362.
37. Miflin, B. J., and Lea, P. J. (1977). *Ann. Rev. Plant. Physiol. 28*: 299-329.
38. Dávila, G., Sánchez, F., Palacios, R., and Mora, J. (1978). *J. Bacteriol. 134*: 693-698.
39. Sánchez, F., Calva, E., Campomanes, M., Blanco, L., Guzmán, J. Saborío, J. L., and Palacios, R. (1980). *J. Biol. Chem., 255*, 2231-2234.
40. Dávila, G., Lara, M., Guzmán, J., and Mora, J. (1980). *Biochem. Biophys. Res. Commun., 92*: 134-140.
41. Espín, G., Palacios, R., and Mora, J. (1979). *J. Gen. Microbiol. 115*: 59-68.
42. Hansberg, W., Espín, G., Palacios, R., Sánchez, F. (1979). *Develop. Biol. 73*: 68-75.
43. Ronzio, R., and Meister, A. (1968). *Proc. Natl. Acad. Sci. U.S.A. 59*: 164-170.
44. Palacios, R., Campomanes, M., and Quinto, C. (1977). *J. Biol. Chem. 252*: 3028-3034.
45. Cooper, A. J. L., and Meister, A. (1977). *CRC. Critical Reviews in Biochemistry 4*: 281-303.
46. Espín, G., and Mora, J. (1978). *J. Gen. Microbiol. 104*: 233-240.
47. Mora, Y., Espín, G., Willms, K., and Mora, J. (1978). *J. Gen. Microbiol. 104*: 241-250.
48. Castañeda, M., Martuscelli, J., and Mora, J. (1967). *Biochim. Biophys. Acta 141*: 276-285.
49. Davis, R. H. (1967). In *Organizational Biosynthesis* (H. J. Vogel, J. O. Lampen, and V. Bryson, eds.) pp. 303-322, Academic Press, New York.
50. Mora, J., Salceda, R., and Sánchez, S. (1972). *J. Bacteriol. 110*: 870-877.
51. Subramanian, K. N., Weiss, R. L., and Davis, R. H. (1973). *J. Bacteriol. 115*: 284-290.
52. Weiss, R. L., and Davis, R. H. (1973). *J. Biol. Chem. 248*: 5403-5408.
53. Weiss, R. L. (1973). *J. Biol. Chem. 248*: 5409-5413.
54. Weiss, R. L. (1976). *J. Bacteriol. 126*: 1173-1179.
55. Legerton, T. L., and Weiss, R. L. (1979). *J. Bacteriol. 138*: 909-914.
56. Martinoia, E., Heck, U., Boller, T. H., Wiemken, A., and Matile, P. H. (1979). *Arch. Microbiol. 120*: 31-34.
57. Cybis, J. J., and Davis, R. H. (1975). *J. Bacteriol. 123*: 196-202.
58. Monder, C., and Meister, A. (1958). *Biochim. Biophys. Acta 28*: 202-203.
59. Streeter, J. G. (1977). *Plant Physiol. 60*: 235-239.
60. Storey, R., and Beevers, L. (1978). *Plant Physiol. 61*: 494-500.

12. The Role of Glutamine in Ammonia Assimilation and Reassimilation in Plants

BENJAMIN J. MIFLIN, PETER J. LEA AND ROGER M. WALLSGROVE

Biochemistry Department
Rothamsted Experimental Station
Harpenden, Herts., U. K.

SUMMARY

Glutamine holds the key position in the nitrogen metabolism of higher plants. It is the portal through which all inorganic nitrogen sources eventually enter into organic combination. Recent research has shown that it is also a central compound in (a) the metabolism of stored-N, in the form of proteins, into nitrogen transport compounds, (b) in the reutilization of those compounds in the recipient plant organ and (c) in the reassimilation of ammonia released in photorespiration. The two enzymes important in the synthesis of glutamine and the transfer of its amide-N into amino acids, glutamine synthetase and glutamate synthase respectively, have been isolated from plants and several of their properties studied. Subcellular localization studies have shown that glutamine synthetase is present in the plastids and the cytosol whilst glutamate synthase is present only in the plastids. It is also probable that glutamine synthetase exists as two forms in separate compartments which may have different regulatory and catalytic properties. Although various studies of possible feedback regulatory mechanisms on glutamine synthetase have been made it is argued that they are unlikely to be of any physiological significance in land-based higher plants. It is suggested that the single factor that is of most significance in the flow of available N into organic combination is the energy status of the plant.

ISBN: 0-12-506040-8

INTRODUCTION

Animals are unable to achieve any net gain of amino-N from inorganic-N but rather must have a source of amino-N (including specific essential amino acids) in their diet. The vast majority of this nitrogen is ultimately derived from higher plants. Consequently the conversion of inorganic-N to amino acid-N in these plants is a topic of considerable importance even though this is not always reflected in the amount of research effort devoted to it. The inorganic sources of nitrogen used by plants are N_2 gas (see below) or combined-N in the form of nitrate, ammonia or urea.

For the majority of the crops harvested by man or his domestic animals combined-N is added as fertilizer at around 40-50 Mt annually. Because of the reactions in the soil the combined-N is chiefly oxidised to NO_3^- before incorporation into the plant. Irrespective of the form of inorganic-N used, it must be reduced to NH_3 before it is assimilated into the organic form. As will be discussed below, this assimilation occurs (in all but the fungi and also bacteria growing on high levels of ammonia; however, see chapter 11, by Mora *et al.*) via the formation of glutamine catalysed by the enzyme glutamine synthetase. Besides this primary assimilation of ammonia it has recently become obvious that considerable recycling of nitrogen out of, and back into, organic combination occurs during its passage through the plant. Evidence will be presented to show that this reassimilation also occurs via glutamine and is catalysed by glutamine synthetase. Thus the conclusion is that glutamine holds a key position in the flow of nitrogen in the biosphere. Although others (18) have claimed that ribulose 1, 5-bisphosphate carboxylase is the protein on which all life ultimately depends, an equally valid case could be made out for glutamine synthetase.

PRIMARY ASSIMILATION OF AMMONIA

The Enzymes of Ammonia Assimilation

Following on from the discovery of NADPH glutamate synthase in bacteria by Tempest, Meers and Brown (82) was the realization that ammonia assimilation via glutamine could be the first step in the formation of amino-nitrogen. Subsequently a similar but subtly different enzyme, ferredoxin-dependent glutamate synthase, was found in the leaves of higher plants (35), which also contain an NAD (P) H-dependent enzyme in non-green tissues (17,52). The flow of NH_3 into amino acids can therefore occur via what others (68) have termed the glutamate synthase cycle (Fig. 1).

(a) Glutamine synthetase

The enzyme has been purified to apparent homogeneity from pea leaves (60, 61), the plant tissue of soyabean root nodules (43) and *Lemna* (69). The enzyme has a mol. wt. in the range of 330,000-376,000 daltons, similar to the mammalian but smaller than the bacterial enzyme, and is probably composed of eight subunits each of a mol. wt. of about 45,000. The subunits in the soyabean enzyme are arranged in two sets of planar tetramers parallel to each other. The enzymes have biosynthetic activity with both Mg^{2+} and Mn^{2+} with respective pH optima of ~9.0 and ~5.0 with little difference in the absolute rates at the two maxima. Studies on the *Lemna* enzyme (69) have shown that the enzyme possesses four SH groups, the number of which varies according to the state of the enzyme.

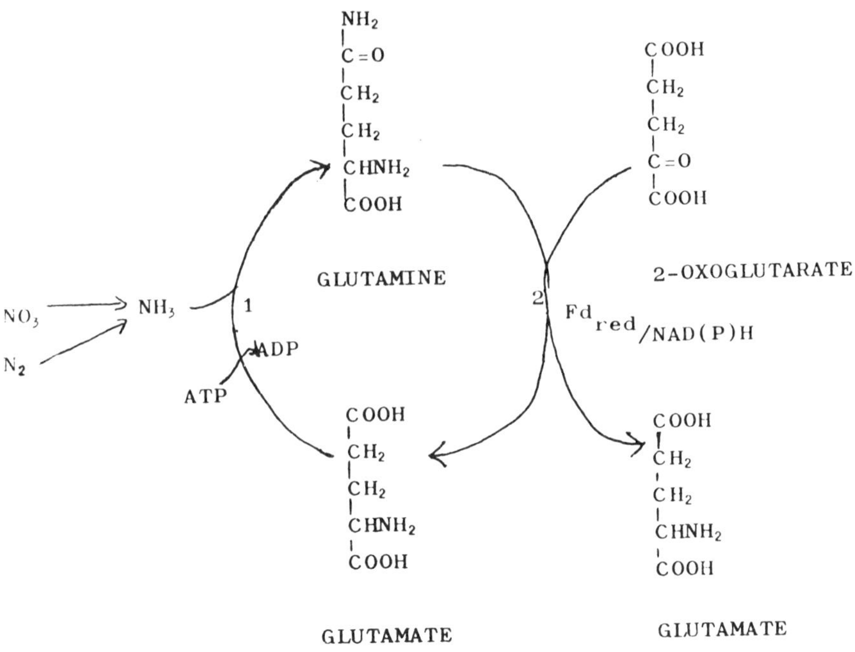

Fig. 1. The glutamate synthase cycle. Enzymes: 1. glutamine synthetase 2. glutamate synthase.

In the presence of enzyme substrates one SH is exposed and this is uniquely associated with enzyme activity.

These studies have all been carried out on a single glutamine synthetase protein. However, recent studies have shown that there are two forms of the enzyme that can be isolated from barley (45) and rice (22) leaves. These enzymes probably have different properties and reflect the chloroplast and cytosol forms of the enzyme. Because the chloroplast form appears to be in the majority and also more stable, it is probably this form that was purified in pea leaves (61). However, until more work has been done in comparing the two forms, it is premature to draw conclusions.

(b) Glutamate synthase

The NADH-dependent enzyme of the plant part of lupin root nodules has been purified to apparent homogeneity (11, 12) and reasonably pure preparations of the *Phaseolus vulgaris* root nodule enzyme (6) and the ferredoxin-dependent leaf enzyme (86) have also been obtained. The molecular weight of the pyridine nucleotide enzyme of 235,000 ± 10% is greater than that of the leaf enzyme (145,000). It is thought that the enzyme consists of one polypeptide chain (12). Various kinetic constants for the enzyme are given in Table I. The enzyme from all the above sources is highly specific in respect of cofactors and substrates with no activity being obtained

TABLE I

Properties of glutamate synthase from different sources

	Source			
	Lupinus angustifolius nodules (12)	*Phaseolus vulgaris* nodules (6)	*Vicia faba* leaves (86)	*E. coli* (57)
Reduced cofactor	NADH	NADH	Ferredoxin	NADPH
Km (μM) for				
2-oxoglutarate	39	18.5	150	7.3
Glutamine	400	222	300	250
Reduced cofactor	1.3	7.1	2.0	7.7
K_i for glutamate	700	2 600	ND	850
pH optimum	9.5	7.6	7.4	7.6
M_r (kilodaltons)	235	222	145	800

with either asparagine or NH_4Cl (6, 12, 86) when they replaced glutamine in the assay. Boland (11) has recently published a kinetic mechanism for NADH-glutamate synthase in which NADH binds first followed by random ordered binding of glutamine and 2-oxoglutarate. Determination of the bound cofactors for this enzyme suggests there are two flavins per mole. There was no evidence for bound flavins or non-haem iron in the partially purified leaf enzyme but this was probably due to the low concentrations of enzyme available (86).

Evidence for the Pathway

The present consensus of the available evidence, which has been reviewed in detail elsewhere (53, 54, 55) is that this pathway is the sole route of ammonia assimilation in higher plants and that there is no role for assimilation via glutamate dehydrogenase. This conclusion is based on the following types of study:

(a) The characteristics of the enzyme

The relative affinities of glutamine synthetase and glutamate dehydrogenase for ammonia are greatly in favour of assimilation occurring via the former. Although there are many potential mechanisms by which this unfavourable situation could be overcome *in vivo* none have yet been demonstrated.

(b) Long term feeding experiments

These have shown that $^{15}NO_3$, [^{15}N]-glutamate and [^{15}N amide]-glutamine are equally available sources for the synthesis of all the amino acids in pea plants (39). This would only be expected if the three sources were interconvertible as shown in Fig. 1.

(c) Kinetic experiments with labelled-N

The majority of the early experiments with ^{15}N were carried out over relatively long times and open to various interpretations (see 53, 55 for discussion). More recently short term exposures to ^{13}N have been used and also detailed kinetic analyses of ^{15}N incorporation. The interpretation of these experiments has been aided by the use of inhibitors, particularly methionine sulphoximine which blocks the action of glutamine synthetase (70) but has no effect upon glutamate dehydrogenase (14, 52, 64).

The radioactive isotope $^{13}N_2$ was first used to show that assimilation of NH_3 formed by nitrogen fixation in the cyanobacterium *Anabaena* took place via the glutamate synthase cycle (92). Subsequent experiments, again over very short time periods (from 30 secs), showed that the same cycle was operating during N_2 fixation in soyabean nodules (47) and nitrate and ammonia assimilation by plant tissue cultures (73), although in the latter case some contribution by glutamate dehydrogenase could not be completely excluded.

The most convincing and complete evidence for the glutamate synthase cycle comes from the recent studies of Rhodes *et al.* (68) of $^{15}NH_4^+$ assimilation in *Lemna* (duckweed) under steady state conditions. Subjecting the results to computer analysis allowed the authors to propose two models which fitted the data. In model A assimilation occurs predominantly via the glutamate synthase cycle in one cell compartment whilst a second site of assimilation via glutamine synthetase occurs in compartment-2. Model B is similar except that the source of glutamate for the second site of glutamine synthesis is via glutamate dehydrogenase whereas in model A it is derived from transport of glutamate produced by glutamate synthase in compartment 1. According to model B the rate of primary assimilation via glutamate dehydrogenase would be 10% of the total. To resolve which model was correct the authors repeated the experiment in the presence of methionine sulphoximine and azaserine arguing that if Model B was correct they should continue to see the 10% of assimilation occurring via glutamate dehydrogenase. However no ^{15}N assimilation occurred and they concluded that the glutamate synthase cycle was the sole route of ammonia assimilation; this conclusion held even under conditions in which the ammonia concentration was increased and the light intensity much decreased (which in theory should favour glutamate dehydrogenase over the glutamate synthase cycle). Their final scheme for N flow in *Lemna* is shown in Fig. 2 which emphasises that the majority of assimilation occurs in compartment-1 which contains both glutamine synthetase and glutamate synthase with ca. 6% occurring in compartment-2 which only has glutamine synthetase.

(d) Distribution of assimilation within the cell

Since early work had shown that glutamine synthetase was in the chloroplasts (49, 59, 71) and glutamate dehydrogenase in the mitochondrion (13, 49) it was of obvious interest to see whether either of these organelles could form organic-N from ammonia. So far the only studies that have demonstrated any net synthesis of gluta-

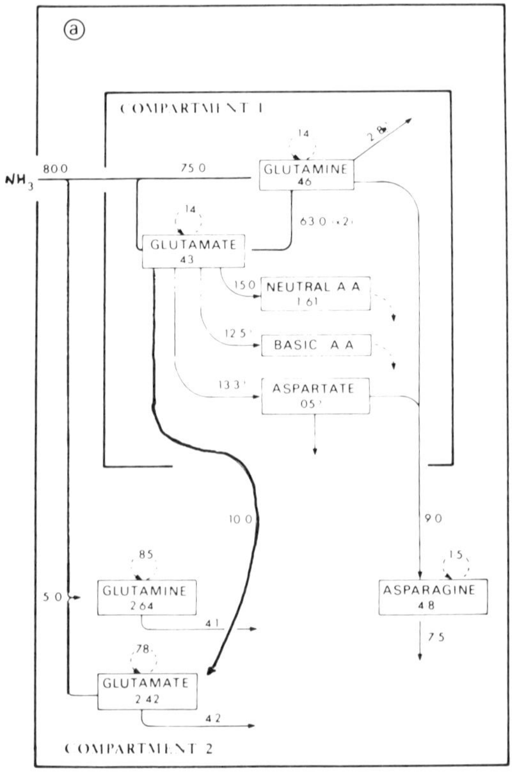

Fig. 2. The pool sizes and rates of metabolism of key N containing intermediates during the assimilation of $^{15}NH_4^+$ by *Lemna*. The numbers in square boxes show the computed pool sizes (mmoles. g^{-1} fr. wt.) and the annotations on the arrows the estimated transfer coefficients (nmole min^{-1} g^{-1} fr. wt.). Quantities marked ? are estimates based on balancing synthesis and evolution rather than isotopic labelling. Arrows with broken lines indicate the rates required to maintain the pools in steady state growth. (Taken from Rhodes *et al.* 68 with permission.)

mate by isolated mitochondria have been carried out under N_2 (16); there is no evidence for ammonia assimilation or glutamate and other amino acid formation by mitochondria incubated under physiological conditions.

In contrast there is plenty of evidence for such reactions occurring in chloroplasts. Isolated chloroplasts will catalyse the light-dependent reduction of nitrite with concomitant formation of both ammonia and amino nitrogen (44, 48). Similarly chloroplasts supplied with glutamate and ammonia in the light make glutamine (58) and chloroplasts provided with glutamine and 2-oxoglutarate catalyse the light-dependent synthesis of glutamate (35). Thus they are capable of all the individual steps of the pathway in figure 1 excluding the reduction of nitrate to nitrite. Studies by Anderson and Done (1, 2, 3) in our laboratory have underlined this conclusion by showing that all the individual steps are linked to the light reactions of photosynthesis and are subject to inhibition by methionine sulphoximine and azaserine as predicted. Although there have been reports of glutamate dehydrogenase in chloroplasts (37, 38), usually at low levels, there is no evidence that it is involved in ammonia assimilation by isolated chloroplasts of higher plants. There is however a possible role for glutamate dehydrogenase in ammonia assimilation by isolated chloroplasts of certain algae which contain a chloroplastic enzyme with a low K_m for ammonia (21, 42).

To further investigate the quantitative location of the various enzymes in the cell, Wallsgrove *et al.* (88) have recently used the technique of isolating organelles from ruptured protoplasts (65) and separating them by non-equilibrium density gradient centrifugation (51). The advantage of this technique is that 80% of the chloroplasts can be recovered intact (compared with 20% by mechanical isolation methods) thus preventing the contamination of the cytosol fraction with large amounts of chloroplast stromal enzymes. This technique has allowed us to propose that nitrite reductase and glutamate synthase are only in one compartment in the cell, namely the chloroplast, whereas glutamine synthetase is present in two, the chloroplast and the cytosol. Note that this conclusion, which was reached independently by different techniques, coincides with the conclusion of Rhodes *et al.* (68) reported above and allows the inference to be drawn that compartment-1 in Figure 2 is the chloroplast and compartment 2 is the cytosol. We are aware that there has been a claim that GS is also associated with the mitochondria (26) but we have carried out further studies (87) and are unable to find any evidence to support such a conclusion.

The subcellular location work discussed above has been drawn almost entirely from studies on leaf tissues; however it is known that a considerable proportion of primary ammonia assimilation can occur in the roots (30, 63). Those studies that have been done on roots or other non-green tissue also suggest that the subcellular distribution of the enzyme is similar (15, 19, 49, 90) and that isolated plastids can carry out certain of the partial reactions discussed above (89).

REASSIMILATION OF AMMONIA

Various findings have led to the realization that the primary assimilation of ammonia may be dwarfed in quantitative terms by various processes that lead to the release of ammonia from organic compounds; if the plant is not to lose all its organic-N in a very short time this ammonia must be reassimilated. Perhaps the most striking of these processes is photorespiration, which for many years was considered to be concerned only with the carbon economy of the plant and not relevant to nitrogen economy; however, ammonia is also released during the catabolism of many organic-N compounds used to transport nitrogen around the plant. These two examples of reassimilation will be discussed separately.

The Photorespiratory Nitrogen Cycle

Most temperate plants form a C_3 compound, phosphoglyceric acid, as the primary product of photosynthesis. They also photorespire i. e., they simultaneously release CO_2 into the atmosphere during the time they are also fixing CO_2 in photosynthesis. This released CO_2 is derived from the oxidation of an intermediate in the photosynthetic process and is chiefly formed from reaction (1) which takes place in the mitochodrion (9, 10).

Reaction 1

$$2\ \mathrm{GLYCINE} + H_2O \rightarrow \mathrm{SERINE} + CO_2 + NH_3 + 2H^+ + 2e^-$$

Much research effort has been expended on this process because it leads to a loss of up to 80 μmoles CO_2 h^{-1} g^{-1} fresh weight (28, 84) which is a substantial proportion of the total carbon fixed. However NH_3 loss from the reaction is equimolar with CO_2 and these rates have to be compared with rates of primary assimilation of between 7-15 μmoles h^{-1} g^{-1} fresh weight. Since, if the plant is to retain its organic-N, reassimilation must take place and can be shown to do so *in vivo* (93) the question is how does this occur? To investigate this Keys *et al.* (27) isolated mitochondria and incubated them with [1-^{14}C, ^{15}N]-glycine and various reactants and measured the loss of $^{15}NH_3$ and $^{14}CO_2$. The results showed that the two compounds were lost stoichiometrically even in the presence of the substrates necessary for reassimilation to occur via either glutamate dehydrogenase or glutamine synthetase. It therefore appeared that the mitochondria did not contain the enzymes necessary to reassimilate the ammonia; however, when isolated glutamine synthetase, partially purified from pea leaves, was added to the incubation mixture in the presence of glutamate the ratio of $^{15}NH_3$:$^{14}CO_2$ released dropped to 0.13. This indicated that the ammonia was being exported from the mitochondria and being reassimilated via the added glutamine synthetase (Fig. 3). Recently however we realised that these experiments might not exactly represent the *in vivo* state of the mitochondria since several workers have suggested that the electron transport chain of the mitochondria is inhibited in the light (see 23). If this occurs the mitochondria are in a state not dissimilar to that induced by anaerobiosis, and since they have been shown to synthesize glutamate under such conditions (16), it may be that glutamate dehydrogenase could be important *in vivo*. We therefore repeated the experiments (87) in the presence of antimycin A, which blocks the reoxidation of NADH and inhibits glycine oxidation, and then tested to see if we could relieve the inhibition with various substrates known to oxidise NADH. Oxaloacetate (presumably via the action of malate dehydrogenase) completely relieved the inhibition but 2-oxoglutarate, which together with the NH_3 produced by glycine oxidation should lead to NADH oxidation via the action of glutamate dehydrogenase, failed to relieve the inhibition. However, some slight alleviation was achieved when large amounts of ammonia were added with 2-oxoglutarate. The conclusion was thus that although glutamate dehydrogenase could work to a small extent under these conditions if sufficient ammonia was supplied it was unlikely to be of any functional significance in ammonia reassimilation and thus the photorespiratory nitrogen cycle (Fig. 3) originally proposed by Keys *et al.* (27) is still considered to be valid. The location of glutamine synthetase and glutamate synthase in this scheme is based on the studies of Wallsgrove *et al.* (88). The significance of the location is that it is likely that the operation of the cycle would not be too energetically wasteful since the major energy requiring step (glutamate synthase) is coupled directly to the light-dependent reduction of ferredoxin.

Reassimilation of Ammonia in Relation to Nitrogen Transport

The major nitrogenous compounds used for transport between different plant organs are glutamate, aspartate, glutamine, asparagine, arginine and the ureides allantoin and allantoic acid. The list is given in order of the decreasing ratio of C:N in the molecules and there appears to be a rough correlation in which the higher the N status of a plant the more the latter compounds in the list are used for transport (see 36 and 63 for discussion). There are also species differences in the use of these compounds; thus many tree species use asparagine and arginine, temperate legumes use asparagine and

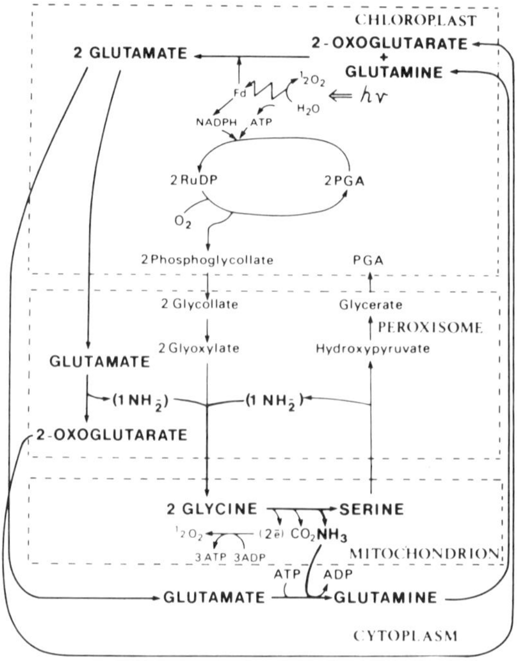

Fig. 3. The photorespiratory nitrogen cycle. Reproduced with permission from (27).

glutamine compared with tropical legumes which use the ureides and cereals in general use predominantly glutamine particularly in the phloem (30, 36, 63, 85).

The synthesis of these compounds from primary assimilation of ammonia must of necessity come through glutamine. Glutamine is also widely involved if synthesis occurs from the breakdown of protein in senescing storage organs (endosperms or cotyledons) or senescing leaves. The pathways that are probably involved are outlined in Fig. 4a. The irreversibility of glutamate synthase suggest that it is unlikely that amino-N can enter glutamine other than via ammonia thus indicating that glutamate dehydrogenase is likely to be important in a catabolic direction. The importance of this enzyme in catabolism is underlined by studies which shows that it increases markedly during leaf senescence (83). The release of N from non-amino N in protein follows many of the same pathways that are outlined below for the breakdown of transport compounds (Fig. 4b).

Once the transport compounds have reached the growing organ they are metabolised. If the 2-amino groups of the above compounds are ignored as they can be simply transferred to other amino acids

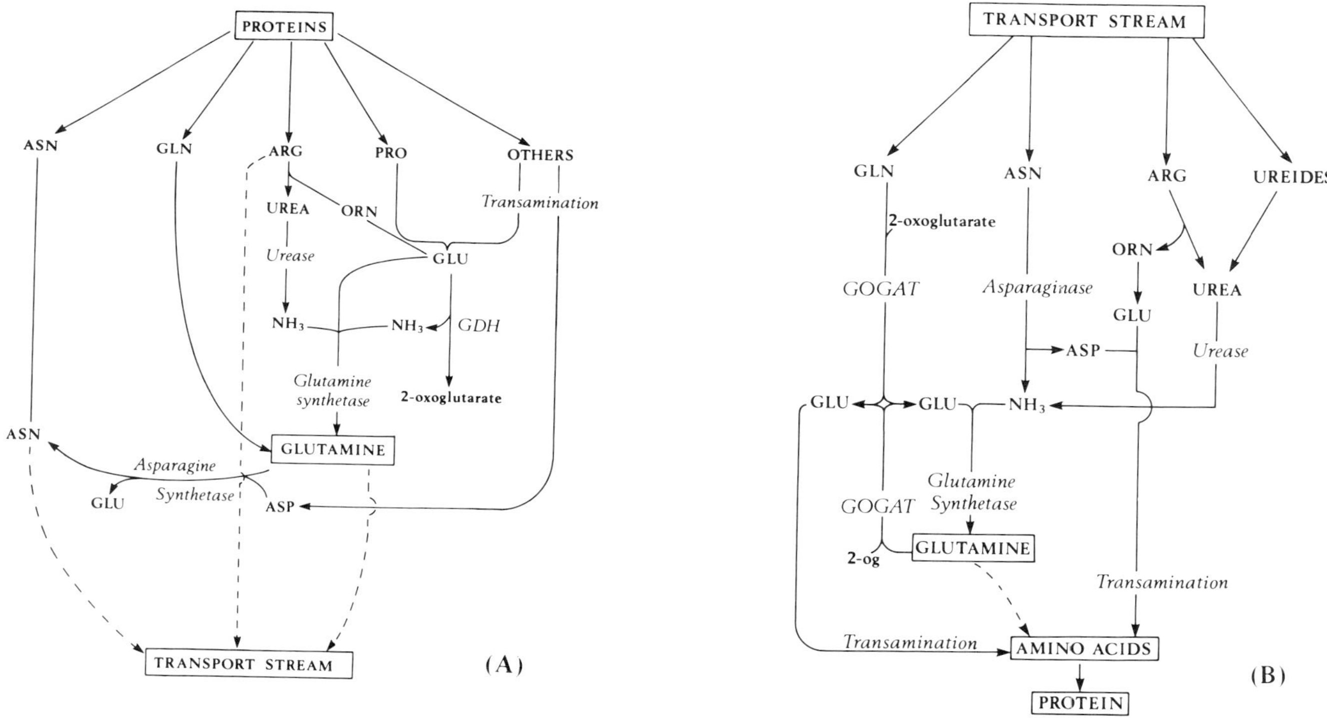

Fig. 4. Pathways for (a) the synthesis from storage protein and (b) the breakdown of nitrogenous transport compounds in plants. GOGAT = glutamate synthase.

by the action of transaminase we are left with the amide-N of asparagine and glutamine, the amidino-N arginine, and the ureides which (apart from direct incorporation of arginine) must be transferred to amino-N if they are to be used in protein synthesis. The mechanism of transfer of these different groups all involve glutamine as discussed below.

(a) Glutamine

Consequent upon the discovery of glutamate synthase a ready mechanism for the transfer of the amide-N to the 2-amino position was available. Subsequent measurements of the enzyme have shown it to be present in organs, especially storage organs, such as developing cereal (56, 75) and legume (32, 78, 79) seeds, carrying out protein synthesis. Although glutaminases and glutamine aminotransferases have been recognized in plants they do not appear to have a physiological role in glutamine utilization.

(b) Asparagine

Although early work (17, 20) suggested that glutamate synthase could use asparagine as an amide donor this was subsequently disproved (52, 75, 78) and so far no convincing evidence exists for the presence of any enzyme able to directly transfer the amide-N of asparagine to the amino position of an amino acid. The two major routes for asparagine breakdown in plants that have been described are reactions 2 and 3.

Reaction 2 Asparaginase [EC 3.5.1.1]

$$\text{ASPARAGINE} + H_2O \rightarrow \text{ASPARTATE} + NH_3$$

Reaction 3 Asparagine aminotransferase

$$\text{ASPARAGINE} + \text{2-OXO ACID} \rightarrow \text{2-OXO SUCCINAMATE} + \text{AMINO ACID}$$

The 2-oxosuccinamate produced by reaction 3 may break down either by reaction 4 a or 4 b – in either case the amide-N of asparagine is eventually liberated as ammonia.

Reaction 4 a

$$\text{2-OXOSUCCINAMIC ACID} + H_2O \rightarrow \text{OXALOACETATE} + NH_3$$

Reaction 4 b

$$\text{2-OXOSUCCINAMIC ACID} \xrightarrow{(2H)} \text{2-HYDROXYSUCCINAMATE}$$

$$\text{2-HYDROXYSUCCINAMATE} \xrightarrow{(H_2O)} \text{MALATE} + NH_3$$

Studies of asparaginase in plants have shown that the enzyme is widespread (5, 33, 74, 80) but that it exists in two forms, one that is K^+-independent and occurs in only relatively few species (33), and one that is K^+-dependent and much more common (74). The latter enzyme is particularly active in developing pea cotyledons which receive a large proportion of their N in the form of asparagine (39, 63). These tissues also contain glutamine synthetase, glutamate dehydrogenase (74, 78) and glutamate synthase (78) and the pattern of activities throughout the development of the cotyledon have been followed. Evidence that the ammonia from asparagine is reassimilated via glutamine synthetase comes from studies with methionine sulphoximine (MSO) (34); isolated cotyledons can synthesize protein at the same rate as those in the plant when they are supplied with asparagine or glutamine; however, MSO inhibits protein synthesis from asparagine by 70% whilst having no effect on glutamine.

Asparagine transaminase activity is found particularly in leaves (40, 80) where labelling studies show that the amide-N of asparagine is actively metabolised (7). Both the patterns of metabolism of ^{15}N and the presence of enzymes in pea leaves (7, 40) and ^{15}N kinetics in analogous tissue in *Lemna* (68) suggest that the ammonia released from asparagine is reassimilated into glutamine and distributed among the various amino acids.

(c) Arginine

Arginine is cleaved by the enzyme arginase to ornithine and urea (Figure 4b). The ornithine can then be transaminated to give glutamic semialdehyde which in turn gives rise to glutamate. The urea formed is hydrolysed to ammonia and CO_2 via urease. Thus of the four nitrogen atoms in arginine two may be salvaged as amino-N and two are released as ammonia. Although as far as we are aware no specific studies on the reassimilation of ammonia released via urease have been carried out, there is no reason to believe that it is metabolised other than through the glutamate synthase cycle.

(d) Ureides

Allantoin is converted to allantoic acid via allantoinase which has been found in leaves, pods and maturing seeds (81). Allantoic acid itself is then cleaved to give glyoxylate and two molecules of urea which, as above, is broken down by urease to give ammonia and CO_2. Thus all of the four nitrogen atoms in the ureides are released as ammonia which presumably enters the general pool for reassimilation. Certainly the cotyledons of soyabean, a plant which exports a high proportion of its nitrogen from the nodule in the form of allantoin and allantoic acid (46)) , contain both glutamine synthetase and glutamate synthase, which are active at the requisite time of the developmental sequence (79).

THE CONTROL OF GLUTAMINE METABOLISM

Many of the basic constraints under which bacteria and higher plants live are entirely different. Consequently it is not surprising that they have evolved different regulatory mechanisms to control their metabolism including that of nitrogen metabolism. However, despite this and probably because of the outstanding work on the regulation of glutamine synthetase in bacteria, much of the work on the plant enzyme has been overly influenced by the bacterial work and attempts made to fit both enzymes into the same mould. Bacteria are often found in an aquatic environment and, have a strong control over what enters the cell and can excrete unwanted products; further their supply of energy is often limited and subject to wide variations.

Higher plants in the main are land based, have no external excretion system and have a ready and constant source of energy from sunlight. In particular plants also appear to have limited control over the uptake of ammonia and in non-agricultural environments are generally limited in their growth by nitrogen availability. Given these background conditions and the fact that ammonia, if it accumulates, is generally toxic to plant cells it is unlikely that a mechanism which *unilaterally* shuts off nitrogen assimilation at the level of glutamine synthetase would have evolved. That this is so is shown by many studies that indicate that when plants are provided with ammonia they take it up rapidly and continue to do so for as long as they have available carbohydrate. Once this is exhausted, the ammonia accumulates and the plants die (e. g. see 29, 66). For plants provided with nitrate the situation is different — there is evidence that nitrate uptake can to some extent be controlled and that its uptake is inhibited in the presence of ammonia (25, 41) ; also once taken into the plant it is non-toxic and can be accumulated in vacuoles to very great concentrations (25). Thus provided plants have mechanisms in which the reduction of nitrate and the assimilation of ammonia are linked to each other and also, preferably, to the supply of energy and carbohydrate, then the plant is unlikely to put itself into the unfavourable position of accumulating ammonia. There are now several lines of evidence that suggest that such mechanisms exist but nevertheless there are also many studies on isolated glutamine synthetase that purport to show that feedback inhibition is an important aspect of the enzyme.

Feedback Control of Glutamine Synthetase

When the level of the enzyme in plants given different nitrogen substrates is measured, there is little convincing evidence for tight controls over the induction and repression of the enzyme. Probably the most complete studies have been done with the aquatic plant *Lemna* (67, 68, 69, 76), which for reasons outlined above may not apply to land plants; these studies show that although the amount of glutamine synthetase is much higher in nitrate than ammonia grown cells, the level of activity in the latter is still more than sufficient to account for ammonia assimilation. Over all the maximum decrease in enzyme activity from the level recorded in the presence of nitrate to that on glutamine (or ammonia) is only 40% (2.5 fold difference) (76) compared with the situation in bacteria in which the enzyme decreases to virtually zero in glucose-limited cultures grown on ammonia (72). The determining factor in these changes appears to be the intracellular concentration of glutamine (67) and the external factors of most importance are the availability of light and ammonia (69).

Studies on the effects of amino acids on isolated plant glutamine synthetase are fairly numerous. In general they show that relative large concentrations of amino acids cause relatively small changes in activity (Table II). The effects of the amino acids on the enzyme from pea leaves (62) and roots (31) are greater when Mn^{2+} is the cofactor rather than Mg^{2+}. These factors generally suggest that the feedback inhibition is unlikely to be of much physiological significance *in vivo*. Of the amino acids measured, alanine and glycine appear to have the major effects which are probably cumulative. There is no suggestion from various studies in a range of organisms that higher plant glutamine synthetase is subject to adenylylation and deadenylylation.

Energy Supply and Glutamine Metabolism

A number of mechanisms exist that relate the activity of glutamine synthetase to the possible energy state of the system. Since the majority of the enzyme involved in primary assimilation is undoubtably in the chloroplast (see above) the properties of the enzyme should be assessed in relation to what we know of the chloroplast environment. When this is done (50), it can be shown that the changes in energy charge, Mg^{2+} concentration and pH that take place on the

TABLE II

Inhibition of glutamine synthetase by various amino acids

	% Inhibition	
	Plant Source	
Amino acid	*Lemna* [1]	*Pisum* [2]
L-Alanine	60	14
Glycine	60	16.7
L-Serine	54	16
Alanine + glycine	70	24
Alanine + serine	70	24
Glycine + serine	67	26.6
Alanine + glycine + serine	77	34.6
L-Ornithine	n.t.	36 (12 mM)
L-Histidine	0 (10 mM)	56 (12 mM)

[1] Assayed by the Mg++ dependent coupled assay involving lactate dehydrogenase and pyruvate kinase (77). Inhibitors added at 25 mM unless stated.

[2] Assayed by the Mn ++ dependent synthetase assay (62). Inhibitors added at 28 mM unless stated.

n.t. — not tested.

transition of chloroplasts from light to dark, all work in such a way that the conditions existing in the light are those which increase the activity of isolated glutamine synthetase (59, 60) whereas those in the dark are unfavourable for activity. When these changes are considered with factors affecting nitrate reduction, it can be seen that the two processes are probably tightly coupled. Thus, since the prior reaction nitrite reductase is dependent on reduced ferredoxin, the rate is likely to be severely decreased in the dark. Similarly in leaves the rate of nitrate reduction is also much less in the dark than the light. Consequently the overall regulating factor determining the potential rate of glutamine synthesis (the actual rate depending on the amount of N available) is likely to be the presence of light. The studies of Rhodes *et al.* (67, 69, 76) have also shown that there are other mechanisms in *Lemna,* grown on ammonia, that bring about a relatively rapid (too rapid to be due to repression) inactivation of glutamine synthetase *in vivo* when the plant is transferred to the dark, such that within 3 hours the activity of GS has dropped to about one third. The activity is regained *in vivo* over a period of 2 h after returning the cultures to the light (76). The inactivated enzyme can also be reactivated *in vitro* by incubating it with various combinations of substrates and ligands that stabilize glutamine synthetase. Rhodes *et al* (69) suggest that such reactivation is spontaneous, provided the enzyme is kept in a stable form, although it is possible to isolate a small heat stable compound which will prevent this occurring. The significance of this mechanism is seen by the authors as preventing excessive chanelling of glutamate into glutamine under conditions in which there is restricted activity of ferredoxin-dependent glutamate synthase.

The reduction of nitrate and assimilation of NH_3 do however go on in the dark as well — both in non-green tissues and leaves (4). Although there is less evidence concerning *in vitro* systems, the major determinant of the rate of the processes *in vivo* can clearly be shown to be the carbohydrate status of the tissue. Thus the amount of nitrate reductase present in response to the addition of nitrate and the *in vivo* assimilation of nitrate and nitrite are stimulated by the provision of carbohydrates to the tissues (see for review 8, 24) and there is also the suggestion that the activity of GS may be regulated by the *in vivo* energy charge in roots (91).

Studies on Glutamate Synthase

All of the results obtained so far indicate that there is little in the way of control mechanisms operating on higher plant glutamate synthase. Probably the only one of importance is the competitive (with 2-oxoglutarate) inhibition of the enzyme by glutamate (12). This would ensure that once sufficient glutamate was available for transmination and glutamine synthesis, the cell's supply of 2-oxoglutarate was not depleted and glutamine could then be accumulated and the synthesis of a storage compound (asparagine or ureide) would take place. If however the cell was rich in carbohydrates with a plentiful supply of ketoacids then any available glutamine could be metabolised to glutamate and lead to protein synthesis. In leaf cells there is evidence that the determining factor in the rate of conversion of glutamine to glutamate is the presence of light which presumably acts on the supply of reduced ferredoxin (see 55 for discussion).

REFERENCES

1. Anderson, J. W., and Done, J. (1977) *Plant Physiol. 60,* 354-359.
2. Anderson, J. W., and Done, J. (1977) *Plant Physiol. 60,* 504-508.
3. Anderson, J. W., and Done, J. (1978) *Plant Physiol. 61,* 692-697.
4. Aslam, M., Huffaker, R. C., Rains, D. W., and Rao, K. P. (1979) *Plant Physiol. 63,* 1205-1209.
5. Atkins, C. A., Pate, J. S., Sharkey, P. J. (1975) *Plant Physiol. 56,* 807-812.
6. Awonaike, K. O. 1980 Ph. D. Thesis, University of London.
7. Bauer, A., Joy, K. W., Urquhart, A. A. (1977) *Plant Physiol. 59,* 920-924.
8. Beevers, L., and Hageman, R. H. (1980) in *Biochemistry of Plants Vol. V 'Amino Acids and their Derivatives'* (B. J. Miflin ed.) pp. 116-168, Academic Press, New York.
9. Bird, I. F., Cornelius, M. J., Keys, A. J., and Whittingham, C. P. (1972) *Biochem. J. 128,* 191-192.
10. Bird, I. F., Cornelius, M. J., Keys, A. J., and Whittingham, C. P. (1972) *Phytochemistry 11,* 1587-1594.
11. Boland, M. J. (1979) *Eur. J. Biochem; 99,* 531-539.
12. Boland, M. J., and Benny, A. G. (1977) *Eur. J. Biochem. 79,* 355-362.
13. Bone, D. H. (1959) *Nature, 184,* 990.
14. Brenchley, J. E. (1973) *J. Bacteriol. 114,* 666-673.
15. Dalling, M. J., Tolbert, N. E., and Hageman, R. H. (1972) *Biochim. Biophys. Acta 283,* 513-519.
16. Davies, D. D. and, Teixeira, A. N. (1975) *Phytochemistry 14,* 647-656.
17. Dougall, D. K. (1974) *Biochem. Biophys Res. Commun. 58,* 639-646.
18. Ellis, R. J. (1979) *Trends in Biochemical Science 4,* 241-244.
19. Emes, M. J., and Fowler, M. W. (1978) *Planta, 144,* 249-253.
20. Fowler, M. W., Jessup, P. W., Stephan-Sarkissian, G. (1974) FEBS *Letts. 46,* 340-342.
21. Gayler, K. R., and Morgan, W. R. (1976) *Plant Physiol., 58,* 283-287.
22. Guiz, C., Hirel, B., Shedlofsky, G., and Gadal, P. (1979) *Plant Science Letts. 15,* 271-274.
23. Heber, U. (1974) in *Proceedings 3rd Int. Congr. Photosynthesis* (M. Avron ed.) pp. 1335-1348, Elsevier, Amsterdam.

24. Hewitt, E. J., Hucklesby, D. B., Notton, B. A. (1976) in *Plant Biochemistry* (Bonner, J. and Varner, J. E. eds.) pp. 633-682, Academic Press, New York.
25. Huffaker, R. C., and Rains, D. W. (1978) in *Nitrogen in the Environment* (Nielsen, D. R., and MacDonald, J. G. eds.) Vol. 2, pp. 1-43 Academic Press, New York.
26. Jackson, C., Dench, J. E., Morris, P., Lui, S. C., Hall, D. O., and Moore, A. L. (1979) *Biochem. Trans. 7,* 1122-1124.
27. Keys, A. J., Bird, I. F., Cornelius, M. J., Lea, P. J., Wallsgrove, R. M., and Miflin, B. J. (1978) *Nature, 275,* 741-743.
28. Keys, A. J., Sampaio, E. V. S. B., Cornelius, M. J., and Bird, I. F. (1977) *J. Exp. Bot. 28,* 525-533.
29. Kirkby, E. A., and Mengel, K. (1970) in *Nitrogen Nutrition of the Plant* (Kirkby, E. A. ed.) pp. 69-77, Univ. of Leeds.
30. Kirkman, M. A., and Miflin, B. J. (1979) *J. Sci. Fd. Agric. 30,* 653-660.
31. Kretovich, W. L. and Estigneeva, Z. G. (1974) in *The Origin and Evolutionary Biochemistry* (Dose, K., Fox, S. W., Deborin, G. A., and Pavlovskaya, eds.) pp. 254-262, Plenum Press.
32. Lea, P. J., and Fowden, L. (1975) *Proc. Royal Soc. London B, 192,* 13-26.
33. Lea, P. J., Fowden, L., and Miflin, B. J. (1978) *Phytochemistry 17,* 217-222.
34. Lea, P. J., Hughes, J., and Miflin, B. J. (1979) *J. Exp. Bot. 30,* 529-537.
35. Lea, P. J., and Miflin, B. J. (1974) *Nature, 251,* 614-616.
36. Lea, P. J., and Miflin, B. J. (1980) in *The Biochemistry of Plants Vol. V 'Amino Acids and their Derivatives'* (B. J. Miflin ed.), pp. 569-608, Academic Press, New York.
37. Lea, P. J., Thurman, D. A. (1972) *J. Exp. Bot 23,* 440-449.
38. Leech, R. M., and Kirk, P. R (1968) *Biochem Biophys. Res. Commun. 32,* 685-690.
39. Lewis, O. A. M., and Pate, J. S. (1973) *J. Exp. Bot. 24,* 596-606.
40. Lloyd, N. D. H., and Joy, K. W. (1978) *Biochem. Biophys. Res. Commun. 81,* 186-192.
41. Lycklama, J. C. (1963) *Acta Bot. Neerl. 12,* 361-423.
42. McKenzie, G. H., Ch'ng, A. L., and Gayler, K. R. (1979) *Plant Physiol. 63,* 578-582.
43. McParland, R. H., Guevara, J. G., Becker, R. R., Evans, H. J. (1976) *Biochem. J. 153,* 597-606.
44. Magalhaes, A. C., Neyra, C. A., Hageman, R. H. (1974) *Plant Physiol. 53,* 411-415.
45. Mann, A. F., Fentem, P. A., Stewart, G. R. (1979) *Biochem. Biophys. Res. Commun. 88,* 515-521.
46. Matsumoto, T., Yatazawa, M., and Yamamoto, Y. (1977) *Plant Cell Physiol. 18,* 613-624.
47. Meeks, J. C., Wolk, C. P., Schilling, N., Shaffer, P. W., Avissar, Y., Chien, W-S (1978) *Plant Physiol. 61,* 980-983.
48. Miflin, B. J. (1974) *Planta 116,* 187-196.
49. Miflin, B. J. (1974) *Plant Physiol. 54,* 550-55.
50. Miflin, B. J. (1978) in *Regulation of Enzyme Synthesis and Activity in Higher Plants* (Smith, H. ed.) pp. 23-40 Academic Press, London.
51. Miflin, B. J., and Beevers, H. (1974) *Plant Physiol. 53,* 870-874.
52. Miflin, B. J., and Lea, P. J. (1975) *Biochem. J. 149,* 403-409.
53. Miflin, B. J., and Lea, P. J. (1976) *Phytochemistry 15,* 873-885.

54. Miflin, B. J., and Lea, P. J. (1977) *Ann. Rev. Pl. Physiol. 28,* 299-329.
55. Miflin, B. J., and Lea, P. J. (1980) in *The Biochemistry of Plants Vol V 'Amino Acids and their Derivatives'* (B. J. Miflin ed.), pp. 169-202, Academic Press, New York.
56. Miflin, B. J., and Shewry, P. R. (1979) in *Cereals* (Laidman, D. and Wyn Jones, R. eds.) pp. 239-273, Academic Press, London.
57. Miller, R. E., and Stadtman, E. R. (1972) *J. Biol. Chem. 247,* 7407-7419.
58. Mitchell, C. A., and Stocking, C. R. (1975) *Plant Physiol. 55,* 59-63.
59. O'Neal, D., and Joy, K. W. (1973) *Nature* (New Biol.) *246,* 61-62.
60. O'Neal, D., and Joy, K. W. (1973) *Arch. Biochem. Biophys. 159,* 113-122.
61. O'Neal, D., and Joy, K. W. (1974) *Plant Physiol., 54,* 773-779.
62. O'Neal, D., and Joy, K. W. (1975) *Plant Physiol., 55,* 968-974.
63. Pate, J. S. (1971) in *Nitrogen-15 in Soil-Plant Studies,* Vienna; IAEA, pp. 165-187.
64. Probyn, T. A., and Lewis, O. A. M. (1979) *J. Exp. Bot. 30,* 299-305.
65. Rathnam, C. K. M., and Edwards, G. E. (1976) *Plant and Cell Physiol. 17,* 177-186.
66. Reisenhauer, H. M. (1978) in *Nitrogen in the Environment* (Nielsen, D. R. and MacDonald, J. G. eds.) Vol. 2, pp. 157-170, Academic Press, New York.
67. Rhodes, D., Rendon, G. A., and Stewart, G. R. (1976) *Planta,* 129, 203-210.
68. Rhodes, D., Sims, A. P., and Folkes, B. F. (1980) *Phytochemistry, 19,* 357-366.
69. Rhodes, D. A., Sims, A. P., and Stewart, G. R. (1979) in *Nitrogen Assimilation of Plants* (Hewitt, E. J., and Cutting, C. V. eds.) pp. 501-520, Academic Press, London.
70. Ronzio, R., Rowe, W., and Meister, A. (1969) *Biochemistry 8,* 1066-1075.
71. Santarius, K. A., and Stocking, C. R. (1969) *Z. Naturforsch.* Teil B246, 1170-1179.
72. Senior, P. J. (1975) *J. Bact. 123,* 407-418.
73. Skokut, T. A., Wolk, C. P., Thomas, J., Meeks, J. C., Shaffer, P. W., Chien, W-S (1978) *Plant Physiol. 62,* 299-304.
74. Sodek, L., Lea, P. J., Miflin, B. J. (1980) *Plant Physiol., 65,* 22-26.
75. Sodek, L., and Da Silva, W. J. (1977) *Plant Physiol. 60,* 602-605.
76. Stewart, G. R., and Rhodes, D. A. (1976) in *Regulation of Enzyme Synthesis and Activity in Higher Plants* (Smith H. ed.), pp. 1-22, Academic Press, London.
77. Stewart, G. R., and Rhodes, D. A. (1977) *New Phytol. 79,* 257-268.
78. Storey, R., and Beevers, L. (1978) *Plant Physiol. 61,* 494-500.
79. Storey, R. and Reporter, M. (1978) *Canad. J. Bot. 56,* 1349-1356.
80. Streeter, J. G. (1977) *Plant Physiol. 60,* 235-239.
81. Tajima, S., Yatazawa, M., and Yamamoto, Y. (1977) *Soil Sci. Plant Nutr. 23,* 225-235.
82. Tempest, D. W., Meers, J. L. Brown, C. M. (1970a) *Biochem. J. 117,* 405-407.
83. Thomas, H. (1978) *Planta 142,* 161-169.
84. Thomas, S. M., Hall, N. P., and Merrett, M. J. (1978) *J. Exp. Bot. 29,* 1161-1168.
85. Tully, R. E., Hanson, A. D. (1979) *Plant Physiol. 64,* 460-466.
86. Wallsgrove, R. M. Harel, E., Lea, P. J., and Miflin, B. J. (1977) *J. Exp. Bot. 28,* 588-596

87. Wallsgrove, R. M., Keys, A. J., Lea, P. J., and Miflin, B. J. (1980) *J. Exp. Bot.,* in press.
88. Wallsgrove, R. M., Lea, P. J., and Miflin, B. J. (1979) *Plant Physiol. 63,* 232-236.
89. Washitani, I., and Sato, S. (1977) *Plant and Cell Physiol. 18,* 1235-1241.
90. Washitani, I., and Sato, S. (1978) *Plant and Cell Physiol. 19,* 43-50.
91. Weissman, G. S. (1976) *Plant Physiol 57,* 339-343.
92. Wolk, C. P., Thomas, J., Shaffer, P. W., Austin, S. M. and Galonsky, A. (1976) *J. Biol. Chem. 251,* 5027-5034.
93. Woo, K. C., Berry, J. C., Osmond, C. B., and Lorimer, G. H. (1977) Abstracts Fourth Int. Congr. on Protosynthesis, Reading, U. K., p. 415.

13. Enterohepatic Aspects of Glutamine Metabolism

HERBERT G. WINDMUELLER

Laboratory of Nutrition and Endocrinology,
National Institute of Arthritis, Metabolism,
and Digestive Diseases
National Institutes of Health, Bethesda, MD (USA) 20205

SUMMARY

The small intestine has been identified as a major site for the metabolism of circulating glutamine. In man and in most laboratory animals, including dogs, cats, hamsters, monkeys and conventional and germfree rats, 20-30% of the total plasma glutamine is extracted in each blood pass through the intestinal vasculature.

Detailed metabolic studies in the rat were performed with (a) isolated intestine vascularly perfused with blood, and (b) jejunal segments *in vivo* having intact arterial blood supply and totally collected venous drainage. The radioactivity of L-[U-^{14}C]glutamine administered vascularly to intestine was released back into blood as CO_2 (64%), organic acids (mostly lactate) (13%), proline (6%), citrulline (5%), glucose (5%), alanine (4%), and glutamate (0.4%) Glutamine nitrogen was accounted for by released ammonia (38%), citrulline (28%), alanine (24%) and proline (7%). Plasma glutamine was the source for 35% of the total CO_2 produced by jejunum of post- absorptive rats. Intestine also metabolized most dietary glutamine during its transport from lumen to blood. The metabolic products were the same as with vascular glutamine.

Glutamine utilization is localized in villus and crypt cells of mucosal epithelium and is initiated by a phosphate-dependent glutaminase with an apparent K_m for glutamine (2.2 mM) that is only one-tenth the K_m of hepatic glutaminase, possibly explaining why intestine is a preferential site for glutamine breakdown. The key enzymes needed to

ISBN: 0-12-506040-8

account for the major end products of intestinal glutamine metabolism have all been demonstrated *in vitro* in cell-free preparations of small intestinal mucosa.

Physiological functions served by intestinal glutamine metabolism include: (a) providing a major source of energy for the intestine; (b) providing the animal indirectly with an endogenous arginine source, by supplying citrulline for arginine biosynthesis in the kidneys and other tissues; (c) providing the liver with nitrogenous precursors for urea biosynthesis. The quantity of glutamine nitrogen processed daily by intestine in fasted rats accounts for more than 30% of the urea nitrogen excreted. Significant net utilization of circulating glutamine by the liver in most animals is observed only when hepatic portal blood glutamine levels are elevated, for example, following the ingestion of a high-protein meal.

It is now evident that intestinal mucosal cells have a significant metabolic function in addition to their better known transport function.

INTRODUCTION

The intracellular metabolic roles of glutamine as a scavenger of ammonia and as a donor of nitrogen in numerous biosynthetic reactions are well-know. Less attention has been directed toward the metabolic fate and physiologic role of the glutamine that circulates in the blood. Glutamine is the most abundant amino acid in the plasma of man and most experimental animals. During periods of metabolic acidosis, glutamine released by muscle, brain and possibly other peripheral tissues is extracted from the blood by the kidney and hydrolyzed as a source of urinary ammonia (1). Following a massive intravenous injection of glutamine into rats, glutamine and glutamate were also observed to accumulate in the liver (2). This and other studies led to the concept that glutamine is a vehicle for transporting nitrogen from peripheral tissues to the liver for urea biosynthesis. Recent evidence indicates, however, that under normal circumstances, in most animal species, the small intestine is a far more important site for the net utilization of circulating glutamine than is the liver. The initial observations were made in fasted dogs by Elwyn *et al.* (3) and by Addae and Lotspeich (4). Arteriovenous difference measurements revealed a net uptake of circulating glutamine by the group of organs drained by the hepatic portal vein — the stomach, small and large intestine, cecum, colon, spleen and pancreas. Similar observations were later made in sheep (5), in rats (6, 7), and in man (8). Subsequent studies, which form the basis for most of this review, have dealt with the quantification, kinetics, and further localization of this uptake, as well as the metabolic fate of the glutamine, its enzymology, and the likely physiologic significance of the process.

ROLE OF THE INTESTINE IN GLUTAMINE METABOLISM

Studies with an Isolated, Perfused Gut Preparation

Our initial studies were done with an isolated, vascularly perfused rat gut preparation where, during unrelated experiments, we observed that glutamine, but no other amino acid, disappeared rapidly from the recycling perfusate (9). The tissue in this gut preparation included the small intestine, cecum, proximal colon, associated mesentery, and part of the pancreas. Perfusate entered the preparation via the superior mesenteric artery and exited from the superior mesenteric vein. The perfusate was composed of heparinized rat blood supplemented with small amounts of norepinephrine (approximately 5×10^{-8}M) and dexamethasone, a synthetic glucocorticoid (6×10^{-7} M). The hormones were needed to control the damaging effects due to denervation: hypermotility, hypersecretion, low vascular resistance, and cellular necrosis (10). The loss of perfusate glutamine was accompanied by a substantial tissue release of alanine, citrulline, proline, and smaller amounts of other amino acids, the last apparently from the digestion of residual protein in the gut lumen (Fig. 1). When sufficient glutamine was infused into the perfusate to maintain the normal blood level, uptake by the tissue was continuous at 74 μmol/h, with 30% of the plasma glutamine being removed in each blood pass through the preparation. The glutamine infusion substantially increased the release of alanine and citrulline (Fig. 1). Results were similar with preparations from conventional and germfree rats. Two lines of reasoning suggested that the utilization of circulating glutamine by the gut represented a quantitatively important pathway: (a) net uptake of glutamine was nearly as large as that of glucose (9) (see also Ref. 11), considered at that time to be the major energy source for intestine; (b) the net uptake rate observed represented a turnover of the rat's entire plasma glutamine pool every 4.6 min and accounted for about 15% of the estimated turnover rate of the total body glutamine pool (12).

Without the addition of glutamine, the perfusate glutamine levels fell rapidly from 600 μM, the normal blood level, to 200 μM, but no lower (Fig. 2). However, the level fell to near zero when the perfusate contained 4 mM L-methionine-SR-sulfoximine, an inhibitor of gluta-

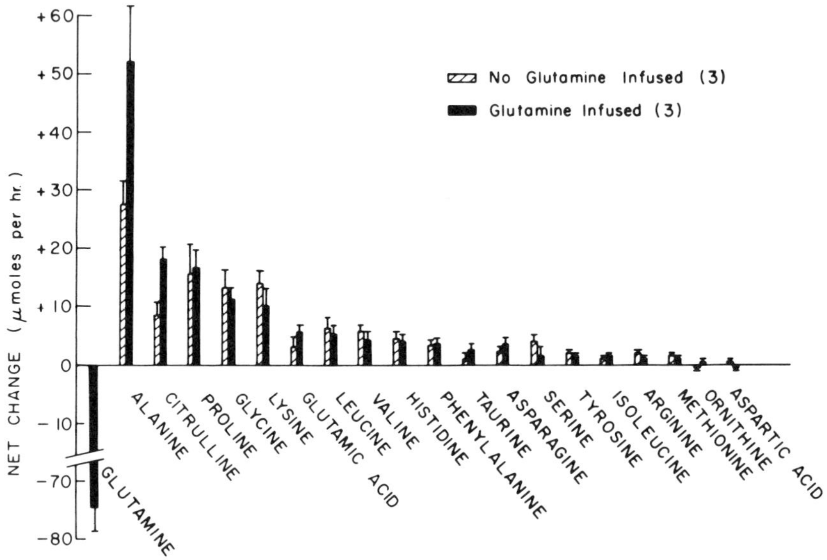

Fig. 1. Net changes in the amino acid content of blood recirculated through an isolated rat gut preparation. The recirculating volume was 120 ml. Glutamine and asparagine were determined enzymatically and other amino acids by ion exchange chromatography. Closed bars: 85 μmoles/h of glutamine were infused continuously into the perfusate; data from one germfree and two conventional rats. Hatched bars: no glutamine supplement; data from three conventional rats. Glutamine data were corrected for the amount added and therefore show net uptake. Each value is the mean ± S. E. Taken from Windmueller and Spaeth (9).

mine synthetase, revealing an ability by some cells in the preparation to make glutamine and release it into the blood (Fig. 2). The rate of glutamine utilization by the perfused gut increased more or less linearly with increasing perfusate concentration and appeared to reach a near maximum rate at about 600 μM, the normal blood level in the rat (Fig. 3) (see also Ref. 11) This concentration dependence can explain the failure of the perfusate glutamine to fall below 200 μM in the absence of the glutamine synthetase inhibitor (Fig. 2). At this concentration, the low production rate and the low utilization rate of glutamine are in equilibrium.

The metabolic fate of glutamine taken up by the gut was determined in two different kinds of studies (9): (a) experiments in which the recycling perfusate contained L-[U-^{14}C]glutamine, and all ^{14}C-labeled products were identified in the perfusate after several hours of perfusion; (b) single-pass experiments, in which the gut was perfused with normal blood containing L-[U-^{14}C]glutamine of known specific radioactivity and the individual ^{14}C-labeled products were quantified in the venous outflow, which was not allowed to recycle.

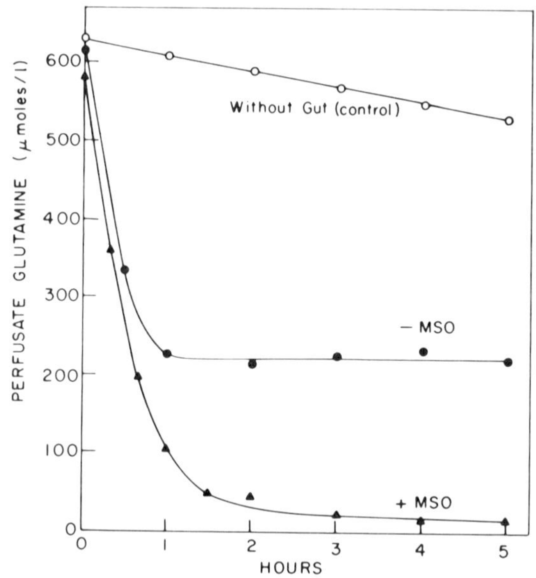

Fig. 2. Glutamine uptake during vascular perfusion of isolated rat gut. Perfusate glutamine values refer to the concentration in the plasma of the recirculated blood. O—O, blood recirculated through the perfusion apparatus in the absence of a gut; ●—●, blood recirculated through gut; △—△, blood containing 4 mM L-methionine-SR-sulfoximine (MSO) recirculated through gut. Values are means from two to four experiments. Taken from Windmueller and Spaeth (9).

Under these latter conditions, the rate of appearance of labeled products in the venous blood reached a steady state in 10-20 min (Fig. 4). The rate of conversion of glutamine carbon to each product was determined from the steady-state rates of appearance of ^{14}C in each product and from the specific radioactivity of the perfusing glutamine (Table I). Fifty-seven percent of the glutamine carbon taken up was released as CO_2. Other important products of glutamine carbon were lactate, citrate, citrulline, and proline. The rate of CO_2 formation from circulating glutamine could account for more than 30% of the total CO_2 produced by the preparation, indicating the importance of glutamine as a respiratory fuel. Glutamine nitrogen taken up could be approximately accounted for by the quantities of citrulline, ammonia, alanine, and proline released back into the perfusate (9). Results were similar with gut preparations from conventional and germfree rats, indicating that all of the products were the result of tissue metabolism and not metabolism by the intestinal

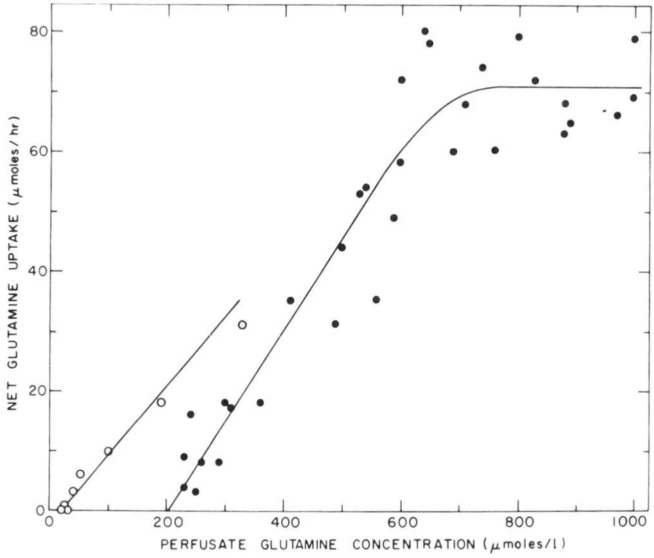

Fig. 3. Concentration dependence of net glutamine uptake by isolated, perfused rat gut. ●, data from 7 experiments in which perfusate glutamine concentration was adjusted by graded additions of glutamine to the perfusing blood; o, data from two experiments in which the perfusate contained 4 mM L-methionine-SR-sulfoximine and no glutamine was added. Net uptake was calculated from the arteriovenous concentration difference across the preparation and the perfusate flow rate. Taken from Windmueller and Spaeth (9).

microflora. In other perfusion studies, Matsutaka *et al.* (13) also identified alanine and ammonia as the main nitrogenous products released when non-hepatic splanchnic organs of conventional or germ-free rats were perfused with very high concentrations of glutamine (16mM).

Whole Animal Studies

Arteriovenous difference measurements in fasted animals indicate that intestinal uptake of circulating glutamine occurs in most laboratory animal species (Table II). Exceptions are the guinea pig and the chicken. In most other animals, there was a net uptake of 20-30% of the plasma glutamine in each blood pass through the tissue. Surgically removing the intestine (13) (Windmueller and Spaeth, unpublished observation) or the liver, spleen, and intestine (15) from the circulation results in a marked increase in the glutamine concentration in the systemic blood. The relative overall importance

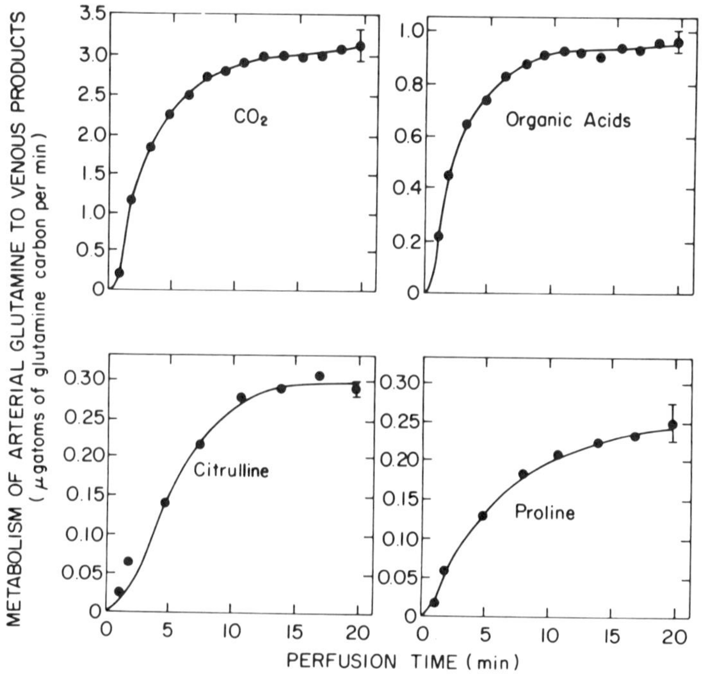

Fig. 4. Metabolism of arterial glutamine to venous products by isolated, perfused rat gut. During the recirculating perfusion of gut preparations with 350 ml blood, 80 μmoles of glutamine per hour were infused continuously into the perfusate to maintain a 600 to 700 μM concentration. At 50 min (zero time), the infusion of nonradioactive glutamine was discontinued, a continuous infusion of L-[U-^{14}C]glutamine (250 mCi/mmole, 9.6 μCi/min) into the arterial cannula was begun, and the entire venous outflow was diverted into stoppered tubes in ice. Over the next 21 min, 14 samples of venous blood were collected at 1.5 min intervals. CO_2, organic acids, and amino acids were isolated and counted for radioactivity. The specific radioactivity of glutamine in the arterial blood was also determined. The rate of appearance of glutamine carbon in the products was calculated for each 1.5 min interval. Only eight of the 14 samples were analyzed for citrulline and proline. Points are the means and vertical lines indicate the range for two experiments. Taken from Windmueller and Spaeth (9).

of the gut as a site of glutamine catabolism is indicated by the finding that evisceration caused a much larger accumulation of plasma glutamine than did nephrectomy (13). Furthermore, arteriovenous measurements made in fasted rats across other large organs or body regions, including liver, kidney, lungs, heart, head, and hindquarters, failed to reveal any appreciable net extraction of glutamine (9).

The site of intestinal glutamine uptake was further localized *in vivo* after infusing a tracer dose of L-[U-^{14}C]glutamine into the superior mesenteric artery of anesthetized rats (Table III). Immediately after a 20s infusion, a total of 26% of the infused dose was

Table I

Rate of metabolism of arterial glutamine by isolated, perfused rat gut

Results are from the two experiments in Fig. 4. The data were calculated from the recovery of ^{14}C in products isolated from the final venous sample, collected 19.5-21.0 min after beginning the arterial infusion of L-[U-^{14}C]glutamine. Tissue acid-insoluble ^{14}C was determined on the whole preparation after the 21-min infusion; incorporation into this fraction was assumed to be linear over the 21-min interval. Taken from Windmueller and Spaeth (9).

Product	Glutamine carbon (μg atoms/h)
In venous blood	
CO_2	187.9
Citrulline	18.2
Proline	15.0
Other amino acids[a]	2.1
Lactate	33.9
Citrate	17.9
Other organic acids[b]	6.2
Tissue, perchloric acid-insoluble	47.3
Total	328.5

[a] Aspartate, ornithine, glutamate, and alanine.
[b] α-Ketoglutarate, malate, pyruvate, and three or more unidentified.

recovered in intestinal tissue, consistent with the earlier results of arteriovenous difference measurements. Most of the dose was recovered in the small intestine, and only there was the glutamine rapidly metabolized. As early as 20 s, 74% of the ^{14}C recovered in the small intestinal mucosa was recovered in metabolic products. These studies pointed to the cells of the small intestinal mucosa as the major site for the glutamine metabolism observed in the whole gut.

Studies with Autoperfused Jejunum in Vivo

An *in situ* autoperfused preparation of rat jejunum was developed to permit studies of glutamine metabolism in the absence of exogenous hormones and also to allow studies with a defined region of small intestine where the fate of luminal as well as arterial glutamine could be quantitatively determined. The preparation has been des-

TABLE II

Arteriovenous difference measurements across intestine and intestinal glutaminase activity[a]

Animal	Plasma Glutamine Concentration[b] (μM) A	Plasma Glutamine Concentration[b] (μM) V-A	Mucosal Glutaminase Activity[c] (μmol·h^{-1}·mg protein^{-1}) Jejunum	Mucosal Glutaminase Activity[c] (μmol·h^{-1}·mg protein^{-1}) Ileum
Rat (conventional)	655	−194	4.8	5.0
Rat (germfree)	596	−196	—	—
Dog	723	−200	5.1	2.8
Cat	602	−129	1.6	4.8
Hamster	741	−153	3.6	4.3
Mouse	—	—	4.9	5.1
Monkey	463	−148	4.7	4.9
Rabbit	433	−40	4.0	5.3
Guinea Pig	247	+14	1.2	1.4
Chicken	330	+53	0.4	0.3

[a] Results taken from Windmueller and Spaeth (9) and Pinkus and Windmueller (14).

[b] Animals were fasted for 24 h and anesthetized. Blood was sampled simultaneously from an artery (A) and from the superior mesenteric vein (V), except in the monkey where venous blood was from an ileal vein.

[c] Phosphate-dependent glutaminase activity was determined in mucosal homogenates prepared from freshly-killed, non-fasted animals.

cribed in detail (16, 17). Briefly, the single vein draining a 2-5 cm jejunal segment of an anesthetized rat is cannulated and all the venous blood from the segment is collected; the arterial blood supply to the segment remains intact. The two ends of the intestinal segment are ligated and may also be cannulated for luminal perfusion. The rat is continuously transfused via a saphenous vein with replacement blood from donor rats. By way of another cannula in the abdominal aorta, arterial blood pressure is constantly monitored and arterial blood samples can be taken periodically.

With autoperfused jejunal segments from rats fasted overnight, the net uptake rate for circulating glutamine was nearly as large as for glucose (Table IV). These results are similar to those obtained earlier in the whole, isolated, perfused gut. Jejunum also took up substantial amounts of ketone bodies. On a molar basis, ammonia was released at about 80% the rate of net glutamine uptake. Large amounts of lactate and alanine were also released, but almost no glutamate. Release of ammonia plus alanine could account for 77%

TABLE III

Prompt metabolism of arterial [^{14}C]glutamine by the gut in vivo

A tracer dose of L-[U-^{14}C]glutamine (20 μCi, 45 μCi/mmol) was infused continuously into the blood stream of the superior mesenteric artery of anesthetized rats for 20 s. Then the gut was excised rapidly, chilled in ice and divided into 4 fractions as indicated. Each fraction was extracted with perchloric acid and the ^{14}C in tissue metabolites determined. Data are means of 4 experiments. Results taken from Windmueller and Spaeth (9).

	Small Intestine Mucosa	Small Intestine Residue	Cecum + Colon	Mesentery
Total ^{14}C (% of dose)	(11)	(6)	(3)	(6)
^{14}C Distribution (% of total)				
Glutamine	20	29	85	87
Glutamate	31	32	12	9
Organic acids	33	26	1	1
Other amino acids	10	9	1	2
Acid-insoluble	6	4	1	1
Total	100	100	100	100

of the glutamine nitrogen taken up. The remaining nitrogen was accounted for in later studies where all amino acids were determined.

To obtain an overview of energy metabolism in rat jejunum and to assess further the relative importance of glutamine as a fuel of respiration, the metabolic fate of each of the substrates taken up from the blood was quantitatively determined (17). In separate experiments, a tracer dose of each substrate, uniformly labeled with ^{14}C, was infused into the arterial blood supply to a jejunal segment, and the labeled metabolic products appearing in the venous blood from the segment were quantified. Therefore, these were single-pass perfusion experiments with normal blood in which one of the blood constituents was labeled at a known and constant specific activity. The rate of appearance of labeled metabolic products in the venous blood reached a steady state in 10-20 min, as in analogous experiments with the isolated gut preparation. And as before, metabolic rates were calculated for each substrate from the steady state production rate of ^{14}C-labeled metabolites and the specific radioactivity of the arterial substrate. Results for glutamine and for glucose are shown in Table V. More than 30% of the vascular glutamine flux through jejunum was taken up by the tissue and metabolized, with

TABLE IV

Net uptake and release of blood metabolites in vivo by jejunum of fasted rats

Net uptake or release was calculated from the blood flow rate through jejunal segments *in vivo* multiplied by the arteriovenous concentration difference in simultaneously collected arterial and venous blood samples. O_2 and CO_2 were determined by Van Slyke analysis and all other metabolites were determined enzymatically. Values are means ± S. E. for 16-48 paired samples from 3 or 4 segments. Results taken from Windmueller and Spaeth (17).

Metabolite	Net Uptake from Blood	Net Release Into Blood
	(nmol · min^{-1} · g intestine^{-1})	
O_2	1950 ± 100	
Glucosé	220 ± 10	
Glutamine	191 ± 2	
Acetoacetate	160 ± 10	
3-Hydroxybutyrate	173 ± 12	
CO_2		1650 ± 88
Ammonia		155 ± 3
Lactate		170 ± 6
Alanine		138 ± 4
Asparagine		4 ± 1
Glutamate		2 ± 2

56% of the carbon appearing in CO_2. About 5% of the glutamine carbon was incorporated into glucose. The products of glutamine carbon in jejunum were otherwise similar to those observed with the whole gut. With arterial glucose, less than 2% of the flux through jejunum was metabolized and only 11% of the carbon appeared in CO_2. The main metabolic products of glucose were lactate and alanine. Although the overall metabolic rates for glutamine and glucose carbon were very similar, the rate of CO_2 production from glutamine was more than five times the rate from glucose. The CO_2 yields from each of the substrates studied are summarized in Table VI. Some arterial lactate was metabolized although there was a net release of lactate into the blood. The combined rate of respiratory CO_2 production from the six arterial substrates listed in Table VI (1704 nmol · min^{-1} · g intestine^{-1}) was very similar to the actual net rate of CO_2 release, as determined by Van Slyke analysis (1650 nmol · min^{-1} · g intestine^{-1}, see Table IV). Thus no additional respiratory fuels of any quantitative importance apparently remain to be identified. Arterial glutamine is clearly an important respiratory fuel for jejunum of fasted rats, accounting for 35% of the CO_2 released.

TABLE V

Metabolism of arterial glutamine and glucose by jejunum of fasted rats

A tracer dose of the substrate, uniformly labeled with ^{14}C, was infused continuously into the arterial blood supplying a jejunal segment, and the steady-state rate of product formation was determined by following the appearance of ^{14}C-labeled products in the segment venous blood for 70-90 min. The specific activity of the labeled substrate in arterial plasma to the segment was determined at 10-min intervals. Values are means ± S. E. for 5-8 steady-state values from each of three experiments. Taken from Windmueller and Spaeth (17).

	Glucose [a]		Glutamine [b]	
Arterial plasma concentration (mM)	7.50 ± 0.18		0.60 ± 0.02	
Flux (μg atom carbon · min^{-1} · g intestine^{-1}) [c]	58.5 ± 3.9		3.50 ± 0.46	
	(ng atoms substrate carbon · min^{-1} · g intestine^{-1})			
Formation rate of metabolic products:				
Released into blood				
CO_2	116 ± 7	(10.8) [d]	597 ± 37	(55.5) [d]
Lactate	463 ± 40	(43.1)	92 ± 20	(8.5)
Citrate			17 ± 5	(1.6)
Other organic acids [e]	51 ± 6	(4.7)	12 ± 4	(1.1)
Alanine	138 ± 13	(12.8)	36 ± 2	(3.3)
Proline	1 ± 0	(0.1)	56 ± 2	(5.2)
Citruline	2 ± 0	(0.2)	50 ± 3	(4.7)
Ornithine	2 ± 0	(0.2)	14 ± 1	(1.3)
Other amino acids [f]	17 ± 1	(1.6)	7 ± 1	(0.7)
Glucose			50 ± 6	(4.7)
Lipids	9 ± 2	(0.8)		
Incorporated into tissue [g]				
Intestine, acid-insoluble	121 ± 20	(11.3)	133 ± 18	(12.4)
Mesentery, lipids	148 ± 22	(13.8)	1 ± 0	(0.1)
Mesentery, lipid-free acid insoluble	6 ± 1	(0.6)	10 ± 1	(0.9)
Total	1074 ± 80	(100.0)	1075 ± 55	(100.0)
Percent of flux metabolized [h]	1.8		30.7	

[a] D-[U—^{14}C]Glucose (300 mCi · mmol^{-1}) infused at 2.0 μCi · min^{-1}.

[b] L-[U—^{14}]Glutamine (40 mCi · mmol^{-1}) infused at 0.4 μCi · min^{-1}.

[c] Flux is the mean flow rate of substrate carbon through the tissue (substrate concentration in plasma · carbon atoms per mol · plasma flow rate).

[d] Percent of total.

[e] Malate, citrate, succinate, α-ketoglutarate, pyruvate.

[f] Glutamate, aspartate, glycine, serine.

[g] Incorporation into tissue was assumed to be linear over the duration of the experiment.

[h] Total rate of product formation · 100/flux.

TABLE VI

Contribution of arterial substrates to respiratory CO_2 in jejunum of fasted rats

The rate of CO_2 production from each arterial substrate was determined during single-pass vascular perfusions with blood containing the ^{14}C-labeled substrate, as described for glutamine and glucose in Table 5. Taken from Windmueller and Spaeth (17).

Substrate	CO_2 Yield	Percent of Total
	($nmol \cdot min^{-1} \cdot g\ intestine^{-1}$)	
Glutamine	597	35.0
3-Hydroxybutyrate	444	26.1
Acetoacetate	410	24.1
Glucose	116	6.8
Lactate	79	4.6
Unesterified fatty acids	58	3.4
Total	1 704	100.0

Intestinal utilization of circulating glutamine is not restricted to fasted animals, however. With jejunal segments from non-fasted rats, the net uptake rate of glutamine from blood was more than twice the rate, on a tissue weight basis, as observed with jejunum from fasted rats (18) (Compare Tables IV and VII). Glucose utilization also increased and the glutamine: glucose utilization ratio remained near one. These measurements were all made after the intestinal lumen was rinsed free of diet residue. The rate of glutamine utilization by jejunum from non-fasted rats is sufficient to account for 46% of the total respiratory CO_2 produced by the tissue (18).

Fate of Glutamine Nitrogen in Intestine

The glutamine nitrogen taken up by jejunal segments from non-fasted rats was fully accounted for by the release rates of ammonia, citrulline, alanine, proline, glutamate, and ornithine (Table VII). The latter two amino acids were very minor products. Small intestine from dog (19, 20), sheep (5), and man (21) also releases these products into the blood.

The metabolic processing by intestine of circulating glutamine nitrogen has some important physiological implications. First, the continous production of metabolic ammonia has clinical significance

TABLE VII

Net uptake and release of blood nitrogen in vivo by jejunum of non-fasted rats

Net uptake and release were determined from arteriovenous difference measurements across autoperfused jejunal segments after rinsing the intestinal lumen free of diet residue. Net uptake of glutamine was equal to 445 nmol · min^{-1} · g^{-1}, equivalent to 890 ng atoms N · min^{-1} · g^{-1}. Results are taken from Windmueller and Spaeth (18).

	Net Release into Blood		
Compound	nmol · min^{-1} · g^{-1}	ng atom N · min^{-1} · g^{-1}	% of Total N
Ammonia	334	334	37.9
Citrulline	81	243	27.6
Alanine	215	215	24.4
Proline	63	63	7.2
Glutamate	16	16	1.8
Ornithine	5	10	1.1
(Total)		(881)	(100.0)

in patients with diseased livers that are inefficient in removing this toxic metabolite from the hepatic portal blood. Therapy in such patients with hepatic encephalopathy has usually included low-protein diets and oral antibiotics, to suppress ammonia production by the intestinal microflora, previously thought to be the principal source of intestinal ammonia. It is now clear that another important source, unaffected by antibiotics, is metabolism of circulating glutamine. Weber and Veach (20) estimate that in fasted dogs, about 50% of the ammonia produced by the entire gut is derived from glutamine metabolized by the jejunum and ileum.

Secondly, the citrulline produced by intestine may contribute significantly to biosynthesis of endogenous arginine by the kidney and also other tissues (22). Arginine is a semi-essential dietary amino acid for most animals, and the kidney is an endogenous source of this amino acid (23). The kidney, however, requires pre-formed citrulline in order to make arginine (24). One source of circulating citrulline may be the liver (25), but the quantitative significance of this source remains uncertain (26). In a 350 g rat, the small intestine would release to the blood daily about 0.5 mmol of citrulline, all potentially available for the synthesis of endogenous arginine. This quantity is approximately equivalent to the quantity of exogenous dietary arginine needed to achieve normal growth in young rats (27), suggesting that the intestinal citrulline contribution could be quantitatively quite significant.

And thirdly, the alanine, ammonia, and proline produced by the intestine are removed efficiently from portal vein blood by the liver, where they contribute importantly as nitrogenous substrates for urea biosynthesis. In fact, all the nitrogenous intestinal products could serve ultimately as urea precursors. The daily urinary output of urea nitrogen in fasted 350 g rats is 9.4 mmol (18). The intestine of such rats metabolizes 3.6 mmol of glutamine nitrogen per day (see Fig. 1). So in fasted rats, more than 30% of the nitrogen utilized for urea biosynthesis in the liver is apparently derived from the nitrogenous products of intestinal glutamine metabolism. Meanwhile, the carbon atoms of the alanine are used for hepatic gluconeogenesis (28). As much as 50% of the alanine utilized for gluconeogenesis in the liver of fasted animals may be produced in the intestine (3, 7).

Metabolism of Luminal Glutamine

After an animal ingests protein, mucosal cells of the small intestine have access to glutamine from the intestinal lumen as well as from the blood. To determine the metabolic fate of luminal glutamine, L-[U-^{14}C]glutamine was administered into the lumen of autoperfused jejunal segments at a concentration of 6 mM, approximating postprandial levels (16). Greater than 94% of the radioactivity appeared in intestinal venous blood within 30 min, but only 34% of the ^{14}C remained in glutamine; the remainder was distributed among a group of metabolic products closely resembling those resulting from the metabolism of arterial glutamine. Thus glutamine was similarly metabolized whether it entered the mucosal cells across the brush border from the lumen or across the basolateral membrane from the blood. When the luminal glutamine concentration was increased to 45 mM, 70% of the glutamine reached the venous blood unchanged. With neither 6 mM nor 45 mM luminal glutamine was there any release of glutamate into the venous blood. Furthermore, when L-[U-^{14}C]glutamate (6 mM) was administered into the lumen, none of it escaped metabolism during transport across the intestine; the ^{14}C-labeled metabolic products were the same as those arising from L-[U-^{14}C]glutamine. These results are consistent with deamidation being the first step in intestinal glutamine catabolism. They also reveal that deamidation is more rate limiting than metabolism of the resulting glutamate. A 40% reduction in the rate of metabolism of arterial glutamine was observed after 6 mM glutamine was administered into the intestinal lumen (16). In other words, the availability of glutamine in the lumen spared its utilization from the blood. The combined rate of metabolism from both sources, however, was 70% greater than the metabolic rate for arterial glutamine alone when the lumen was empty. A similar result was obtained in an isolated intestinal preparation by Hanson and Parsons (11). These studies further underscore the large capacity of the small intestine for glutamine catabolism.

Studies with Intestinal Preparations In Vitro

The exogenous energy source added by most investigators during incubation studies *in vitro* with preparations of small intestine is glucose. Evidence exists, however, for the preferential utilization of glutamine. Slices of rabbit ileum preferred glutamine to glucose (29), and glutamine was more effective than glucose in stimulating O_2 uptake by rabbit ileal mucosa (30). With isolated rat enterocytes, glucose plus glutamine stimulated O_2 uptake more than either substrate alone (31). Meanwhile, Matsutaka *et al.* (13) found that in rat ileal slices, glutamine was hydrolyzed to ammonia and glutamate but there was little evidence for further metabolism of the glutamate. With tissue slices, rings, everted sacs, and cells prepared from intestine, the full range of metabolic activity may not always be demonstrable due to leakage of cellular adenine nucleotides and other metabolites during incubations *in vitro* (see Ref. 31).

Enzymology of Intestinal Glutamine Metabolism

In order to define the major pathway for glutamine metabolism in the small intestine, the activities of several glutamine degrading enzymes were measured and data for additional enzymes was obtained from a survey of scattered reports in the literature (14). Homogenates prepared from mucosal scrapings were found to contain an active phosphate-dependent glutaminase (3-6 $\mu mol \cdot h^{-1} \cdot mg$ $protein^{-1}$), confirming previous reports (32-35). Relative to the activity of this enzyme, there is very low tissue activity of phosphate-independent glutaminase, glutamine transaminase, and several glutamine amidotransferases (carbamoyl phosphate synthetase, phosphoribosyl pyrophosphate amidotransferase, cytidine triphosphate synthetase, and fructose-6-phosphate aminotransferase). Subject to the uncertainties associated with the determination of enzyme activities in crude tissue preparations, the results indicate that phosphate-dependent glutaminase (referred to hereafter as glutaminase) is by far the most active glutamine degrading enzyme in small intestine and the only one with sufficient activity *in vitro* to account for the rates of glutamine metabolism observed *in vivo*.

Glutaminase was found in the small intestinal mucosa of all animals surveyed (see Table II). With several exceptions, the specific activities are similar in jejunum and ileum. Guinea pig and chicken intestine have relatively low activity; it is noteworthy that in these animals the arterial plasma glutamine concentration is also low, and, from arteriovenous difference measurements across the intestine, a net release rather than a net uptake of glutamine was found.

The localization of glutaminase in rat intestine was studied in some detail (14). The specific activity is similar in mucosa from duodenum, upper and lower jejunum, and upper and lower ileum (3.5-4.2 $\mu mol \cdot h^{-1} \cdot mg\ protein^{-1}$). Much lower activity was found in mucosa from the stomach, cecum, and colon. In the small intestine, virtually all the glutaminase is present in mucosal epithelial cells; the muscularis is nearly devoid of activity. Thus, glutaminase is concentrated in those regions of the small intestine found to be most active in metabolizing circulating glutamine (see Table III). The enzyme specific activity is about the same in rapidly-dividing crypt cells and in more mature, non-dividing villus cells, where most intestinal absorption takes place. Virtually all intestinal glutaminase is membrane-bound, and when cell homogenates were centrifuged, both glutaminase and the mitochondrial marker enzyme, cytochrome oxidase, were enriched in a 15,000 x g pellet. Following sucrose-density centrifugation of this pellet, both enzymes were found in a band corresponding to the density of mitochondria. Therefore, like glutaminase in a number of other tissues, the glutaminase of small intestinal mucosa appears to be a mitochondrial enzyme.

Some properties of glutaminase were determined in mitochondria from rat intestine (14). The enzyme has a rather broad pH optimum between pH 7.6 and pH 9.2. The apparent K_m for glutamine at pH 8.1 was 2.2 mM. This is lower by a factor of ten than the K_m reported for glutaminase from liver (28 mM) (36) and may explain the 30-fold lower tissue levels of glutamine found in intestinal mucosa (0.2 μmol/g) compared with liver (5.9 μmol/g) (9). It may also help to explain why intestine is quantitatively so important as a site for the metabolism of circulating glutamine. It is interesting to note that in intestinal mucosa the tissue:plasma concentration ratio for glutamine is about 0.3 while in liver it is near 10.

The apparent $K_{0.5}$ for activation by phosphate was 22 mM. However, in the absence of phosphate or with sub-optimal concentrations, glutaminase was activated by a number of other compounds, including ammonium sulfate, glucose-6-phosphate, maleate, ammonium bicarbonate, oxalate, pyrophosphate, citrate, and arsenate. The enzyme did not hydrolyze D-glutamine or L-asparagine, explaining why asparagine, unlike glutamine, is transported across the rat intestine without being metabolized (37).

Glutaminase is inactivated by the sulfhydryl inhibitor p-hydroxymercuri-benzoate. It is irreversibly inhibited by two analogs of glutamine, 6-diazo-5-oxo-L-norleucine (DON) and L-2-amino-4-oxo-5-chloropentanoic acid (chloroketone); glutamine offers significant protection, suggesting that the analogs inhibit by reaction with the glutamine binding site(s) of the enzyme. Results of experiments with these analogs *in vivo* are consistent with the proposed role of the enzyme in intestinal glutamine metabolism. Chloroketone administered luminally into an autoperfused jejunal segment preparation resulted in 50% inhibition of the whole tissue glutaminase and totally inhibited the hydrolysis of glutamine subsequently absorbed from the lumen. The hydrolysis of glutamine taken up from arterial blood was not affected. Meanwhile, DON administered arterially inhibited glutaminase almost completely and inhibited the hydrolysis of both luminal and arterial glutamine (14).

Intestinal glutaminase activity in rats could not be increased by greatly increasing the glutamine content of the diet. There was also little effect of age on the level of activity. The intestinal glutaminase appears similar in many of its properties to the glutaminase of kidney and brain and, in fact, it cross reacts immunochemically with the kidney enzyme (see Ref. 14). However, metabolic acidosis that resulted in a 4-fold increase in kidney glutaminase activity had no effect on the level of activity in the intestine (38, 14). Likewise, acidosis did not increase the rate of glutamine utilization by isolated, perfused rat intestine (11) or by intestinal slices incubated *in vitro* (38).

The proposed metabolic scheme in Fig. 5 can account for all of the observed products of intestinal glutamine metabolism. The reactions that are numbered have all been demonstrated during incubation studies with cell free intestinal mucosal preparations (Table VIII). Glutaminase (Reaction 1) appears to be the main initiating step, as already discussed. Analysis of metabolites in intestinal venous blood indicates that about 80% of the glutamine amide nitrogen is released into the portal vein as ammonia (see Table 7). The remaining ammonia is apparently converted to carbamoyl phosphate

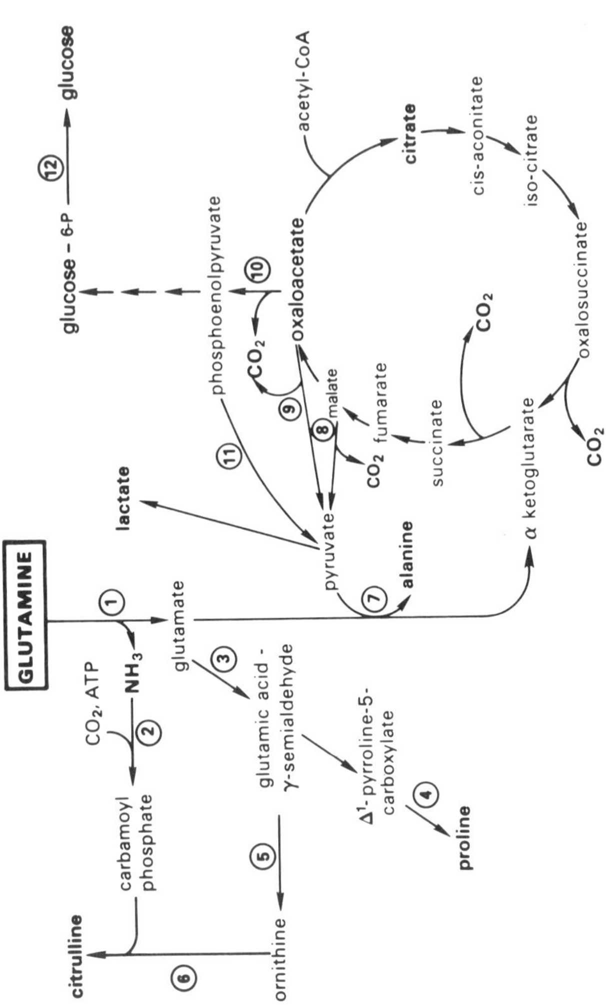

Fig. 5. Pathways suggested for glutamine metabolism by rat intestinal mucosa. Bold print indicates metabolic products released into intestinal venous blood. Numbered reactions have all been demonstrated in rat intestinal mucosa *in vitro* (see Table VIII).

TABLE VIII

Enzymes present in small intestinal mucosa and relevant to the metabolism of glutamine

Reaction Catalyzed[a]	Enzyme	EC Number	References
1	Glutaminase	3.5.1.2	14, 32-35
2	Carbamoyl-phosphate synthase (ammonia)	2.7.2.5	39-41
3	(enzymology remains uncertain)[b]	——	42
4	Pyrroline-5-carboxylate reductase	1.5.1.2	43
5	Ornithine-oxo-acid aminotransferase	2.6.1.13	44, 42
6	Ornithine carbamoyltransferase	2.1.3.3	41, 45-47
7	Alanine aminotransferase	2.6.1.2	48, 49
8	Malate dehydrogenase (decarboxylating) ($NADP^+$)	1.1.1.40	50, 31
9	Oxaloacetate decarboxylase	4.1.1.3	31
10	Phosphoenolpyruvate carboxykinase (GTP)	4.1.1.32	51, 31
11	Pyruvate kinase	2.7.1.40	50
12	Glucose-6-phosphatase	3.1.3.9	52

[a] Numbers refer to metabolic reactions in Fig. 5.
[b] In homogenates, the reaction requires ATP, Mg^{2+}, and NADPH and, in analogy to the pathway in E. coli, probably proceeds through the intermediate formation of γ-glutamyl phosphate.

(Reaction 2). In addition to carbamoyl phosphate synthetase, rat intestine has been shown to contain N-acetylglutamate, a cofactor for the enzyme, and is able to synthesize it from glutamate *in vitro* (53). Some of the glutamate formed in Reaction 1 is reduced to glutamic acid γ-semialdehyde (Reaction 3), which can be further reduced to proline (Reaction 4) or transaminated to ornithine (Reaction 5). Glutamate can serve as amino donor in Reaction 5. Ornithine plus carbamoyl phosphate yields citrulline (Reaction 6). We have failed to find any evidence for the conversion of citrulline to arginine (9). Some arginase is present (37) but apparently not the complete urea cycle. The end-product of this pathway appears to be citrulline. Additional glutamate can be transaminated with pyruvate to give alanine and α-ketoglutarate (Reaction 7). As suggested by Watford *et al.* (31), providing the large amounts of pyruvate needed for transamination may be viewed as a physiological explanation for the high rates of aerobic glycolysis in the intestine. Several pathways are present to explain the appearance of glutamine carbon in 3-carbon products: "malic enzyme" (Reaction 8), oxaloacetate decarboxylase

(Reaction 9), and phosphoenolpyruvate carboxykinase plus pyruvate kinase (Reactions 10 and 11). Reactions 10 and 12 and the presence of the glycolytic pathway explain the recovery of glutamine carbon in glucose.

ROLE OF THE LIVER IN GLUTAMINE METABOLISM

During the post-absorptive period between meals, the glutamine concentration in the blood perfusing the liver (about 75% portal blood and 25% arterial blood) will be lower than the glutamine concentration in the arterial blood perfusing other body tissues. Under these conditions, there appears to be little or no net uptake or release of circulating glutamine by the liver of rats (7, 9, 54) or dogs (55). Rat hepatocytes incubated in normal blood concentrations of glutamine also fail to utilize the amino acid (54), although the liver cell membrane is readily permeable to it (56). Some net uptake by liver has been observed in fasted sheep (57). Postprandially, after a high-protein meal, however, the portal blood glutamine levels become elevated. Under these conditions, the entire net output of glutamine from the intestine is immediately extracted by the liver (3, 58). Conversely, when the liver is perfused with solutions containing lower than normal glutamine concentrations, a net hepatic release of glutamine is observed (15). Thus, the liver appears to play a regulatory role in restoring blood glutamine levels to normal when they are either elevated or depressed. The liver is amply supplied with glutamine synthetase (15, 59, 60) and also with glutaminase (35, 36), but the regulation of their activity *in vivo* is poorly understood. In normal animals in the post-absorptive state, both enzymes are presumably low-functioning. Evidence *in vivo* (54, 57) and from liver perfusion experiments (61, 62) suggests that during metabolic acidosis, when the glutamine demand by the kidney is increased, the net hepatic production of glutamine is also increased, at the expense of urea production.

REFERENCES

1. Van Slyke, D. D., Phillips, R. A., Hamilton, P. B., Archibald, R. M., Futcher, P. H., and Hiller, A. (1943), *J. Biol. Chem. 150,* 481-482.
2. Waelsch, H. (1952), *Adv. Enzymol. 13,* 237-319.
3. Elwyn, D. H., Parikh, H. C., and Shoemaker, W. C. (1968), *Am. J. Physiol. 215,* 1260-1275.
4. Addae, S. K., and Lotspeich, W. D. (1968), *Am. J. Physiol. 215,* 269-277.
5. Wolff, J. E., Bergman, E. N., and Williams, H. H. (1972), *Am. J. Physiol. 223,* 438-446.

6. Ishikawa, E., Aikawa, T., and Matsutaka, H. (1972), *J. Biochem. (Tokyo) 71,* 1097-1099.
7. Aikawa, T., Matsutaka, H., Yamamoto, H., Okuda, T., Ishikawa, E., Kawano, T., and Matsumura, E. (1973), *J. Biochem. (Tokyo) 74,* 1003-1017.
8. Felig, P., Wahren, J., and Räf, L. (1973), *Proc. Nat. Acad. Sci. U. S. A. 70,* 1775-1779.
9. Windmueller, H. G., and Spaeth, A. E. (1974), *J. Biol. Chem. 249,* 5070-5079.
10. Windmueller, H. G., Spaeth, A. E., and Ganote, C. E. (1970), *Am. J. Physiol. 218,* 197-204.
11. Hanson, P. J., and Parsons, D. S. (1977), *Biochem. J. 166,* 509-519.
12. Duda, G. D., and Handler, P. (1958), *J. Biol. Chem. 232,* 303-314.
13. Matsutaka, H., Aikawa, T., Yamamoto, H., and Ishikawa, E. (1973), *J. Biochem. (Tokyo) 74,* 1019-1029.
14. Pinkus, L. M., and Windmueller, H. G. (1977), *Arch. Biochem. Biophys. 182,* 506-517.
15. Lund, P. (1971), *Biochem. J. 124,* 653-660.
16. Windmueller, H. G., and Spaeth, A. E. (1975), *Arch. Biochem. Biophys. 171,* 662-672.
17. Windmueller, H. G., and Spaeth, A. E. (1978), *J. Biol. Chem. 253,* 69-76.
18. Windmueller, H. G., and Spaeth, A. E. (1980), *J. Biol. Chem. 255,* 107-112.
19. Weber, F. L., Jr., Maddrey, W. C., and Walser, M. (1977), *Am. J. Physiol. 232,* E263-E269.
20. Weber, F. L., Jr., and Veach, G. L. (1979), *Gastroenterology 77,* 235-240.
21. Felig, P., and Wahren, J. (1971), *J. Clin. Invest. 50,* 2703-2714.
22. Ratner, S. (1973), *Adv. Enzymol. 39,* 1-90.
23. Featherston, W. R., Rogers, Q. R., and Freedland, R. A. (1973), *Am. J. Physiol. 224,* 127-129.
24. Borsook, H., and Dubnoff, J. W. (1941), *J. Biol. Chem. 141,* 717-738.
25. Drotman, R. B., and Freedland, R. A. (1972), *Am. J. Physiol. 222,* 973-975.
26. Hems, R., Ross, B. D., Berry, M. N. and Krebs, H. A. (1966), *Biochem. J. 101,* 284-292.
27. Milner, J. A., and Visek, W. J. (1978), *J. Nutr. 108,* 382-391.
28. Ross, B. D., Hems, R., and Krebs, H. A. (1967), *Biochem. J. 102,* 942-951.
29. Neptune, E. M., Jr. (1965), *Am .J. Physiol. 209,* 329-332.
30. Frizzell, R. A., Markscheid-Kaspi, L., and Schultz, S. G. (1974), *Am. J. Physiol. 226,* 1142-1148.
31. Watford, M., Lund, P., and Krebs, H. A. (1979), *Biochem. J. 178,* 589-596.
32. Richterich, R., Goldstein, L., and Cohen, E. (1958), Hoppe-Seyler's *Z. Physiol. Chem. 312,* 45-50.
33. Spencer, R. P., and Zamcheck, N. (1961), *Gastroenterology 40,* 423-426.
34. Horowitz, M. L., and Knox, W. E. (1968), *Enzym. Biol. Clin. 9,* 241-255.
35. Katunuma, N., Katsunuma, T., Towatari, T., and Tomino, I. (1973), in *The Enzymes of Glutamine Metabolism* (Prusiner, S., and Stadtman, E. R., eds.) pp. 227-258, Academic Press, New York.

36. Huang, Y-Z., and Knox, W. E. (1976), *Enzyme 21,* 408-426.
37. Windmueller, H. G., and Spaeth, A. E. (1976), *Arch. Biochem. Biophys. 175,* 670-676.
38. Anderson, N. M., Bennett, F. I., and Alleyne, G. A. O. (1976), *Biochim. Biophys. Acta 437,* 238-243.
39. Hall, L. M., Johnson, R. C., and Cohen, P. P. (1960), *Biochim. Biophys. Acta 37,* 144-145.
40. Jones, M. E., Anderson, A. D., Anderson, C., and Hodes, S. (1961), *Arch. Biochem. Biophys. 95,* 499-507.
41. Tatibana, M., and Shigesada, K. (1972), *Adv. Enz. Regul. 10,* 249-271.
42. Ross, G., Dunn, D., and Jones, M. E. (1978), *Biochem. Biophys. Res. Commun. 85,* 140-147.
43. Herzfeld, A., Mezl, V. A., and Knox, W. E. (1977), *Biochem. J. 166,* 95-103.
44. Herzfeld, A., and Knox, W. E. (1968), *J. Biol. Chem. 243,* 3327-3332.
45. Lowenstein, J. M., and Cohen, P. P. (1956), *J. Biol. Chem. 220,* 57-70.
46. Reichard, H. (1959), *J. Lab. Clin. Med. 53,* 417-425.
47. Raijman, L. (1974), *Biochem. J. 138,* 225-232.
48. Brin, M., and McKee, R. W. (1956), *Arch. Biochem. Biophys. 61,* 384-389.
49. Volman-Mitchell, H., and Parsons, D. S. (1974),*Biochim. Biophys. Acta 334,* 316-327.
50. Tyrrell J. B., and Anderson, J. W. (1971), *Endocrinology 89,* 1178-1185.
51. Anderson, J. W. (1970), *Biochim. Biophys. Acta 208,* 165-167.
52. Anderson, J. W. (1974), *Am. J. Physiol. 226,* 226-229.
53. Shigesada, K., and Tatibana, M. (1971), *J. Biol. Chem. 246,* 5588-5595.
54. Lund, P., and Watford, M. (1976), in *The Urea Cycle* (Grisolia, S., Báguena, R., and Mayor, F., eds.) pp. 479-488, John Wiley & Sons, New York.
55. McMenamy, R. H., Shoemaker, W. C., Richmond, J. E., and Elwyn, D. (1962),*Am. J. Physiol. 202,* 407-414.
56. Pardridge, W. M. (1977), *Am. J. Physiol. 232,* E492-E496.
57. Heitmann, R. N., and Bergman, E. N. (1978), *Am. J. Physiol. 234,* E197-E203.
58. Yamamoto, H., Aikawa, T., Matsutaka, H., Okuda, T., and Ishikawa, E. (1974), *Am. J. Physiol. 226,* 1428-1433.
59. Tate, S. S., and Meister, A. (1973), in *The Enzymes of Glutamine Metabolism* (Prusiner, S., and Stadtman, E. R., eds.) pp. 77-127, Academic Press, New York.
60. Deuel, T. F., Lerner, A., and Albrycht, D. (1973), in *The Enzymes of Glutamine Metabolism* (Prusiner, S., and Stadtman, E. R., eds.) pp. 129-144, Academic Press, New York.
61. Lueck, J. D., and Miller, L. L. (1970), *J. Biol. Chem. 245,* 5491-5497.
62. Oliver, J., Koelz, A. M., Costello, J., and Bourke, E. (1977), *Europ. J. Clin. Invest. 7,* 445-449.

14. Glutamine Metabolism in Skeletal Muscle

ALAN J. GARBER

Internal Medicine Department
Division of Endocrinology and Metabolism
Baylor College of Medicine
Houston, Texas, USA

SUMMARY

Although glutamine comprises only 3% of the amino acid residues in skeletal muscle protein, it accounts for nearly 30% of amino acid efflux. Amino acids, particularly branched chain amino acids, stimulate the synthesis and release of glutamine from skeletal muscle. This may reflect a precursor role for these amino acids since they may contribute carbon and nitrogen for net glutamine synthesis. Amino acid precursor for glutamine synthesis may derive from the degradation of muscle proteins. Modulation of glutamine synthesis and release is achieved in part by modifications of protein synthesis and protein degradation. Insulin stimulates protein synthesis and simultaneously inhibits protein degradation. Both actions divert amino acid precursors from net glutamine synthesis towards net protein accumulation. Thyroid hormone stimulates proteolysis and thereby increases precusor availability for glutamine formation. Hormonal agonists linked through cyclic nucleotides also modulate muscle protein degradation. Adrenergic agonists, acting through a β-adrenergic receptor, a hormone sensitive adenylyl cyclase and increased cAMP levels inhibit proteolysis and the synthesis and release of glutamine in skeletal muscle. Similarly, serotonergic agonists, acting through a D-serotonergic receptor, a hormone sensitive adenylyl cyclase and increased cAMP levels, also inhibit proteolysis and glutamine formation and release. Cholinergic agonists, acting through a nicotinic cholinergic receptor and increased cyclic GMP levels, accelerate muscle proteolysis and

ISBN: 0-12-506040-8

thereby increase the synthesis and release of glutamine. Disordered regulation of cyclic nucleotide mechanisms controlling protein homeostasis in skeletal muscle are associated with concomitant derangements in the regulation of muscle glutamine metabolism.

INTRODUCTION

Skeletal muscle represents the principal site of glutamine production in man and other mammals. This tissue releases comparatively large quantities of glutamine and to a lesser extent of alanine (1-4). Indeed, the release of these two amino acids together accounts for more than half of the total α-amino nitrogen release from skeletal muscle under basal circumstances (1, 5). Paradoxically, however, both glutamine and alanine provide not more than 8% of the amino acid residues found in mammalian skeletal muscle protein (6). The metabolic basis for this relatively disproportionate release of glutamine and alanine from skeletal muscle as well as the physiologic significance of this finding with regard to whole body protein, amino acid and carbohydrate homeostasis is the subject of this chapter.

RESULTS AND DISCUSSION

As shown in Fig. 1, the spontaneous release of glutamine from the isolated intact epitrochlaris preparation of rat skeletal muscle, which has been developed and validated for studies of amino acid metabolism *in vitro* (5), vastly predominates over the release of all other amino acids. In fact, glutamine release exceeds by 50% the release of alanine, the latter being the only other amino acid released to a substantial extent from skeletal muscle. On the other hand, a large number of amino acids such as aspartate and the branch-chain amino acids are released to only a negligible extent compared to the release of glutamine and alanine. However, the content of these amino acids in rat skeletal muscle protein equals or exceeds that of alanine and glutamine. Similar, disproportionate patterns of amino acid release have been observed in other preparations of rat skeletal muscle and in arteriovenous catheterization studies of human extremities (2-4). Of note, however, rat skeletal muscle releases small quantities of glutamate whereas human skeletal muscle appears to take up small quantities of this amino acid (1, 5).

It is possible that this disproportionate release of glutamine and alanine from rat skeletal muscle may be the result of a relatively

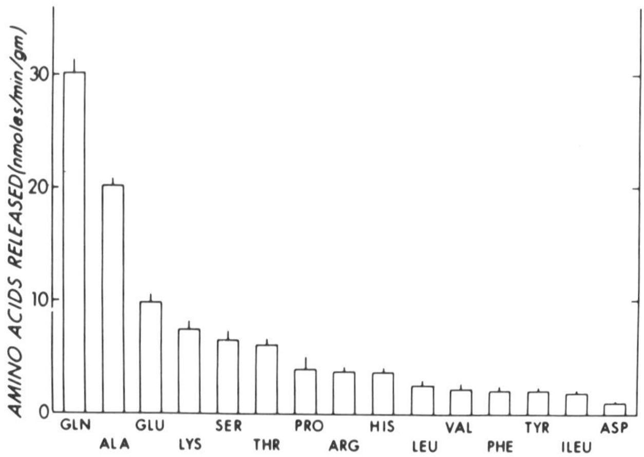

Fig. 1. Profile of amino acid release from epitrochlaris preparations of rat skeletal muscle.

rapid degradation of one or more classes of proteins having extremely high glutamine and alanine enrichment, relative to other classes of protein comprising rat skeletal muscle. This hypothesis has been evaluated *in vitro* by a number of investigators using inhibitors of glutamine synthetase and alanine aminotransferase such as methionine sulfoximine and aminooxyacetic acid (1, 5). As shown in Table I, these studies show clearly that the largest portion of alanine and glutamine released from rat skeletal muscle most probably derives from the net synthesis of these two amino acids since the disproportionate release of either alanine or glutamine is obliterated by inhibition of alanine aminotransferase and glutamine synthetase, respectively. Similar results have also been obtained using cycloserine as an inhibitor of alanine aminotransferase (7). It is also interesting to note that inhibition of glutamine output by methionine sulfoximine results in augmentation of alanine output from rat skeletal muscle, whereas aminooxyacetate inhibition of alanine synthesis results in a concomitant increase in glutamine output from rat skeletal muscle. These observations suggest the possibility that the pathways and mechanisms producing the selective synthesis of alanine and glutamine in skeletal muscle are interrelated. On the other hand, the release of other amino acids such as aspartate was increased when there was inhibition of either alanine aminotransferase or glutamine synthetase, The significance of these observations will be explored in greater detail below.

Several mechanisms have been proposed to account for the apparent selective synthesis and subsequent release of glutamine and

TABLE I

Effects of methionine sulfoximine and aminooxyacetate on amino acid and pyruvate release from epitrochlaris preparations of rat skeletal muscle

Addition	Glutamine	Alanine	Glutamate Release	Aspartate	Pyruvate
			nmoles/min/g muscle		
None	30.1 ± 2.0	20.3 ± 1.1	10.2 ± 0.6	0.8 ± 0.1	19.3 ± 2.0
Aminooxyacetate	39.2 ± 1.4	7.1 ± 1.0	8.5 ± 0.6	5.0 ± 0.9	31.6 ± 2.7
Methionine Sulfoximine	18.0 ± 2.0	30.2 ± 2.2	11.9 ± 0.9	2.2 ± 0.2	12.9 ± 1.8

alanine from skeletal muscle. Because alanine is also the predominant amino acid substrate for hepatic gluconeogenesis in man (8, 9), and because this amino acid also produces relatively high rates of glucose production in the perfused liver (10), a glucose-alanine cycle has been hypothesized (10, 11). In the formulation of this cycle, glucose released by the liver is taken up by skeletal muscle and pyruvate is then generated by way of anaerobic glycolysis. This glucose-derived pyruvate is then transaminated with glutamate to form α-ketoglutarate and alanine, the latter being released from skeletal muscle and returned to the liver for subsequent reconversion to glucose. Such a cycle, although attractive as a mechanism for muscle nitrogen disposal by way of hepatic urea synthesis, does not provide a mechanism for the net transfer of carbon into the total body alanine pool. As a consequence, this cycle cannot account for the transfer of carbon derived from protein into glucose as is known to occur *in vivo* during periods of starvation or uncontrolled diabetes, because alanine is the predominant amino acid substrate for hepatic gluconeogenesis. Furthermore, a glucose-alanine cycle does not explain the undiminished rates of alanine and glutamine output from skeletal muscle in the absence of glucose in the media during *in vitro* studies with isolated muscles (5), and the absence of net glycogenolysis in these preparations. Lastly, studies *in vitro* have failed to demonstrate any relationship between rates of glucose uptake and rates of alanine and glutamine output from skeletal muscle (5, 12, 13), and inhibition of glucose utilization by iodoacetate failed to diminish alanine and glutamine output from skeletal muscle (Table II). Thus, it appears clear that no clearcut relationship can be demonstrated between net glucose utilization by skeletal muscle and net alanine and glutamine synthesis. Similar results have also been reported using fluoride inhibition of glycolysis (13), although 2-deoxyglucose apparently causes an inhibition of alanine output coordinate with inhibition of glucose utilization (14). Nevertheless, direct or indirect effects of 2-deoxyglucose on adenine nucleotide and high energy phosphate metabolism were not excluded.

Studies using the direct addition of pyruvate, glutamate, or ammonium chloride to skeletal muscle incubations have suggested that the addition of these compounds produces an augmented output of alanine from skeletal muscle. With pyruvate addition, a concomitant decrease in glutamate and glutamine output was observed, whereas with glutamate addition, alanine and glutamine output were augmented but pyruvate release was decreased (Table III). Ammonium chloride increased both glutamine and alanine synthesis. These findings suggest that an increased delivery of either substrate

TABLE II

Effects of iodoacetate on glucose and amino acid metabolism in rat skeletal muscle

Addition	Alanine	Glutamine Release	Pyruvate	Glucose Uptake
		nmoles/min/g muscle		
Control	20.6 ± 1.7	30.3 ± 1.9	19.7 ± 1.8	119.5 ± 21.2
Iodoacetate	29.7 ± 1.3	34.5 ± 3.5	6.2 ± 2.9	8.1 ± 4.2

TABLE III

Effect of pyruvate, ammonium chloride and glutamate addition on amino acid and pyruvate release from rat skeletal muscle

Addition	Glutamine	Glutamate	Alanine	Pyruvate
		Release		
		nmoles/min/g muscle		
Control	29.9 ± 2.0	10.1 ± 1.0	19.8 1.6	20.2 1.8
Pyruvate	20.3 ± 1.8	5.8 ± 0.8	28.8 2.3	– 1.8
Glutamate	52.1 ± 3.6	–	29.5 ± 2.1	13.1 2.0
Ammonium	40.1 ± 2.4	15.3 ± 0.8	30.2 2.4	11.9 0.8

alone for alanine formation (pyruvate or glutamate) increases net alanine synthesis in skeletal muscle, but that compensatory decreases in the output of other, related substrates must occur because the latter are not supplied simultaneously in an increased amount. Although glucose itself produces little or no augmentation of alanine and glutamine output from skeletal muscle, [^{14}C]-label from glucose can be found in alanine released from muscle, suggesting that this pyruvate derived from glucose mixes with much of the pyruvate pool (s) within the skeletal muscle cell prior to pyruvate transamination to form alanine (14).

In view of the markedly reduced release of branch-chain amino acids and aspartate from skeletal muscle, and in view of the observation that inhibition of glutamine and alanine synthesis produces a compensatory increase in the release of other, more poorly released amino acids such as aspartate, we have investigated the effects of added amino acids on alanine and glutamine synthesis and release from skeletal muscle (15). As shown in Figures 2-4, a number of amino acids stimulate the synthesis and release of alanine and glutamine. Four different kinds of results may be discerned. Certain amino acids such as the branch-chain amino acids appear to stimulate the synthesis of both glutamine and alanine whereas amino acids such as aspartate stimulate the release of only alanine but not glutamine. On the other hand, a third group of amino acids may stimulate

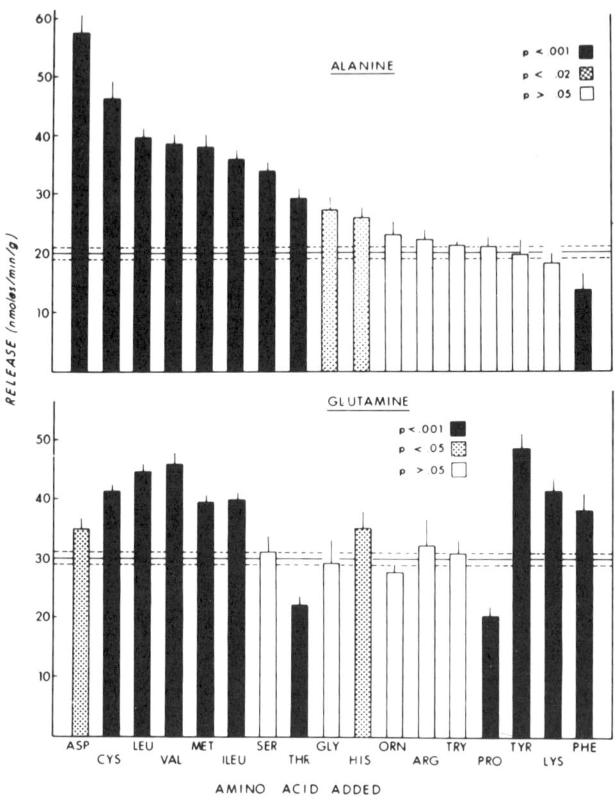

Fig. 2. The effect of added, exogenous amino acids on alanine and glutamine synthesis and release from rat skeletal muscle preparations *in vitro*.

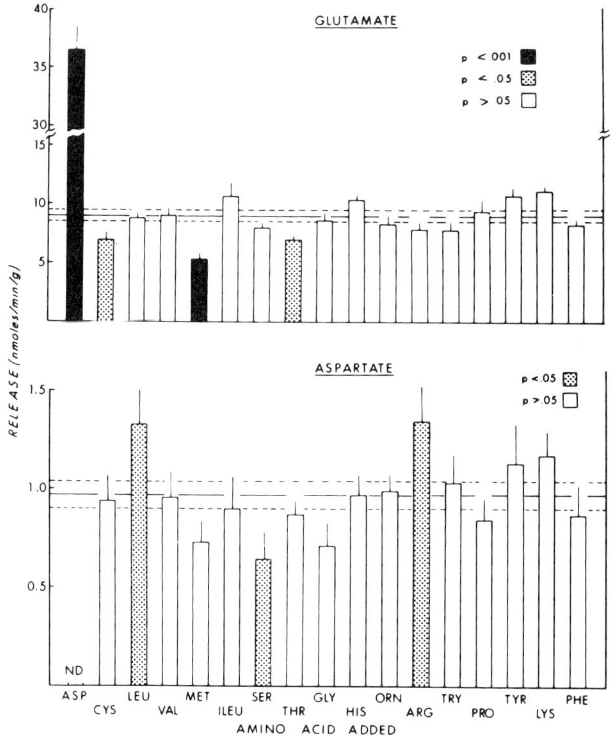

Fig. 3. The effect of added amino acids on glutamate and aspartate synthesis and release from rat skeletal muscle preparations *in vitro*.

the release of glutamine alone whereas a fourth group of amino acids such as the urea cycle amino acids appear not to stimulate either alanine or glutamine output from skeletal muscle. These increased rates of skeletal muscle alanine and glutamine production occur without changes in the rate of glucose utilization or without reciprocal decreases in either glutamate or pyruvate output. Thus, it seems likely that added amino acids must stimulate alanine and glutamine synthesis by mechanisms other than by influencing the supply of only a single precursor such as pyruvate or glutamate alone, because reciprocal decreases in the release of other precursors were not observed. The latter would have been anticipated based on studies with added pyruvate or glutamate alone. It is therefore possible to conclude that these added amino acids most likely stimulate alanine and glutamine synthesis by supplying both the amino groups as well as a portion of the carbon skeletons required for

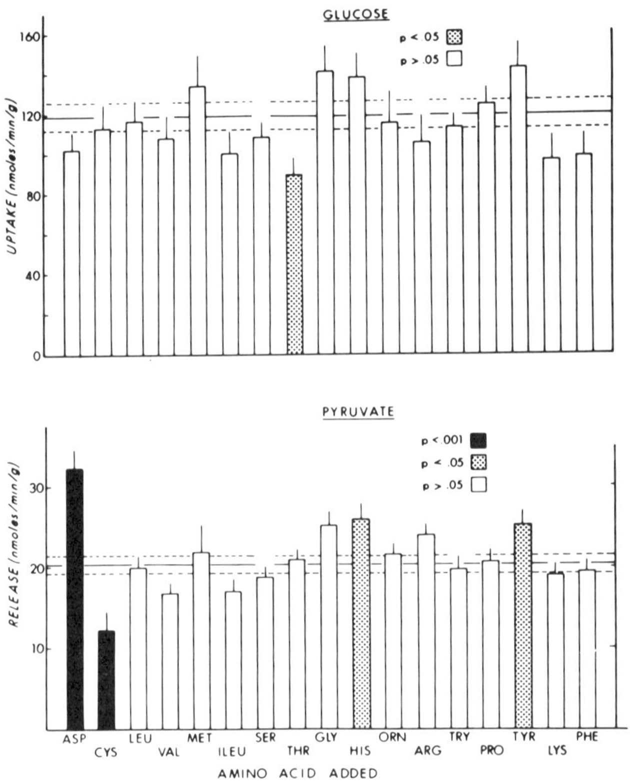

Fig. 4. The effect of added amino acid on glucose uptake and pyruvate formation in rat skeletal muscle *in vitro*.

ongoing alanine and glutamine synthesis in skeletal muscle. These results obtained in epitrochlaris preparations of rat skeletal muscle have also been observed by other investigators using perfused hemicorpuses and hindquarters of rats as well as rat diaphragm preparations (7, 12, 13). These data do not, however, provide an exact mechanism for the stimulation of alanine and glutamine synthesis by each of the amino acids added. It is possible that pyruvate carbon required for ongoing alanine synthesis as well as glutamine synthesis in skeletal muscle is derived from the diversion of the citric cycle intermediates. α-Ketoglutarate production by citric acid cycle flux may be directed by transamination or reductive amination to glutamate, which in turn is then aminated to form glutamine. However, the situation for potential net production of pyruvate carbon from citric acid cycle intermediates is less clear since at least two pathways may be invoked to explain such a phenomenon. As shown in Fig. 5, malate carbon may be converted directly to pyruvate by the action of NAD+ or NADP+-malate dehydrogenase (decarboxylating). This enzyme exists in relatively high concentration in rat skeletal muscle (greater than one unit/gm muscle, wet weight) and may therefore account for maximal rates of ongoing alanine synthesis in the absence of glucose from added amino acids such as aspartate (15, 16). On the other hand, other investigators have proposed that oxalacetate carbon may be converted to phosphoenolpyruvate (Fig. 6) by the action of cytosolic and/or mitochondrial phosphoenolpyruvate carboxykinase (13). This phosphoenolpyruvate may be dephosphorylated by pyruvate kinase to form pyruvate. Skeletal muscle phosphoenolpyruvate carboxykinase has approximately 1/10 to 1/100 the activity of NAD (P) + malate dehydrogenase (decarboxylating) as assayed *in vitro*. Nonetheless, the action of either enzyme could explain net pyruvate formation from citric acid cycle intermediates, and might therefore account for the finding of labeled bicarbonate

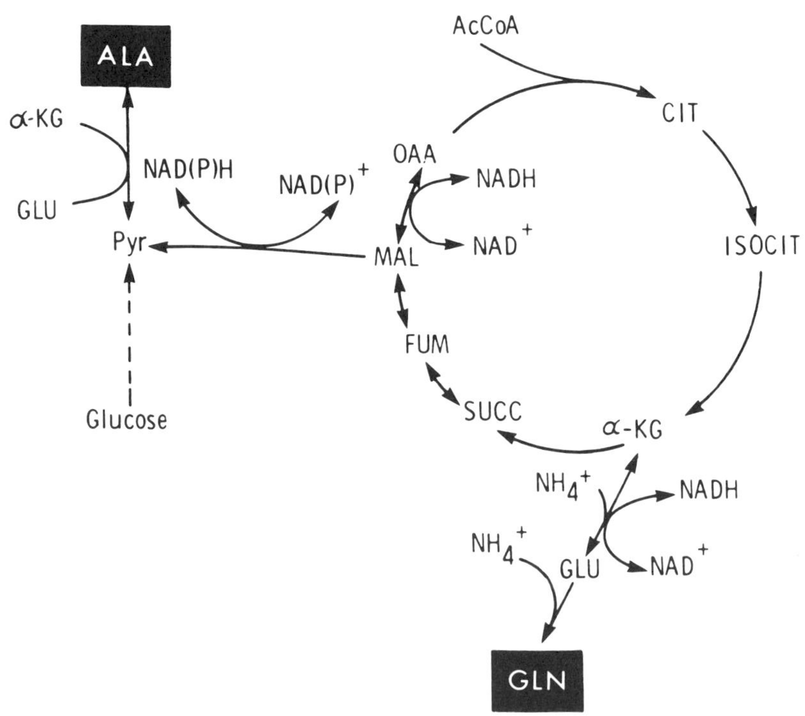

Fig. 5. The potential role of NADP+-malate dehydrogenase in the diversion of citric acid cycle intermediates for net pyruvate synthesis in skeletal muscle.

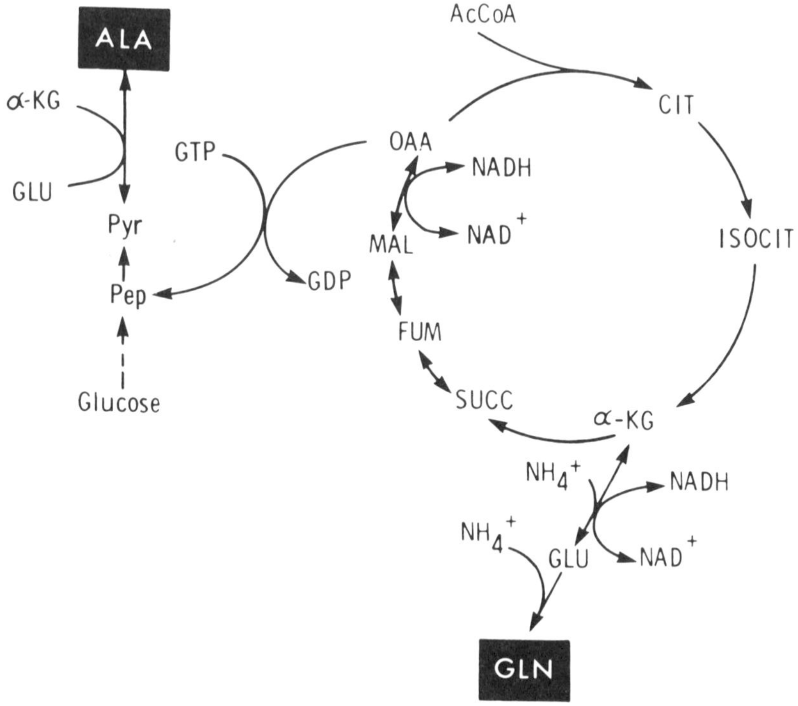

Fig. 6. The potential role of phosphoenolpyruvate carboxykinase in the diversion of citric acid cycle intermediates for net pyruvate synthesis in skeletal muscle.

incorporation into citric acid cycle precursors as observed by Davis and coworkers (16).

In the course of our studies on the relationship between glucose metabolism and alanine and glutamine synthesis, we found that glycogenolytic agents such as epinephrine inhibited rather than augmented alanine and glutamine output from skeletal muscle (17). This effect of epinephrine was found with physiologic concentrations of the agonist (10^{-9} - 10^{-7} M). As shown in Figure 7, concentrations of epinephrine activating glycogenolysis (10^{-8} M - 10^{-5} M), as judged by increased lactate and pyruvate production from skeletal muscle, also inhibited alanine and glutamine output from skeletal muscle. This effect was reproduced by isoproterenol (10^{-6} M) but not by phenylephrine (10^{-6} M), suggesting the participation of a β-adrenergic receptor for this effect. Further support for this concept is provided by the findings that α-adrenergic antagonists such as phentolamine do not block these effects of epinephrine, but

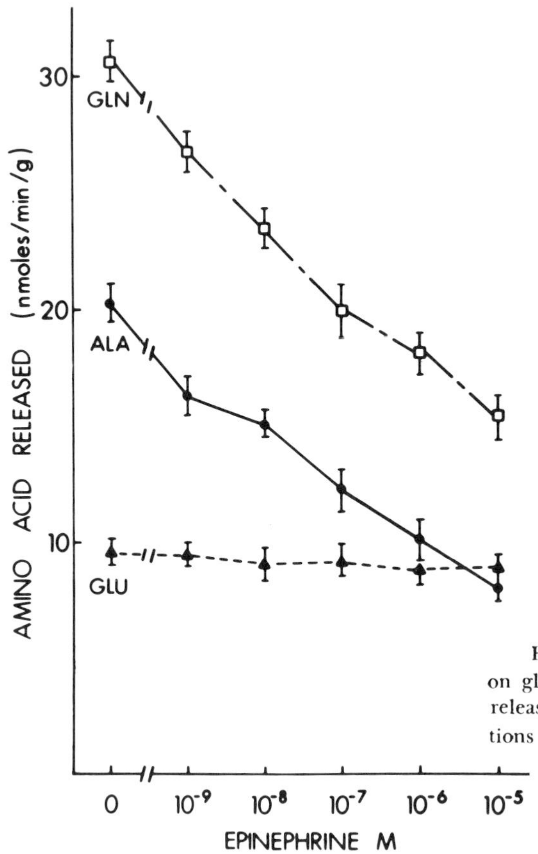

Fig. 7. The effect of epinephrine on glutamine, alanine and glutamate release from skeletal muscle preparations *in vitro*.

that β-adrenergic antagonists such as propranolol completely block adrenergic inhibition of muscle glutamine formation and release (18). The finding of β-adrenergic receptor participation in adrenergic inhibition of glutamine output from skeletal muscle is similar to the receptor requirement for adrenergic regulation of muscle glycogenolysis. Therefore, the skeletal muscle adenylyl cyclase-cAMP system may also participate in this inhibition. Concentrations of epinephrine causing inhibition of glutamine production from skeletal muscle and activation of muscle glycogenolysis also activated skeletal muscle membrane adenylyl cyclase activity and caused the accumulation of increased intracellular levels of cAMP. Maximal stimulation of adenylyl cyclase activity was found at 10^{-5} M epinephrine, a concentration producing maximal inhibition of glutamine and alanine output from skeletal muscle. These observations support the concept that the β-adrenergic mechanism regulating muscle glycogen homeostasis also participates in the mechanism of adrenergic inhibition of glutamine and alanine synthesis and release from skeletal muscle.

In view of these findings, other agents including such autacoids as serotonin were studied for potential effects on glutamine formation and release in skeletal muscle. Serotonin, an important autacoid influencing smooth muscle and platelet function, was also found to have direct effects on skeletal muscle protein and amino acid metabolism (19). As shown in Table IV, serotonin produces a concentration-dependent inhibition of glutamine as well as alanine output from skeletal muscle. These effects of serotonin appear to be mediated by a specific D-serotonergic receptor since the inhibition of glutamine production is blocked by cyproheptadine (10^{-6} M) as well as methysergide but is not influenced by morphine sulfate and tetradotoxin. The former two inhibitors define D-serotonergic receptor effects such as those receptors found in the gastrointestinal tract. The inhibition of skeletal muscle glutamine synthesis produced by serotonin is accompanied by a serotonin-induced, concentration-dependent increase in muscle cAMP levels (Fig. 8). Serotonin appears to increase skeletal muscle cAMP levels not by indirect actions on such enzyme systems as cAMP phosphodiesterase but instead by a direct effect on skeletal muscle adenylyl cyclase, which is stimulated approximately two-fold by serotonergic agonists *in vitro*. As is the

TABLE IV

Effects of serotonin and serotonergic antagonists on amino acid release from skeletal muscle

Addition		Inhibitor	Alanine Release	Glutamine Release
			nmoles min/g muscle	
Control		—	20.1 ± 1.2	31.4 ± 2.3
Serotonin	10^{-8}		17.9 ± 1.6	27.1 ± 2.3
	10^{-6}		14.8 ± 2.0	23.0 ± 2.2
	10^{-4}		12.3 ± 1.1	19.8 ± 2.0
	10^{-4}	Methysergide	21.2 ± 1.3	30.8 ± 2.6
	10^{-4}	Cyproheptidine	20.8 ± 1.8	31.7 ± 3.0

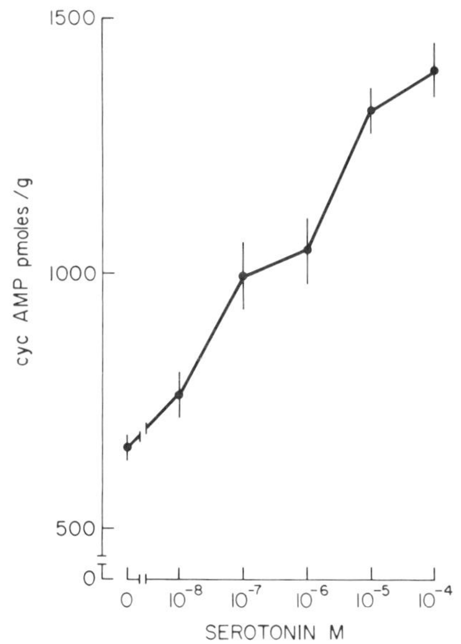

Fig. 8. The effect of serotonin on muscle cyclic AMP levels *in vitro*.

case with epinephrine, serotonin does not alter glutamine formation from added amino acids, suggesting that the mechanism of adrenergic as well as serotonergic inhibition of glutamine synthesis in muscle must involve an inhibition of precursor availability for glutamine synthesis rather than an inhibition of enzymes involved in the conversion of precursor amino acids to form the carbon and nitrogen requirements for net glutamine synthesis in skeletal muscle. Support for this concept is provided by the fact that coordinate inhibition of alanine synthesis is also achieved in skeletal muscle by adrenergic as well as serotonergic agonists. In the case of alanine synthesis, little overlap exists in the pathways required for alanine as opposed to glutamine synthesis in muscle. In light of these considerations, direct effects of adrenergic and serotonergic agonists, perhaps mediated by increased intracellular levels of cyclic AMP, on skeletal muscle proteolysis appears to be the most likely explanation for the inhibition of glutamine synthesis in muscle produced by these autacoids. Direct effects of serotonin on skeletal muscle proteolysis have been assessed *in vitro* using muscle preparations prelabeled *in vivo* using [guanido-^{14}C]arginine. On incubation of these preparations *in vitro* in media containing [^{3}H]arginine to quantitate reincorporation of label into muscle protein and in the presence of cycloheximide, serotonin was found to diminish the degradation of muscle proteins as indicated by a decreased rate of appearance of [^{14}C]label in trichloroacetic acid soluble fractions (Table V).

Because of these previously unsuspected effects of cyclic AMP related agonists on skeletal muscle protein and glutamine metabolism, the effects of agonists linked to cyclic GMP have also been studied. In cardiac muscle and in gastrointestinal smooth muscle, muscarinic cholinergic agonists are associated with marked elevations of intracellular cGMP levels (20, 21). In skeletal muscle, acetylcholine in combination with neostigmine or carbamylcholine increased the net synthesis and release of glutamine as well as alanine (22). This effect is dependent upon the concentration of carbamylcholine or other cholinergic agonists studied. Concentrations as low as 10^{-10} M carbamylcholine significantly augment the output of glutamine and alanine from skeletal muscle preparations (Fig. 9). This increased amino acid output from skeletal muscle appears to be due in part to a facilitation of amino acid transport by cholinergic agonists since glutamine levels in skeletal muscle decline in a concentration-dependent fashion according to the amount of cholinergic agonist added. Nevertheless, the amount of the increased glutamine output produced by cholinergic agonists derived from increased transport and disposal of the intracellular glutamine pool does not account for more than 39% of the augmented glutamine release from muscle. On the other hand, more than 61% of the increased glutamine release produced by cholinergic agonists appears to be due instead to an

TABLE V

Effect of serotonin on the distribution of [^{14}C]arginine label in skeletal muscle preparations

Treatment	Muscle Acid Insoluble	Muscle Acid Soluble	Released to Medium		Increment in Soluble Label		Decrement in Insoluble Label	
	DPM/mg muscle							
Freeze-Clamped *in situ*	168.2 ± 3.1	57.2 ± 1.6	—		—		—	
Incubated:								
Control	142.5 ± 3.7	28.0 ± 3.0	54.1	3.2	24.9	2.0	25.8	2.3
Serotonin	158.4 ± 2.4	29.6 ± 2.0	39.8	3.7	12.2	1.6	10.1	1.1

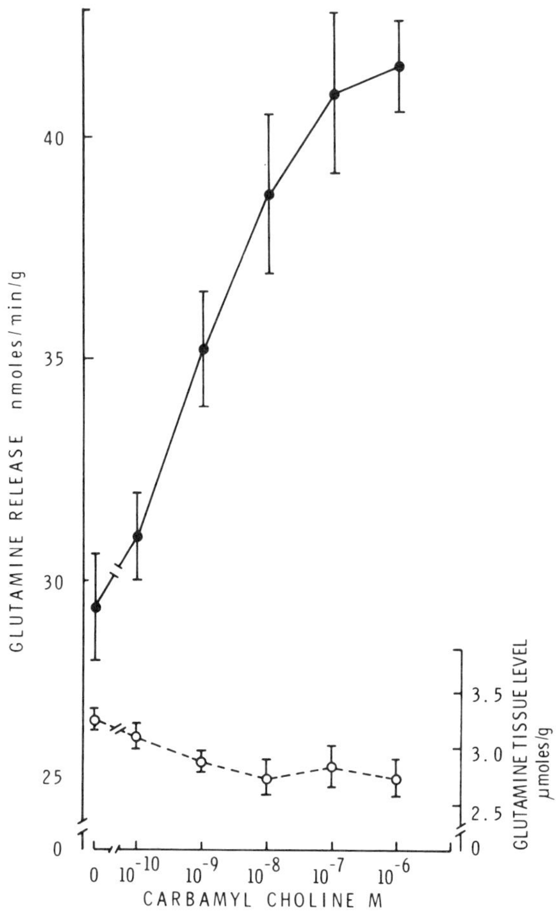

Fig. 9. The effect of carbamylcholine on skeletal muscle glutamine synthesis and release *in vitro*.

increased net synthesis of this amino acid as well as to an increased net synthesis of alanine. These effects of cholinergic agonists are not due to an inhibition of amino acid oxidation in the citric acid cycle or to an augmentation of muscle contraction by the cholinergic agonist. Support for the latter concept is provided by observations that muscle glycogen levels are unchanged in the presence of any concentrations of cholinergic agonists studied in this system, thereby indicating that significant contraction did not occur.

Glutamine and alanine synthesis from added precursor amino acids was not further augmented by cholinergic agonists suggesting that the mechanism of cholinergic action on skeletal muscle glutamine synthesis was not due to a direct activation of the enzymes for glutamine formation in skeletal muscle. Instead, a more likely hypothesis is that cholinergic action provides an increased availability of precursor amino acids for glutamine synthesis. Support for this concept is provided by the observation that cholinergic agonists facilitate alanine output, presumably by supplying an increased availability of other precursors which are different from those used for glutamine synthesis. Thus, cholinergic agonists may accelerate one or more steps in the pathway of intracellular proteolysis in skeletal muscle. Using prelabeled skeletal muscle preparations to assess the degradation of endogenous muscle proteins, cholinergic agonists were found to greatly facilitate the transfer of amino acid label from protein to acid soluble cellular and extracellular compartments during *in vitro* incubation studies. Thus, it seems likely that cholinergic agonists act oppositely with respect to glutamine synthesis in skeletal muscle as do adrenergic and serotonergic agonists.

However, unlike other muscle systems, a muscarinic cholinergic receptor could not be implicated in the mechanism of cholinergic action on amino acid metabolism. Instead, good evidence has been obtained to demonstrate that a nicotinic cholinergic receptor appears to be the most probable mechanism by which cholinergic agonists augment glutamine synthesis and release in skeletal muscle (23). For example, carbamylcholine inhibition of glutamine synthesis was antagonized by hexamethonium or decamethonium but not by atropine. Similarly, the action of carbamylcholine was reproduced by tetramethylammonium hydroxide but not by pilocarpine. Thus, it seems clear that the cholinergic receptor participating in the augmentation of glutamine synthesis in muscle is a nicotinic rather than a muscarinic receptor. Although this is an unusual receptor requirement for cholinergic action in skeletal as compared to smooth muscle, cholinergic agonists did produce a concentration-dependent increase in skeletal muscle cGMP levels. The lowest concentration of carbamylcholine producing significant increments in cGMP levels was 10^{-9} M and the highest increments observed were found at 10^{-5} M carbamylcholine (Fig. 10). Thus, concentrations of cholinergic agonists augmenting glutamine synthesis in muscle also produce significant increments in cGMP levels. The association between increased cGMP levels and increased glutamine synthesis in muscle is strengthened by the results of studies with cGMP phosphodiesterase inhibitors such as methylisobutylxanthine or theophylline. Both inhibitors, at low concentrations, do not affect either glutamine synthesis or basal cGMP levels. However, the addition of submaximal

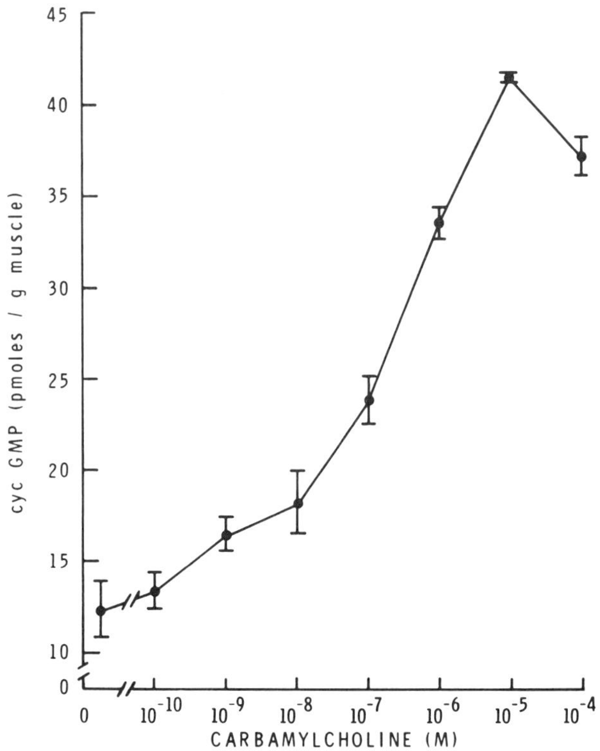

Fig. 10. The effect of carbamylcholine on steady-state levels of intracellular cGMP in skeletal muscle.

concentrations of cholinergic agonists result in augmented cGMP responses and increased stimulation of glutamine synthesis when phosphodiesterase inhibitors are included. Furthermore, nicotinic cholinergic antagonists block both the increment in cGMP levels and the increment in glutamine synthesis whereas nicotinic agonists such as tetramethylammonium also increase cGMP levels as well as facilitate glutamine synthesis. Muscarinic cholinergic agonists have no effect on cGMP levels nor do they augment glutamine synthesis in skeletal muscle. The direct addition of cGMP derivatives to skeletal muscle reproduces the effects of cholinergic agonists on glutamine synthesis and release from muscle and on amino acid transport.

As a result of these studies, it can be demonstrated therefore that two coordinated and presumably independent systems associated with cyclic nucleotide metabolism influence and modulate skeletal muscle protein and amino acid metabolism. Cyclic AMP and agonists generating increased cyclic AMP levels in skeletal muscle produce a marked inhibition of both proteolysis and the synthesis and release of glutamine. On the other hand, cGMP and agonists generating increased cGMP levels in skeletal muscle produce a marked stimulation of muscle proteolysis and of glutamine synthesis and release. The dependence or independence of these two pathways and the precise localization of their mechanism of action remain to be elucidated.

The physiologic significance of these cyclic nucleotide associated mechanisms controlling glutamine synthesis and release from skeletal muscle can be assessed in part by investigations in pathologic models of muscle wasting such as chronic renal failure and uncontrolled diabetes mellitus. In skeletal muscle from these pathologic models, defective cyclic nucleotide regulation of muscle protein and amino acid metabolism appears to play an important role not only in the accelerated glutamine production and release observed but also in the muscle wasting and muscle dysfunction. In skeletal muscle preparations from rats made experimentally uremic by unilateral nephrectomy and subtotal infarction of the contralateral kidney, there is an increased release of glutamine, alanine and most other amino acids (Table VI). This increased glutamine and alanine output

TABLE VI

Amino acid release from skeletal muscle of control and uremic rats

Treatment	Alanine	Aspartate	Glutamate	Glutamine
	Release			
	nmoles/min/g muscle			
Control	20.4 ± 1.6	0.7 ± 0.2	10.4 0.7	30.3 2.6
Uremic	29.8 ± 2.2	2.1 ± 0.9	16.8 1.2	41.2 3.1

from uremic muscle does not derive from diminished amino acid reutilization via oxidation to CO_2 or reincorporation into muscle protein (24). The latter pathways are quantitatively insignificant in muscle compared to the net rate of glutamine output and release. This increased glutamine synthesis and release in uremic muscle is not due to an increased maximal capacity of enzymes associated with glutamine formation in skeletal muscle, since added amino acids did not stimulate total rates of glutamine and alanine formation to any extent greater than that observed in control muscle. Instead, the accelerated muscle proteolysis observed in uremic muscle is associated with and may be responsible for the increased rates of glutamine formation since increased muscle proteolysis provides an increased availability of carbon and nitrogen precursors for net glutamine synthesis.

The accelerated glutamine synthesis and release owing to accelerated muscle proteolysis in chronically uremic skeletal muscle does not appear to be due to primary abnormalities of insulin action on uremic muscle protein and amino acid metabolism (25). Instead, glutamine synthesis and release showed marked insensitivity to suppression by adrenergic agonists such as epinephrine. As shown in Fig. 11, the dose response curve of glutamine release to added

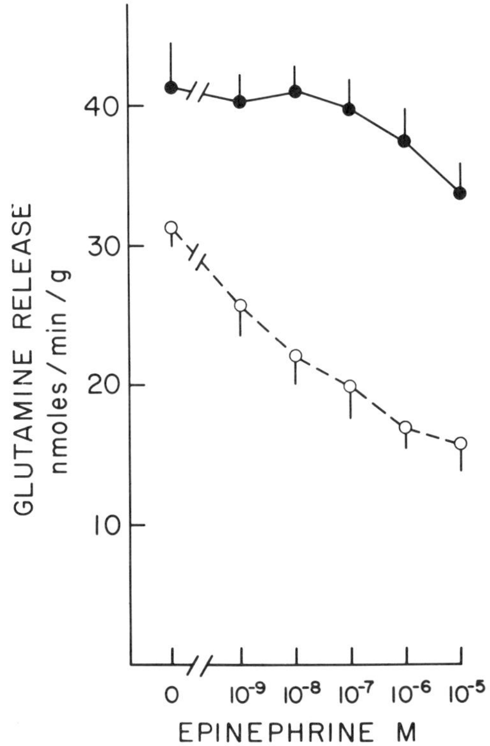

Fig. 11. Suppression of glutamine formation and release by epinephrine from skeletal muscle preparations of control (O —— O) and uremic (● —— ●) rats.

epinephrine was shifted not only upwards but to the right by approximately three log unit intervals in epinephrine concentration. Similar results were also obtained for alanine release from skeletal muscle of uremic as compared to control rats. A virtually identical shift in the dose response curve of glutamine release to added serotonin was also found (Fig. 12). Inhibition of glutamine release

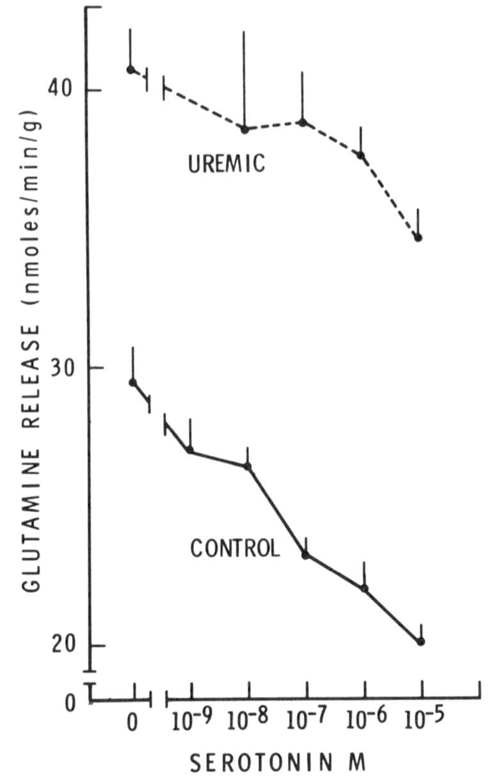

Fig. 12. The effect of serotonin on glutamine synthesis and release in skeletal muscle preparations of control and uremic rats.

by concentrations of serotonin less than 10^{-5} M were not observed with uremic muscle, whereas concentrations of serotonin as low as 10^{-8} M clearly produced inhibition of glutamine output from control muscle. This loss of adrenergic and serotonergic action on glutamine synthesis and release in uremic muscle was not due to a loss of cyclic AMP responsiveness since dibutyryl cyclic AMP derivatives maintained their inhibitory effect on glutamine synthesis in uremic muscle. Instead, the metabolic basis for the insensitivity to adrenergic and serotonergic action in uremic skeletal muscle alanine synthesis appears to be due to a primary failure of cAMP generation in uremic muscle. As shown in Table VII, at any

TABLE VII

Effect of epinephrine and serotonin on cAMP levels in skeletal muscle of control and uremic rats

Addition	Concentration	cAMP Level Control	Uremic
		pmoles/g muscle	
Control	–	609 ± 39.5	543 ± 30.9
Epinephrine	10^{-8}	726 ± 32.1	595 ± 40.1
	10^{-6}	1710 ± 71.2	794 ± 61.3
	10^{-5}	2246 ± 95.2	1091 ± 84.0
Serotonin	10^{-7}	995 ± 65	649 ± 40
	10^{-5}	1381 ± 45	734 ± 18

concentration of epinephrine or serotonin studied, lower levels of cAMP were produced in uremic as compared to control muscle. This lack of stimulation to cAMP levels from adrenergic and serotonergic agonists results from a diminished activity of the hormone sensitive adenylyl cyclase in uremic muscle. Although basal and fluoride stimulated cyclase activities were preserved, epinephrine stimulated adenylyl cyclase activity in skeletal muscle membrane preparations of chronically uremic rats showed approximately 50% of the activity observed in membrane preparations of control rats.

A similar loss of responsiveness to adrenergic inhibition of muscle proteolysis and hence of glutamine synthesis and release from skeletal muscle was also observed in muscle preparations obtained from streptozotocin-diabetic rats (26). As shown in Fig. 13, glutamine synthesis and release was increased by 30% in muscle of diabetic as compared to control rats, and alanine synthesis and release was increased by more than 50%. An insensitivity to adrenergic suppression of glutamine output from skeletal muscle of diabetic rats was observed. However, no insensitivity or loss of responsiveness to serotonergic

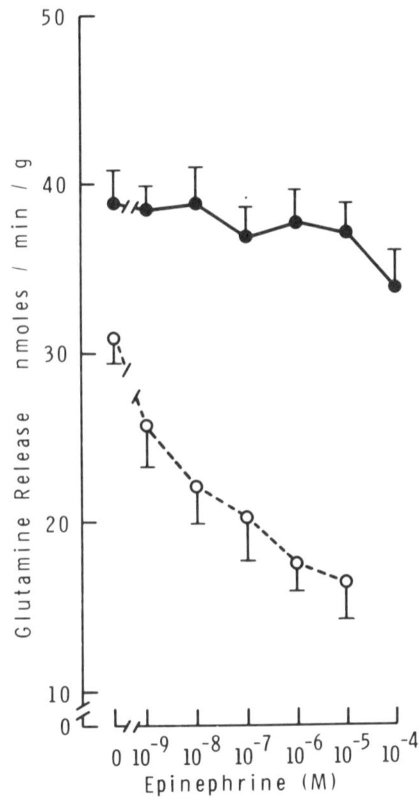

Fig. 13. The effect of epinephrine on glutamine synthesis and release from skeletal muscle preparations of control (O——O) and streptozotocin-diabetic (●——●) rats.

agonists were observed in diabetic muscle. As shown in Fig. 14, glutamine synthesis and release from diabetic muscle was suppressed to the same extent as in control muscle, although at a higher basal rate in diabetic muscle. The insensitivity to adrenergic action in diabetic muscle was due in part to a diminished activity of the epinephrine stimulated adenylyl cyclase in membrane preparations of diabetic as compared to control muscle. It is interesting to note that not only was basal and fluoride stimulated adenylyl cyclase activity preserved in diabetic muscle, but that serotonin stimulated adenylyl cyclase was entirely normal in diabetic muscle. As a consequence, cyclic AMP levels in diabetic muscle were reduced by 50-60% with epinephrine as agonist but were unchanged from normal in studies of serotonergic action on muscle cyclic AMP levels.

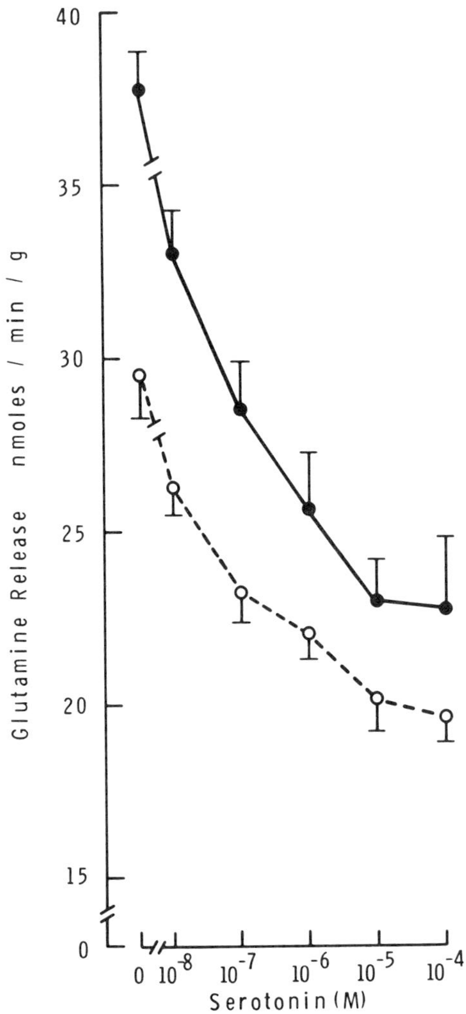

Fig. 14. The effect of serotonin on glutamine synthesis and release from skeletal muscle preparations of control (O——O) and diabetic (●——●) rats.

Although these studies do not provide a clear mechanism by which to account for the loss of adrenergic responsiveness in these pathologic models, it seems clear that defective regulation of skeletal muscle protein and amino acid metabolism by cyclic nucleotide associated mechanism can result in severe muscle wasting disorders. Thus, these cyclic nucleotide mechanisms may play a potentially important physiologic role in the regulation of protein balance *in vivo* in mammalian skeletal muscle. As a consequence of this regulation, glutamine synthesis in skeletal muscle may be modulated by mechanisms such as those primarily influencing the rate of muscle proteolysis (Fig. 15) since it is the latter process which in part controls the availability of at least some precursors for glutamine synthesis.

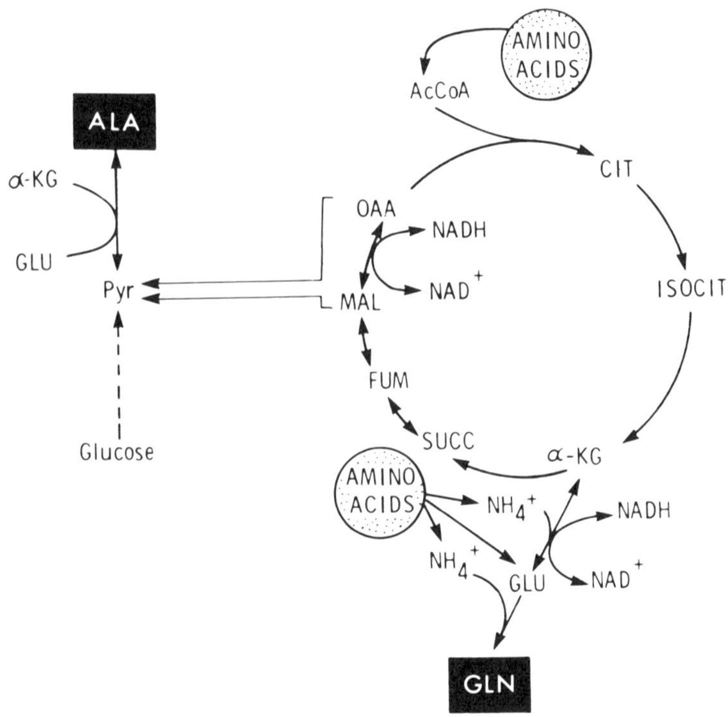

Fig. 15. Schematic representation of the regulation of glutamine synthesis in skeletal muscle by the delivery of precursor amino acids secondary to skeletal muscle proteolysis.

REFERENCES

1. Ruderman, N. B. and Lund, P. (1972) *Israel J. Med. Sci. 8,* 295-302.
2. Pozefsky, T., Felig, P., Tobin, J. D., Soeldner, J. S. and Cahill, G. F., Jr. (1969) *J. Clin. Invest. 48,* 2273-2382.
3. Pozefsky, T. and Tancredi, R. G. 1972 *J. Clin. Invest. 51,* 2359-2369.
4. Ahlberg, G., Felig, P., Hagenfeld, L., Hendler, R., and Wahren, J. (1974) *J. Clin. Invest. 53,* 1080 1090.
5. Garber, A. J., Karl, I. E., and Kipnis, D. M. (1976) *J. Biol. Chem. 251,* 826-835.
6. Kominz, D. R., Hough, A., Symond, P., and Loki, K. (1954) *Arch. Biochem. Biophys. 50,* 148-159.

7. Goldstein, L. and Newsholme, E. A. (1976) *Biochem. J. 154,* 555-558.
8. Garber, A. J., Menzel, P. H., Boden, G., and Owen, O. E. (1974) *J. Clin. Invest. 54,* 981-989.
9. Snell, K. (1979) *Trends. Biochem. Sci. 4,* 124-128.
10. Mallete, L. E., Exton, J. H., and Park, C. R. (1969) *J. Biol. Chem. 244,* 5713-5723.
11. Felig, P., Pozefsky, T., Marliss, E., and Cahill, G. F., Jr. (1970) *Science 167,* 1003-1004.
12. Ruderman, N. B. and Berger, M. (1974) *J. Biol. Chem. 249,* 5500-5506.
13. Snell, K. and Duff, D. A. (1977) *Biochem. J. 62,* 399-403.
14. Chang, T. W. and Goldberg, A. L. (1978) *J. Biol. Chem. 263,* 3677-3684.
15. Garber, A. J., Karl, I. E., and Kipnis, D. M. (1976) *J. Biol. Chem. 251,* 836-843.
16. Lee, S-H., and Davis, E. J. (1979) *J. Biol. Chem. 254,* 420-430.
17. Garber, A. J., Karl, I. E., and Kipnis, D. M. (1976) *J. Biol. Chem. 251,* 851-857.
18. Li, J. B. and Jefferson, L. S. (1977) *Amer. J. Physiol. 252,* 243-249.
19. Garber, A. J. (1977) *Mol. Pharmacol. 13,* 640-651.
20. Van Sande, J., DeCoster, C., and Dumont, J. E. (1975) *Biochem. Biophys. Res. Commun. 62,* 168-175.
21. Yamashita, K. and Field, J. B. (1972) *J. Biol. Chem. 247,* 7065-7066.
22. Garber, A. J., Harari, Y., and Entman, M. L. (1978) *J. Biol. Chem. 253,* 7918-7923.
23. Garber, A. J., Entman, M. L., and Birnbaumer, L. (1978) *J. Biol. Chem. 253,* 7924-7930.
24. Garber, A. J. (1978) *J. Clin. Invest. 62,* 623-632.
25. Garber, A. J. (1978) *J. Clin. Invest. 62,* 633-641.
26. Garber, A. J. (1980) *J. Clin. Invest., 65,* 478-487.

15. Glutamine Metabolism in Brain

RICARDO TAPIA

Departamento de Neurociencias
Centro de Investigaciones en Fisiología Celular
Universidad Nacional Autónoma de México
México, D. F.

SUMMARY

The function of synapses is possibly the crucial event in the physiology of the nervous system. Although we are still far from knowing the basic mechanisms responsible for synaptic transmission, the progress made in the last 30 years in the chemistry, physiology and cytology of the brain has established that the biochemical events taking place at the presynaptic terminals are an essential part of the system. Thus neurochemistry has become a specialized discipline of biochemistry, which studies not only the general reactions occurring in nervous tissue, but rather attempts to discover how the biochemical pathways are linked to the specific functions and mechanisms of the brain. In this respect the metabolism of glutamine is an example particularly interesting. This amino acid constitutes not only the key metabolite for the elimination of ammonia in nervous tissue, but also, as indicated by recent evidence from studies on metabolic compartmentation, it may be an important precursor of the pools of both glutamate and γ-aminobutyric acid (GABA) specifically related to their role as synaptic transmitters in the central nervous system. This evidence can be summarized as follows: 1) After the administration of certain labeled precursors, but not with others, the specific radioactivity of glutamine is higher than that of its immediate precursor, glutamate; this has been observed both *in vivo* and *in vitro*, in nervous tissue but not in other tissues. Thus, at least two pools of glutamate have been metabolically characterized: a "small" pool, which is rapidly converted into glutamine, and a "large" pool. 2) Glutamine synthetase has been located mainly in the glial cells, whereas glutaminase

ISBN: 0-12-506040-8

is relatively more concentrated in nerve endings. 3) In radioautographic experiments in dorsal root ganglia, the precursors which feed the small pool labeled preferentially the glial cells, whereas those entering the large pool labeled mainly the neurons. 4) Depolarization of brain slices notably stimulated the release of labeled glutamate and GABA derived from labeled glutamine, in a calcium-dependent manner. These data give support to the hypothesis that glutamine is synthesized mainly in glial cells, for later transportation into the nerve endings to serve as a precursor of the neurotransmitter pools of glutamine and GABA. This role of glutamine illustrates the importance of metabolic reactions in the basic physiological mechanisms of nervous activity.

INTRODUCTION

The metabolism of glutamine in brain can be considered from at least three points of view. The first one is simply the description of its synthetic and degradative reactions, which are not very different from those in other tissues, and of the characteristics of the synthesizing and degradative enzymes in nervous tissues. The second is related to the role of glutamine synthesis in the elimination of ammonia, whose toxic effect of the nervous system has long been known. The third is the relationship of glutamine to the specialized mechanisms responsible for the particular function of the brain, namely the initiation, propagation and transmission of nerve impulses in neuronal structures. The present article will refer mostly to the latter aspect of glutamine metabolism, since at the present time the biochemistry of an organ as complex as the brain becomes much more interesting and challenging when considered in relation to the neurophysiological and structural parameters. Of course, this consideration would be rather difficult to apply if the current knowledge of the role of glutamine in brain function were nonexistent or limited. Fortunately, the studies of a number of laboratories, beginning with the discovery by Waelsch and his collaborators of glutamate-glutamine metabolic compartmentation in brain tissue (1-4), have established that glutamine may be an important metabolic precursor of the pools of two amino acids which play a fundamental role in the transmission of nerve impulses across synapses in the central nervous system, namely γ-aminobutyric acid (GABA) and glutamate.

COMPARTMENTATION OF GLUTAMATE METABOLISM IN BRAIN TISSUE

The previously mentioned pioneering work by Waelsch and his coworkers established that when isotopically labeled glutamate was injected into the brain, the specific radioactivity of glutamine was higher than that of glutamate itself after short periods of time (Table I), a behavior that deviates from the usual precursor-product

TABLE I

Specific radioactivity of glutamine relative to that of glutamate in brain, after intracisternal administration of ^{14}C glutamic acid to rats.

Time (min)	Glutamine sp. act. / Glutamate sp. act.	total cpm (glutamine + glutamate)
0.25	0.43	92,000
1	1.3	84,000
	3.8	91,000
2[a]	cerebellum 5.3	
	pons-medulla 5.0	
	anterior-superior 5.4	
	posterior-superior 3.2	
	anterior-inferior 4.8	
	posterior-inferior 3.9	
5	5.2	97,000

[a] The regions indicated at 2 min refer to the cerebral hemispheres cut in 4 parts. Data from reference (3).

relationship. Since the total counts of glutamine + glutamate was constant, this result cannot be explained by a rapid metabolic degradation of labeled glutamate to other products. Later work by the same group and by others demonstrated that a similar result could be obtained with other precursors such as ammonia (labeling of the α-amino group of glutamine), bicarbonate, acetate, butyrate, citrate and several amino acids, but not with glucose, pyruvate, lactate or glycerol. Thus, a list of compounds which give specific radioactivity ratios glutamine/glutamate greater than 1.0 was compiled, as opposed to other precursors giving ratios lower than 1.0. Since the only known route of glutamine synthesis is from glutamate, the explanation for these findings was that at least two metabolic pools of glutamate exist in nervous tissue: a "large" pool, and a "small" pool in which glutamate is rapidly converted into glutamine before mixing with the "large" pool; the precursors giving a specific activity ratio glutamine/glutamate greater than 1.0 are feeding the "small" pool, whereas the other precursors feed mainly the "large" pool (Table II).

TABLE II

Glutamate-glutamine compartmentation in brain

Small glutamate compartment (precursors giving a specific activity radio gln/glu > 1.0)	Large glutamate comparment (precursors giving a specific activity ratio gln/glu < 1.0)
Glutamate, aspartate, GABA	Glucose
leucine, glycine, serine, proline	β-Hydroxybutyrate
Acetate, butyrate	Acetoacetate
Bicarbonate	Pyruvate
Ammonia (α-amino)	Lactate
Citrate	Glycerol

Subsequent studies (for reviews see references 5 and 6) demonstrated that this compartmentation of glutamate metabolism was: 1) a particular property of the nervous system, since it was not observed in other organs; 2) present in all regions of the brain; 3) present in all layers of the cerebral cortex after topical application of precursors; 4) present in several mammalian species; and 5) demonstrable also *in vitro,* in slices of brain cortex or other regions.

Since the carbon atoms of all precursors have to be channeled through the citric acid cycle in order to be converted to glutamate and subsequently to glutamine, it has been concluded that at least two metabolically different citric acid cycles exist in brain: one mainly for energy purposes, which gives rise to the large pool of glutamate, and a second one for synthetic purposes, which is related to the small pool of glutamate rapidly converted into glutamine and probably also to GABA (5, 7, 8).

STRUCTURAL CORRELATES OF THE GLUTAMATE-GLUTAMINE COMPARTMENTATION

For several years the obvious question relating to the possibility of different morphological structures accounting for the different compartments of glutamate metabolism remained unanswered.

However, in the last few years a picture seems to be emerging in which the two main types of cells in the nervous system, the neurons and the glial cells, appear to be responsible for the large and the small glutamate pools, respectively. This hypothesis had been previously suggested by some authors on the basis of the effect of metabolic inhibitors on compartmentation (9-11), and it received direct support in the elegant study by Minchin and Beart, published in 1975 (12), using the dorsal root ganglion as an experimental model. This ganglion does not contain nerve endings nor synapses, but only neuronal cell bodies and their satellite glial cells, and both cell populations are easily distinguishabe by means of the light microscope. Minchin and Beart first demonstrated that the ganglion incubated with labeled acetate or glucose for 15-90 min did show the compartmentation of glutamate metabolism discussed above: with acetate as a precursor the specific radioactivities ratio glutamine/glutamate was higher than 1.0, at all times studied, whereas with glucose it was only 0.3-0.7 during the experimental period. They then carried out light microscope autoradiography of ganglia incubated with labeled glucose or labeled acetate, and clearly demonstrated that the large pool, labeled by glucose, was associated with the neuronal bodies, whereas the small pool, labeled by acetate, was located within the satellite glial cells.

In further support for this direct evidence, it was later found that glutamine synthetase (GS), studied immunohistochemically both by light microscopy (13) and electronmicroscopically (14), was located exclusively in the glial cells, and more specifically in the astrocyte cytoplasm and in its neuropil processes.

Other authors have recently employed biochemical techniques to study the distribution of GS in bulk isolated neuronal bodies, glial cells and nerve endings (synaptosomes) (15). The results of these experiments indicate that, although glial cells seem to be enriched in GS, both sinaptosomes and neuronal bodies, particulary the latter, possess GS in fairly large amounts. Methodological problems related to the isolation of the cell populations might be the explanation for the discrepancy between these results and the immunohistochemical observations, but is is also difficult to conceive that neurons completely lack GS.

On the other hand, in the same study (15) it was shown that glutaminase is clearly concentrated in synaptosomes, as compared to neurons or glial cells. This result thus confirms a previous report (16) in which a large proportion of glutaminase was found in the synaptosomal fraction, as well as in mitochondria. As will be discussed

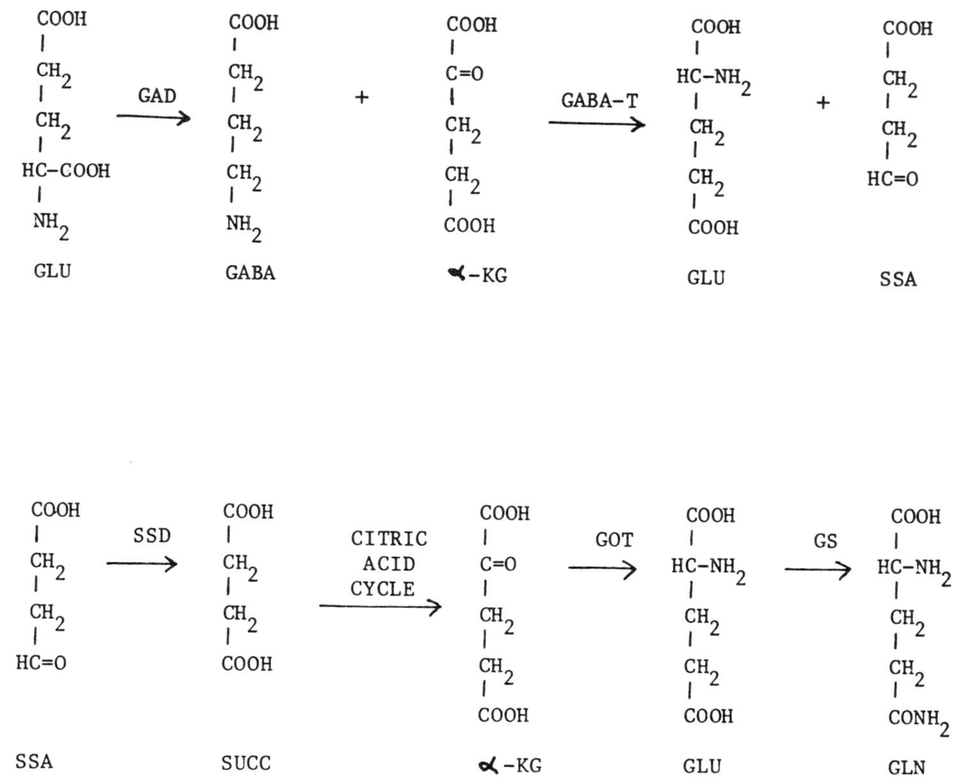

Fig. 1. Conversion of the carbon skeleton of GABA into glutamine. GABA is synthesized by α-decarboxylation of glutamate (GLU), catalyzed by L-glutamate decarboxylase (GAD). It is transaminated with α-ketoglutarate (α-KG) by GABA-transaminase (GABA-T), to yield succinic semialdehyde (SSA) and glutamate. SSA is subsequently oxidized by succinic semialdehyde dehydrogenase (SSD) to succinate, which through the reactions of the citric acid cycle is converted to α-KG. Transamination with aspartate, catalyzed by glutamic-oxalacetic transaminase (GOT), forms glutamate, which is converted to glutamine (GLN) by glutamine synthetase (GS).

in the next section, the localization of glutaminase in the nerve endings may be of physiological importance at the level of glutamatergic and GABAergic synapses.

PHYSIOLOGICAL CORRELATES OF GLUTAMINE METABOLISM IN BRAIN; GLUTAMATERGIC AND GABAERGIC SYNAPTIC TRANSMISSION

As mentioned in the Introduction, the basic mechanisms of brain function are the initiation, propagation and transmission of nerve impulses. Whereas the former two events are electrical phenomena, accounted for by the movements of ions across the axonal membrane, the transmission of the nerve impulse from one neuron to another is essentially a chemical event which takes place at the specialized structures known as chemical synapses. At any given synapse the arrival of an action potential (depolarization) to the presynaptic axon terminal induces a Ca^{2+}-dependent release of a specific metabolite, which acts as neurotransmitter by combining with a specific receptor molecule located in the postsynaptic membrane of the second neuron. The interaction of the neurotransmitter with its receptor results in a change of permeability to different ions: Na^+ if the synapse is excitatory, with a consequent depolarization, or Cl^- if the synapse is inhibitory, with a consequent hyperpolarization.

Some amino acids seem to act as neurotransmitters in the central nervous system. Among them, glutamic acid appears to be an excitatory synaptic transmitter whereas GABA — the product of α-decarboxylation of glutamate, catalyzed by glutamate decarboxylase — is considered the major inhibitory synaptic transmitter in the mammalian brain. After these two amino acids have been released from presynaptic endings and have combined with their respective postsynaptic receptors, they are taken up by the presynaptic ending, the postsynaptic neuron or the adjacent glial cell processes, by a high affinity mechanism dependent on Na^+ and Cl^- (17, 18).

From the compartmentation evidence reviewed in the previous sections (9-11) it had been hypothesized that glutamate released at glutamatergic synapses might be taken up by the glial cells and converted to glutamine by the high concentration of GS present in such cells. Similarly, the GABA released at GABAergic synapses may be taken up by glial cells and its carbon skeleton converted there to glutamine via the well known metabolic reactions shown in Fig. 1: transamination to succinic semialdehyde, oxidation of this semialdehyde to succinic acid, conversion into α-ketoglutarate through the citric acid cycle reactions, transamination with aspartate to yield glutamate and finally glutamine synthesis by GS. (GABA is a precursor of the "small" glutamate pool, see Table II). The glutamate resulting directly from transamination of GABA in glia could be also converted into glutamine by GS. According to the hypothesis, glutamine thus synthesized would be transported into nerve endings, where glutaminase would convert it to glutamate. In glutamatergic synapses this glutamate would be available for release, whereas in GABAergic synapses it would be decarboxylated to GABA, to a pool from where it could be released by depolarization of the nerve ending.

This idea that glutamine may be a precursor of the releasable pool of glutamate and GABA has been strongly supported by several recent findings which demonstrate that labeled glutamate and labeled GABA synthesized from labeled glutamine are released in a Ca^{2+}-dependent manner by depolarization of brain slices. Fig. 2 shows the results of experiments carried out in our laboratory (19) on the spontaneous release of labeled GABA derived from [^{14}C] glutamate or [^{14}C] glutamine in mouse brain cortex slices; Fig. 3 shows the release induced by depolarization by high K^+ concentration in the pre-

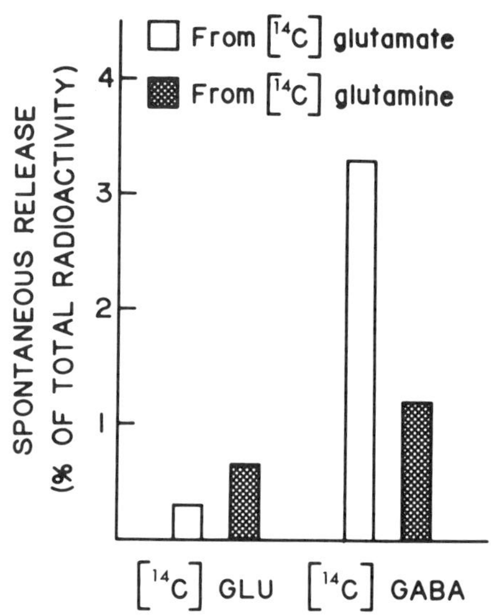

Fig. 2. Spontaneous release (before depolarization) of [^{14}C] GABA and [^{14}C] glutamate derived from [^{14}C] glutamate or [^{14}C] glutamine, in mouse brain cortex slices. The release is expressed as percent radioactivity in each amino acid released per min, considering as 100% the total radioactivity (total released + tissue radioactivity at the end of the experiment). Data from reference (19).

sence of Ca^{2+}. As can be seen in these figures, the spontaneous release of labeled GABA derived from glutamate is notably higher than that of labeled GABA derived from glutamine, whereas the opposite results were obtained in the case of the depolarization-induced release. Very similar results, also shown in Fig. 3, were reported by Reubi *et al.*, in pigeon optic tectum slices (20). In contrast to these findings with GABA, the spontaneous release of labeled glutamate was similar whether the amino acid was derived from glutamine or taken up by the tissue (Fig. 2). In addition, glutamate release stimulated by depolarization was very high both when glutamate was taken up from the medium or derived from glutamine (Fig. 3).

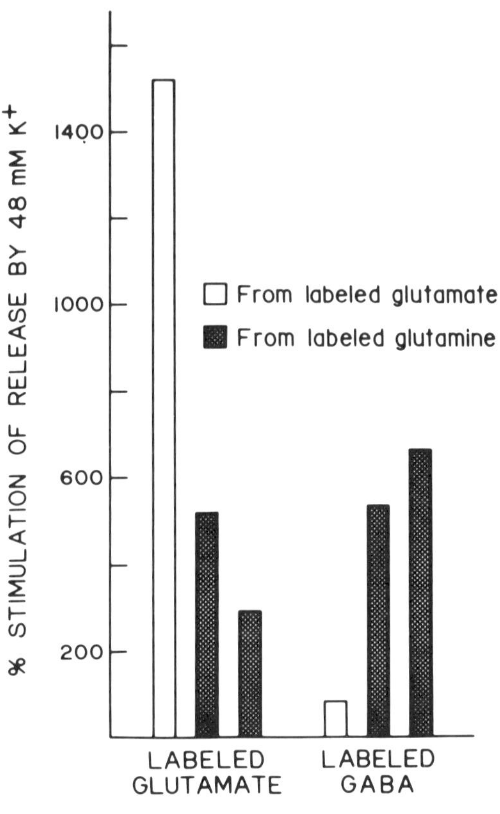

Fig. 3. K^+-depolarization-induced release of labeled glutamate and GABA derived from labeled glutamate or glutamine, in mouse brain cortex slices (first two bars in each set of three) and in pigeon optic tectum slices (last bar in each set). The release is expressed as percent stimulation by 48 mM K^+ over the control pre-stimulation value. Data from references (19) and (20).

The above results have been interpreted as an indication that the glutamate taken up by the tissue is probably a better precursor for the spontaneous release of GABA than the glutamate derived from glutamine, whereas the reverse is true for the K^+-depolarization-induced release; this interpretation implies that there are two pools of glutamate which mix with one another comparatively very slowly. On the other hand, since both the glutamate derived from glutamine and that taken up from the medium are released by K^+-depolarization, it is possible that the two pools of glutamate are available for depolarization-induced release.

Although from the above discussion it seems clear that glutamine may be an important precursor of the glutamate and GABA available for depolarization-induced release, similar experiments with pyruvate or glucose as precursors indicate that the pools of these two transmitter amino acids derived directly from the citric acid cycle can also be released by depolarization (21-23). Part of this work is summarized in Fig. 4 for glutamate (23). It can be seen that the release of glutamate derived from glucose, pyruvate or glutamine is similarly stimulated by K^+-depolarization in the presence of Ca^{2+}.

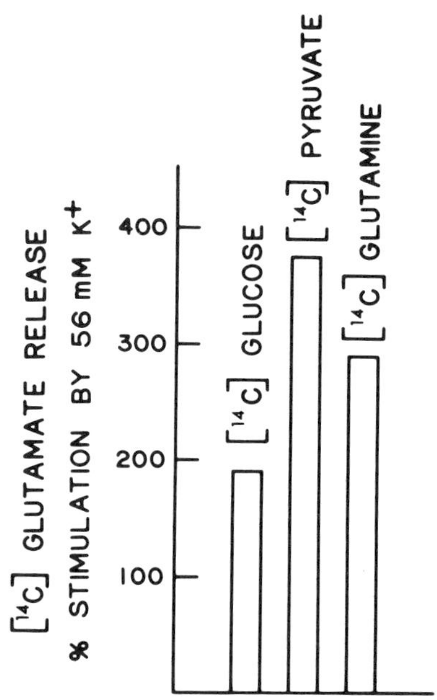

Fig. 4. K^+-depolarization-induced release of [^{14}C] glutamate derived from labeled glucose, pyruvate and glutamine, in rat hippocampus slices. The release is expressed as in Fig. 3. Data from reference (23).

From the available evidence discussed in the above paragraphs the metabolic-physiological relationships shown in Figs. 5 and 6 can be postulated for glutamatergic and GABAergic synapses. In the case of the inhibitory GABA synapses it is possible that GABA derived from the citric acid cycle can be released also by depolarization. However, in Fig. 6 the role of glutamate as a precursor of the spontaneous release of GABA is emphasized, because of the important physiological role that this tonic release seems to have in the regulation of cerebral excitability through the maintenance of a continuous inhibitory control or neuronal activity (24, 25).

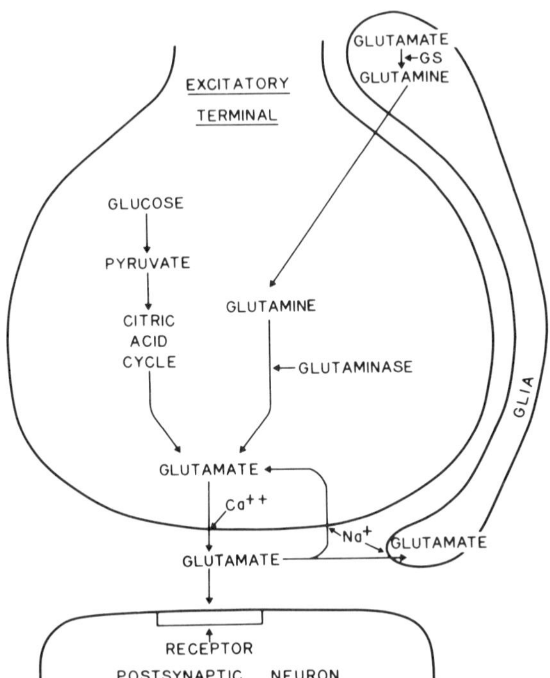

Fig. 5. Metabolic-physiological relationships in an excitatory glutamatergic synapse. Glutamate synthesized from glucose, pyruvate or glutamine is released by depolarization in the presence of Ca^{2+}. After interaction with its postsynaptic receptor it is taken up by the nerve ending or the adjacent glial cells by a Na^{+}-dependent mechanism. The glutamate entering the glia is converted to glutamine by glutamine synthetase (GS) and glutamine is eventually transported to the presynaptic ending to serve as a precursor of the releasable glutamate. This scheme is based on many of the references mentioned in the text, particularly (12-17, 19, 22 and 23).

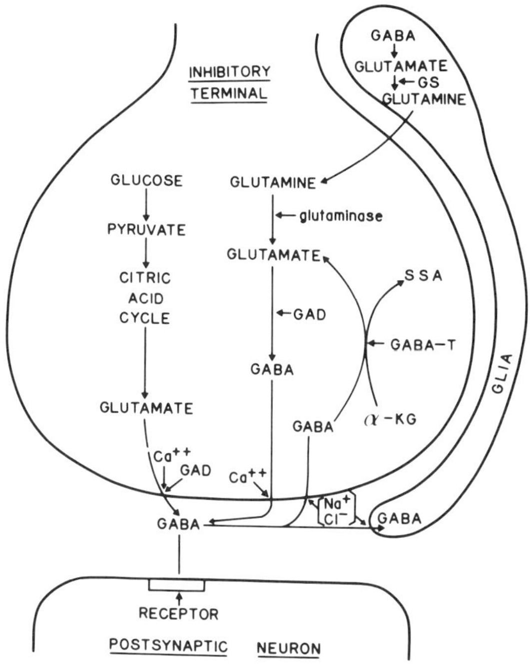

Fig. 6. Metabolic-physiological relationships in an inhibitory GABAergic synapse. GABA is synthesized from two different pools of glutamate by glutamate decarboxylase (GAD). The pool of glutamate derived from glucose or pyruvate, or taken up by the terminal, gives rise to a GABA that is preferentially released in a process coupled to its synthesis. This release occurs in the absence of depolarization (spontaneous, tonic release), mediated by a population of GAD that binds to the presynaptic membrane in the presence of Ca^{2+}. The other pool of glutamate, derived from glutamine, is decarboxylated in the cytoplasm of the nerve ending and the GABA thus synthesized is preferentially released by depolarization, in the presence of Ca^{2+}. After interaction with its postsynaptic receptor, GABA is taken up by the nerve ending or the adjacent glial cells by a Na^{+}- and Cl^{-}-dependent mechanism. In glia, GABA is transformed into glutamine through the reactions shown in Fig. 1, and this glutamine is transported to the nerve ending and converted to glutamate by glutaminase. In the nerve ending the GABA taken up may be metabolized by GABA-T. Abbreviations are defined in the legends to Fig. 1 and 5. This scheme is based on many of the references cited in the text, particularly (12-21 and 24-26).

REFERENCES

1. Waelsch, H. (1960) in *Structure and Function of the Cerebral Cortex* (Tower, D. B., and Schadé, J. P., eds.), pp. 313-326, Elsevier, Amsterdam.
2. Lajtha, A., Berl, S., and Waelsch, H. (1960) in *Inhibition in the Nervous System and Gamma-Aminobutyric Acid* (Roberts, E., Baxter, C. F., van Harreveld, A., Wiersma, C. A. G., Adey, W. R., and Killam, K. F., eds), pp. 460-467, Pergamon, Oxford.
3. Berl, S., Lajtha, A., and Waelsch, H. (1961) *J. Neurochem.* 7, 186-197.

4. Takagaki, G., Berl, S., Clarke, D. D., Purpura, D. P., and Waelsch, H. (1961) *Nature (Lond.) 189*, 326.
5. Berl, S., and Clarke, D. D. (1969) in *Handbook of Neurochemistry* (Lajtha, A., ed.) Vol. 2, pp. 447-472, Plenum, New York.
6. Berl, S., Clarke, D. D., and Schneider, D. (eds.) (1975) *Metabolic Compartmentation and Neurotransmission*, 721 pp., Plenum, New York.
7. Van der Berg, C. J., Krzalic, L. J., Mela, P., and Waelsch, H. (1969) *Biochem. J. 113*, 281-290. W
8. Berl, S., Nicklas, W. J., and Clarke, D. D. (1970) *J. Neurochem. 17* 1009-1015.
9. Balázs, R., Machiyama, Y., Hammond, B. J., Julian, T., and Richter, D. (1970) *Biochem. J. 116*, 445-467.
10. Benjamin, A. M., and Quastel, J. H. (1972) *Biochem. J. 128*, 631-646.
11. Van den Berg, C. F., Matheson, D. F., Ronda, G. Reijnierse, G. L. A., Blokhuis, G. G. D., Kroon, M. C., Clarke, D. D., and Garfinkel, D. (1975) in *Metabolic Compartmentation and Neurotransmission* (Berl, S., Clarke, D. D., and Schneider, D., eds.), pp. 515-543, Plenum, New York.
12. Minchin, M. C. W., and Beart, P. M. (1975) *Brain Research 83*, 437-449.
13. Martínez-Hernández, A., Bell, K. P., and Norenberg, M. D. (1977) *Science (N. Y.) 195*, 1356-1358.
14. Norenberg, M. D., and Martínez-Hernández, A. (1979) *Brain Research 161*, 303-310.
15. Weiler, C. T., Nyström, B., and Hamberger, A. (1979) *Brain Research 160*, 539-543.
16. Bradford, H. F., and Ward, H. K. (1976) *Brain Research 110*, 115-125.
17. Iversen, L. L. (1971) *Brit. J. Pharmacol. 41*, 571-591.
18. Kuhar, M. J., and Zarbin, M. A. (1978) *J. Neurochem. 31*, 251-256.
19. Tapia, R., and González, R. M. (1978) *Neurosci. Letters 10*, 165-169.
20. Reubi, J.-C., Van den Berg, C., and Cuénod, M. (1978) *Neurosci. Letters 10*, 171-174.
21. Gauchy, C. M., Iversen, L. L., and Jessell, T. M. (1977) *Brain Research 138*, 374-379.
22. Hamberger, A., Chiang, G., Nylén, E. S., Scheff, S. W., and Cotman, C. W. (1978) *Brain Research 143*, 549-555.
23. Hamberger, A., Chiang, G. H., Nylén, E. S., Scheff, S. W., and Cotman, C. W. (1979) *Brain Research 168*, 513-530.
24. Tapia, R. (1974) in *Neurohumoral Coding of Brain Function* (Myers, R. D., and Drucker-Colín, R. R., eds.) pp. 3-26, Plenum, New York.
25. Tapia, R., Sandoval, M. E., and Contreras, P. (1975) *J. Neurochem. 24*, 1283-1285.
26. Covarrubias, M., and Tapia, R. (1978) *J. Neurochem. 31*, 1209-1214.

16. Regulatory Aspects of the Induction of Glutamine Synthetase in Embryonic Neural Retina

ARON A. MOSCONA, PAUL LINSER, PAULA MAYERSON AND MALKA MOSCONA

Laboratory for Developmental Biology, Cummings Life Science Center University of Chicago, Chicago, Illinois 60637

SUMMARY

The results described here demonstrate that the cortisol-elicited induction of GS in the embryonic neural retina occurs in Müller glia cells. They indicate that in addition to the hormonal inducer, contact-dependent interactions between Müller cells and retina neurons are involved in the mechanism of GS induction. The possible nature of these interactions is discussed. BrdU-elicited malformation of the neural retina is described and the non-inducibility of GS in this situation is discussed.

INTRODUCTION

Glutamine synthetase, to be referred to as GS, is found throughout the nervous system in higher vertebrates, as first discovered by Krebs (1). There is good evidence that it is one of the key enzymes in the so-called "small glutamate compartment" in neural tissues (2). One of the functions of this compartment is to salvage neuronally released neurotransmitter metabolites, such as glutamate, and to convert them into glutamine which is then made again available to

ISBN: 0-12-506040-8

neurons (3, 4). Because of the importance of GS in this cycle, and the involvement of GS in the synthesis of mucopolysaccharides and nucleic acids (5), its role in the biology of neural tissues is of considerable and growing interest (6). Measurements of GS in the brain of various animals (7) and in various parts of the nervous system have shown that the neural retina contains the highest level of this enzyme (8). In the neural retina of the chicken the level of GS is particularly high, while in the retina of the chick embryo its level is very low. Some years ago we became interested in this difference (9) and began a series of studies that brought us face to face with fundamental questions concerning the control of this enzyme and its involvement in the embryonic differentiation of the neural retina (10).

GS INDUCTION IN THE RETINA

The retina represents, developmentally and functionally, an extension of the brain. It arises in the embryo as a vesicle from the wall of the midbrain. It starts out as a relatively simple neuroepithelium and follows a program of growth and differentiation that turns it into a system specialized for reception, integration and transmission of light signals. One of the characteristic biochemical features of this developmental program of the retina is the induction of GS.

In the retina of the early chick embryo the level of GS is low and increases very slowly. However, on the 16th day of incubation, GS in the retina begins to rise sharply, increases 100-fold in 5 to 6 days, then plateaus at this high level. This sudden, fast and large rise of GS is unique to the embryonic retina. It begins several days after growth and cell replication in the retina had virtually ceased (Fig. 1), and it is triggered by the elevation of adrenal corticosteroids in the embryo that takes place one or two days earlier.

However, what makes this system especially interesting is the fact that GS in the retina can be induced precociously long before the time of its normal rise in the embryo (as early as in retina of 8-day embryo). The precocious induction can be accomplished *in vivo*, by injecting cortisol into eggs (11); and *in vitro* in organ cultures of retina tissue isolated from 9 to 15-day embryos by adding cortisol (or related 11-β-hydroxycorticosteroids) to the culture medium. Organ cultures of retina tissue have been more widely used in these studies because they lend themselves better to controlled biochemical investigations. Thus, when retina tissue isolated from 10-day chick embryos is incubated in cortisol-containing synthetic culture medium, *de novo* enzyme synthesis begins after a short lag period and its rate of accumulation increases rapidly, as demonstrated by immunotitration of the radioactively labeled nascent enzyme (12, 13); in 48 hrs, the tissue contains 10 to 30 times the original level of GS.

The induction of GS involves differential gene expression elicited by the hormone. The steroid rapidly enters the cells and binds to cytoplasmic receptors; the steroid-receptor complex is translocated into nuclei (14, 15, 16) where it associates with chromatin and elicits transcription of mRNA for GS, resulting in enzyme synthesis and accumulation (17-20). Actinomycin D blocks GS induction at the transcriptional level; cycloheximide and puromycin inhibit GS synthesis and thereby prevent the expression of GS induction (21). It is noteworthy that only 11-β-hydroxycorticosteroids elicit the gene expression required for GS induction in the neural retina (22); and that only in the neural retina (and to a lesser extent in the optic tectum (23)) is GS inducible by this class of hormones.

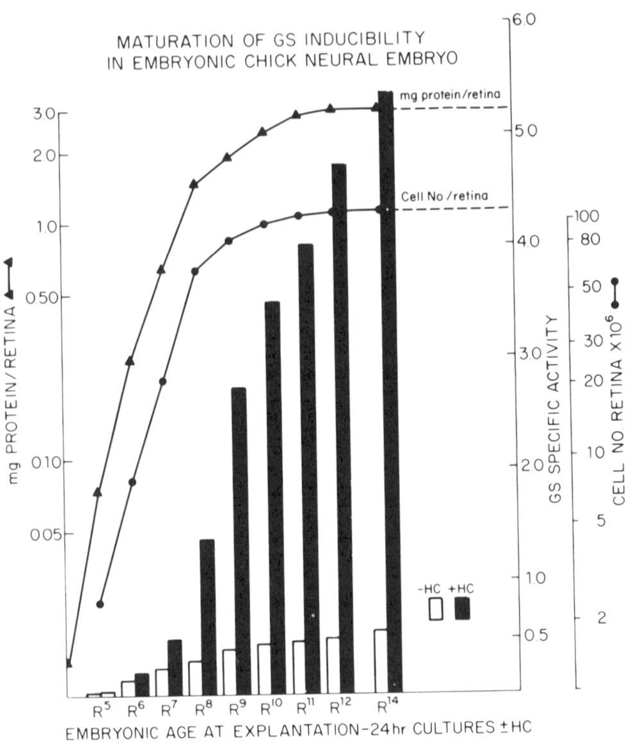

Fig. 1. Development of inducibility for GS in the neural retina of the chick embryo between 5 and 14 days of incubation: relation to embryonic age and to changes in cell number and in total protein per retina. Retinas dissected from embryos were cultured for 24 h in medium with cortisol (+HC) or without it (−HC). The *black bars* show the levels of GS activity induced in retinas of different embryonic ages. The *white bars* show the levels of GS in the absence of the steroid inducer (From 24).

Detailed studies on retinas of various embryonic ages (24), summarized in Fig. 1, have shown that competence for GS induction in the retina of the chick embryo is detectable by the 7th day, and that this competence develops during the 5 to 7th day; during the precompetence period, GS is not inducible, or only minimally inducible. After the 7th day, inducibility of GS increases towards a maximum on the 16th day.

In addition to the crucial role of the specific hormonal stimulus in GS induction, still another control mechanism appears to be implicated in this process; this mechanism involves cell interactions and will be discussed below.

CHARACTERISTICS OF RETINAL GS

GS from chicken retina has been purified (25); upon electrophoresis in SDS-polyacrylamide gels the purified preparation yields a single band in a region corresponding to M. W. 42,000 ± 2,000. Examination by electron microscopy has shown that the native enzyme is a rectangular particle consisting of eight subunits. The molecular weight of the octamer is in the order of 340,000. The amino acid composition of chicken retina GS shows considerable similarity with that of sheep brain GS and bacterial GS (26, 27). The purified chicken retina enzyme is antigenic in rabbits and antibodies highly specific for GS have been obtained (28). Immunological studies revealed no differences between GS of mature chicken retinas and GS precociously induced in embryonic retina (29). By using the specific antiserum, the nascent (radioactively labeled) GS peptide was immunoprecipitated with polysomes isolated from induced embryonic retina (28). Extension of this work led to isolation of GS mRNA and its activity was demonstrated in an *in vitro* translation system (20; work in progress).

LOCALIZATION OF GS IN MÜLLER CELLS

The neural retina consists of five major types of neural cells, and of glia cells known as Müller fibers (cells). An essential step in the analysis of GS induction has been to determine in which of these cell types the enzyme is induced. This question was investigated by immunohistochemical methods. Histological sections of retinas were reacted with the anti-GS antiserum in order to determine (by indirect immunofluorescence) the distribution of the GS antigen in the different types of cells (29).

Localization of the enzyme was studied, first, in mature retinas (from 6-week old chicken) because of their very high content of GS. The enzyme was detected only in Müller cells (Fig. 2); these are the only glia elements known to be present in the chicken retina and they are readily identifiable by their characteristic morphology. It is of interest that also in adult rat retina GS is confined to Müller cells (30, 31). The location of GS in glia cells agrees with the view that they are the site of the "small glutamate compartment" (32, 33) in which GS is assumed to be an important constituent.

Next, we determined that also precociously induced GS is localized in Müller cells. Retinas from 11-day embryos were cultured for 48 hrs in cortisol-containing medium to induce the enzyme. GS was also induced prematurely in retinas *in vivo* by administering cortisol to 11-day embryos (22). Histological sections of these retinas (and of appropriate control retinas) were immunostained with anti-GS antiserum; their examination revealed that induced GS was invariably localized only in Müller cells (Fig. 2). In non-induced control retinas Müller cells showed a very weak reaction with the antiserum, consistent with the low level of GS activity in the non-induced tissue.

Similar studies performed on embryonic retinas of different embryonic ages confirmed that there was a direct correspondence between the level of GS induction (assayed biochemically) and the intensity of immunostaining of Müller cells. The conclusion derived from these results is that the cortisol-induced precocious accumulation of GS in the embryonic retina is confined to Müller cells; hence, these glia elements represent the predominant, and possibly the only cellular location of GS in the embryonic-induced retina, as well as in the adult retina.

Evidence from still other studies further confirms the localization of GS in Müller cells. These studies were based on earlier findings by Olney *et at.* (34) and Pedersen and Karlsen (35) who found that DL-alpha aminoadipic acid caused cytotoxic changes in Müller cells in infant mice and adult rats, without noticeably affecting retina neurons. We examined if alpha-aminoadipic acid prevented the induction of GS in chick embryo retina (Linser and A. A. Moscona, in preparation); such a result would be expected, assuming that this amino acid is cytotoxic for Müller cells also in chick embryo retina, and that these cells are actually involved in GS induction.

Retinas from 11-day chick embryos were cultured in medium with cortisol, with and without the addition of alpha-aminoadipic acid (200 μg/ml). After 4 days GS specific activity was assayed. In the cortisol-induced retinas, GS activity increased 400% above non-induced control level; however, in cultures which contained alpha-

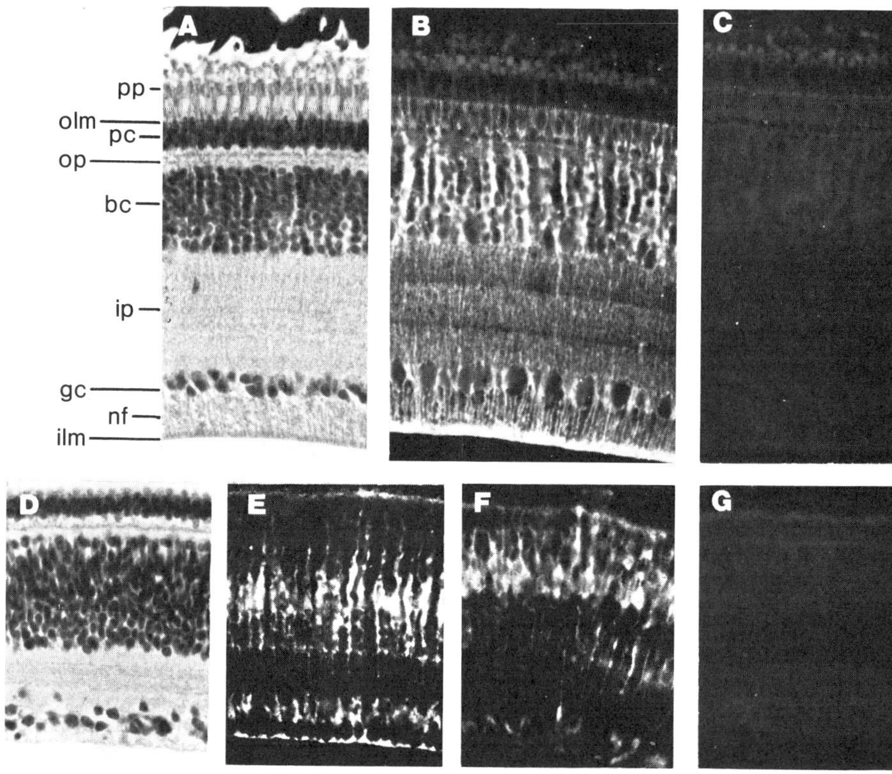

Fig. 2. Cellular localization of GS in mature chicken retina (A-C) and in cortisol-induced embryonic chicken retina (D-G) by immunostaining and indirect immunofluorescence. (A) Section of mature retina (6-week-old chicken) stained with hematoxylin and eosin. pp, Photoreceptor processes; olm, outer limiting membrane; pc, photoreceptor cell layer; op, outer plexiform layer; bc, bipolar cell layer; ip, inner plexiform layer; gc, ganglion cell layer; nf, nerve fiber layer; ilm, inner limiting membrane. (B) Section similar to that in A, immunostained with anti-GS-antiserum and FITC-GAR. Light areas represent immunofluorescence in Müller fibers. (C) Control section treated with nonimmune rabbit serum and FITC-GAR. (D) Section of 13-day embryo retina stained with hematoxylin and eosin. (E) Thirteen-day retina induced *in vivo* with cortisol. Section treated with anti-GS-antiserum and FITC-GAR. Fluorescence is localized to Müller fibers. (F) Section of 13-day retina induced in organ culture with cortisol; immunofluorescence shows localization of induced GS in Müller fibers. (G) Section of control, noninduced 13-day retina treated as in E and F; no immunofluorescence. All magnifications were X350. (From 29).

aminoadipic acid in addition to cortisol there was no GS induction and, in fact, the enzyme level was slightly below that in non-treated controls. Histological sections of retinas treated with alpha-amino-adipic acid showed cytotoxic changes only in Müller cells. Furthermore, when thus treated retinas were dissociated into single cells and these were plated in monolayer cultures (Fig. 3), only neuronal cells were present and no glial cells were found. In contrast, cultures derived from retinas that were not treated with alpha-aminoadipic acid contained, in addition to neurons also numerous large epithelioid glia cells which are known from previous studies (36) to be derived from Müller fibers (Fig. 3). We conclude that alpha-amino-adipic acid prevents GS induction in chick embryo retina because it is selectively cytotoxic for Müller cells; this supports a direct correlation between GS induction in the retina and localization of the enzyme in Müller cells.

Taken as a whole, the above results demonstrate that GS in the retina is a characteristic differentiation product and a biochemical

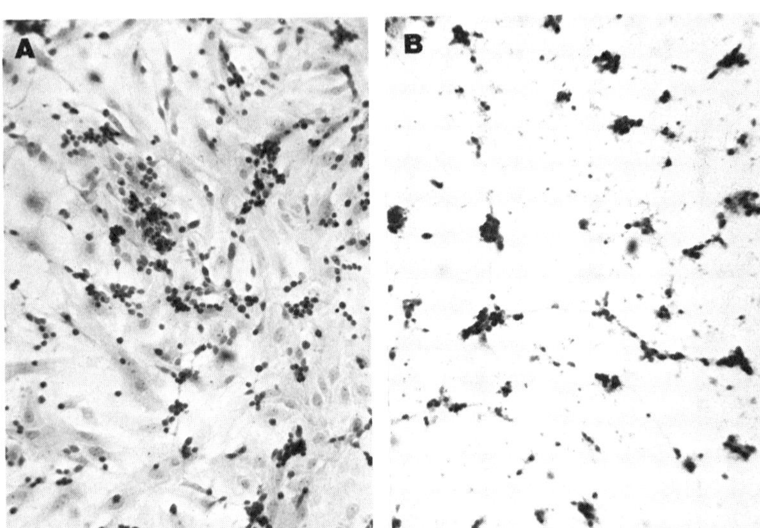

Fig. 3. Monolayer cultures of retina cells; the cells were obtained by dissociation of retinas that were originally isolated from 11-day embryos, and were then organ-cultured for 4 days. The organ cultures were maintained either in normal medium (controls), or in medium with alpha-aminoadipic acid. A: Cell culture obtained from control retinas; note the presence of dark-nucleated neurons and of large epitheliocytes derived from Müller cells. B: Cell culture obtained from retina that was treated with alpha-aminoadipic acid; note absence of epitheliocytes.

marker of the Müller glia cells and that it is inducible specifically in these cells by cortisol and related adrenocortical steroids. This cellular location of GS agrees with the view that the "small glutamate compartment" of which GS is a part, is located in Müller cells. At the present time, it is unknown whether other enzymes associated with this compartment also are hormonally inducible coordinately with GS induction, or if they are independently controlled.

CELL INTERACTIONS AND GS INDUCTION

Experiments begun some years ago suggested that, in addition to the hormone, still another control mechanism may be involved in the induction of GS in the retina (21, 37, 38). These experiments examined whether tissue integrity and specific cell contacts were essential for GS induction, or if the enzyme could also be induced in disassociated and dispersed retina cells. The fundamental question was whether cell interrelationships, structural and functional, of the kind that exist within intact retina tissue are involved in the mechanism of the hormonal induction of GS.

To examine this question, retina tissue from 10-day embryos was dissociated (by mild trypsinization) (39) into cell suspensions, and these were plated in culture dishes at low densities which favored the cells remaining dispersed and forming a monolayer. Cortisol was added at plating time; GS specific activity was measured daily over a period of several days. There was no measurable GS induction in these cultures of dispersed cells (Fig. 4). It was determined that the steroid inducer was taken up into the cells, and there was no enzyme loss from cells into the medium; yet, the level of GS specific activity did not increase. Since in retina tissue GS is confined to Müller cells, special attention was paid to them. As mentioned above, Müller cells rapidly assume in such monolayer cultures the shape of large flattened epithelioid cells (36). Immunostaining of cortisol-treated monolayer cultures with anti-GS antiserum failed to detect at any time GS induction in these cells, or in any type of cell present in these cultures.

These results were consistent with the assumption that histological cell associations, possibly specific juxtapositions and interactions between Müller cells and neurons, are essential for the mechanism of hormonal induction of GS. An alternative possibility was that the process of dissociating retina tissue into single cells might have rapidly and permanently damaged the ability of Müller cells to undergo GS induction. These possibilities were examined by reag-

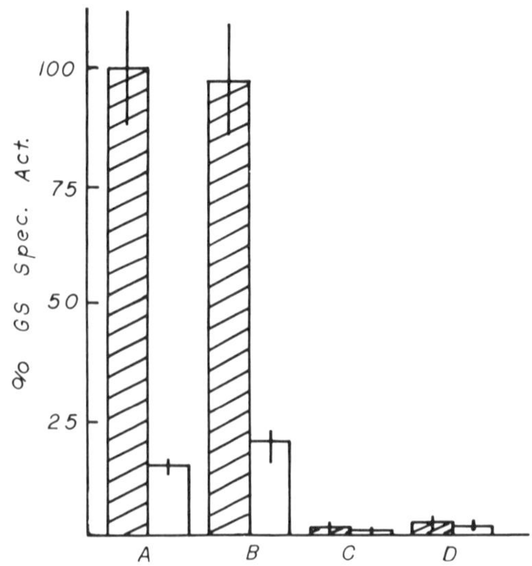

Fig. 4. Specific activity of GS (expressed as percentage of bars A) induced by cortisol in: organ cultures of embryonic retina tissue (bars A); cell aggregates (bars B); monolayer cell cultures (bars C and D). Hatched bars, cortisol present; open bars, cortisol absent. Bars A: GS activity in retina from 11-day embryo cultured for 2 days. Bars B: GS activity in aggregates of 6-day retina cells cultured for 5 days and then induced with cortisol. Bars C and D: monolayer cultures of retina cells from 6-day (C) and 10-day (D) embryos. (From 29).

gregating dissociated retina cells into multicellular aggregates; in such aggregates the cells reestablish retinotypic contacts and reconstruct within 2 days their characteristic tissue pattern (38, 39). We examined if GS could be induced in such cell aggregates, and if it was localized in Müller cells (29).

Cell aggregates obtained from suspensions of cells dissociated from 8 to 10-day retinas were found to be inducible for GS (Fig. 4). In fact, their inducibility increased coordinately with the progress of histological organization of the cells. Fig. 4 compares the levels of GS induction in organ cultures of retina tissue, in cell aggregates and in monolayer cell cultures. Histologically, the cell aggregates consisted of characteristic retinal rosettes (Fig. 5). Immunostaining with anti-GS antiserum of histological sections of cortisol-treated cell aggregates showed that GS was localized in cells identifiable as Müller fibers (Fig. 5).

Therefore, the non-inducibility of GS in fresh monolayer cultures of dispersed retina cells is not due to a rapid and permanent damage

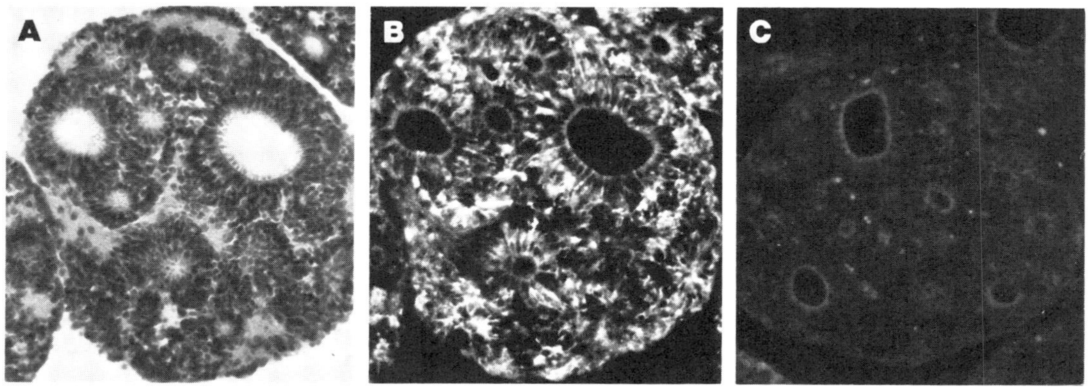

Fig. 5. Cellular localization of GS induced in retina cell aggregates (see text). (A) Section stained with hematoxylin and eosin showing retinal rosettes. (B) Section similar to A immunostained for GS; immunofluorescence shows localization of induced GS in cells identified as Müller fibers. (C) Section of noninduced aggregate immunostained for GS. All magnifications were X225. (From 29)

to Müller cells resulting in loss of their competence for induction. The more likely explanation is that the absence of specific intercellular contacts causes a failure at some step in the mechanism required for GS induction. Accordingly, our working hypothesis is that cell associations of the kind present in the tissue are essential for cell responsiveness to the hormonal induction of GS.

The nature and functional significance of such cell associations is presently being investigated. They may involve specific contacts between Müller cells and retina neurons. Such cell contacts could be necessary to maintain a particular molecular organization of the cell surface, or to enable interactions between specific cell-cell ligands that are involved in histotypic cell associations (38, 40). There is much evidence from various systems that conditions and changes at the cell surface can generate signals which are relayed to within the cells and affect cytoplasmic constituents and gene expression. GS induction involves differential gene expression mediated by hormone-activated cytoplasmic receptors; therefore, it could be susceptible to signals from the cell surface and might be prevented by modification of the cell surface resulting from cell separation. Restitution of histotypic cell contacts by reaggregation of the cells would restore conditions at the cell surface that are permissive for cell responsiveness to GS induction by the hormone. One attractive possibility is that the stability of cytoplasmic receptors for the steroid hormone, or their functional characteristics, are modified as a result of changes at the cell surface due to cell separation; this could interfere with transport of the hormone into the nucleus, or with its binding to the specific regions of DNA involved in GS induction. Another possibility is that cell separation results in a change in the transcriptional activity of these DNA regions, or in the processing of the primary transcripts.

Still another possibility is that histotypic cell contacts are necessary for intercellular transfer of metabolites which may participate in the regulation of GS induction. For example, the transfer of neuronally released substances into the "small glutamate compartment" in Müller cells could play a regulatory role in GS induction; if such transfer requires specific associations between Müller cells and neurons, cell separation would prevent GS induction.

These are presently largely hypothetical considerations; however, they suggest directions and guidelines for further exploration of the significance of cell associations in the mechanism of GS induction. The possibility of a causal link between histotypic cell associations and differential gene expression has potentially very important and broad implications. Various developmental abnormalities are initiated by disorganization of cells due to genetic or environmental factors. Loss of specific cell contacts might modify cell responsiveness to hormonal and other inducers and thus start a cascade of developmental defects rooted in the initial failure of cell associations. The results described in the next section relate, in part, to this problem.

THE EFFECTS OF BrdU

The possible dependence of GS induction in the retina on histotypic cell associations is being examined from yet another direction. It has been found that the cellular organization of retina tissue can be grossly modified without resorting to tissue dissociation and cell separation. This has been accomplished by briefly treating retinas of 5-day chick embryos with BrdU (5-bromo-2′-deoxyuridine), a thymidine analogue that is incorporated into DNA of replicating retina cells.

Retina tissue isolated from 5-day chick embryos and maintained in organ culture continues its development and morphogenesis, and attains in a few days a histological structure closely resembling that of a retina *in vivo* of a corresponding embryonic age. It also develops competence for GS induction (see Fig. 1). However, if retinas isolated from 5-day embryos are treated for 24 h with BrdU, then washed and further cultured in the absence of this agent, a dramatic result is obtained (Fig. 6). Instead of orderly histological development, the cells become disorganized, misaligned and misconnected, and progressively form a chaotic tissue mass (38, 41). The cells proliferate, but their normal spatial relationships become drastically altered (for detailed description see 41).

Particularly significant is the fact that in such malformed retinas GS is not inducible, or its inducibility is greatly reduced (24). In fact, there exists a close correlation between the extent of cell disorganization and the reduction of GS inducibility. Similar treatment with BrdU of older retinas (from 8-day or later embryos) that already had attained their characteristic histological organization does not cause disruption of tissue architecture and does not reduce the inducibility of GS (24).

The analysis of this BrdU effect is still in progress and it follows two working hypotheses. Both assume that the effect of BrdU is due to its incorporation into specific regions of DNA in retina cells. In fact, there is evidence for its incorporation into moderately repetitive DNA sequences (42). One possibility is that BrdU modifies in retina cells the function of genes that control the mechanism of morpho-

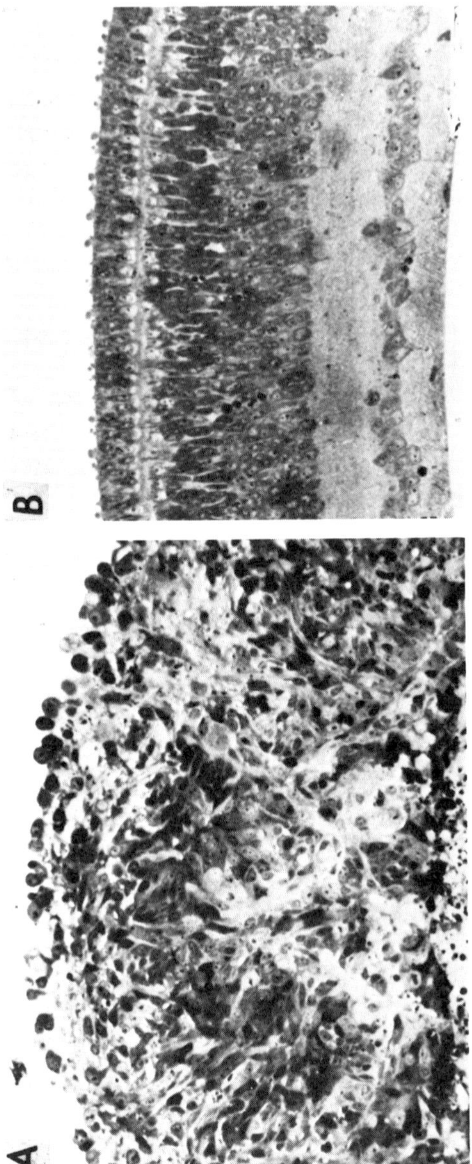

Fig. 6. Effect of treatment with BrdU on histogenesis of retina from 5-day embryo. A: Retina isolated from 5-day embryo was treated with BrdU for 24 hrs, then rinsed and organ-cultured for additional 6 days; note misorganization of the cells. B: Normal control; 12-day retina showing normal cell organization. (From 41).

genetic cell associations and thereby causes malformation of the retina. GS is not inducible because the misorganized neurons and Müller cells are unable to interact effectively. In other words, the situation here may be essentially similar to that proposed above for disassociated retina cells, except that the effect of BrdU is irreversible. Another possibility is that BrdU irreversibly modifies also the function of genes involved in GS induction; hence, that its effects on retina morphogenesis and on GS inducibility are coincidental.

These hypothetical suggestions are experimentally testable. The results would provide further information relevant to several important issues: the mechanisms that control GS induction in the retina; the specific nature and role of contact-mediated cell interactions in the induction and regulation of this enzyme in Müller cells and in retina differentiation; and the significance of GS in the development and metabolism of this neural tissue. Considering that the retina is a relatively "simplified" extension of the brain, it is not unreasonable to expect that clarification of the above issues might shed light also on related problems in the developmental physiology of the central nervous system.

Acknowledgments. This work is part of a research program supported by grant HD01253 from the National Institute of Child Health and Human Development (to A. A. M.), and by a postdoctoral fellowship (to P. L.) from Training grant GM7542 from the National Institutes of Health to the University of Chicago.

REFERENCES

1. Krebs, H. A. (1935) *Biochem. J. 29,* 1951-1969.
2. Van den Berg, C. J. (1970) in *Handbook of Neurochemistry* (Lajtha, A., ed.) Vol. III, pp. 355-379, Plenum Press, New York.
3. Kennedy, A. J., Voaden, M. J., and Marshall, J. (1974) *Nature 252,* 50-52.
4. Starr, M. S. (1974) *J. Neurochem. 23,* 337-344.
5. Meister, A. (1968) *Adv. Enzymol. 31,* 183-205.
6. Norenberg, M. D. (1979) *J. Histochem. Cytochem. 27,* 756-762.
7. Wu, C. (1963) *Comp. Biochem. Physiol. 8,* 335-351.
8. Rudnick, D., and Waelsch, H. (1955) *J. Exp. Zool. 129,* 309-326.
9. Moscona, A. A., and Hubby, J. L. (1963) *Develop. Biol.* 7, 192-206.
10. Moscona, A. A. (1972) in *Biochemistry of Cell Differentiation.* FEBS Proceed. (Monroy, A., and Tsanev, R., eds.) Vol. 24, pp. 1-23, Academic Press, London.
11. Moscona, A. A., and Piddington, R. (1966) *Biochim. Biophys. Acta 121,* 409-411.
12. Alescio, T., and Moscona, A. A. (1969) *Biochem. Biophys. Res. Commun. 34,* 176-182.

13. Moscona, M., Frenkel, N., and Moscona, A. A. (1972) *Develop. Biol. 28*, 229-241.

14. Sarkar, P. K., and Moscona, A. A. (1975) *Am. Zool. 15*, 241-247.

15. Koehler, D. E., and Moscona, A. A. (1975) *Arch. Biochem. Biophys. 170*, 102-113.

16. Sarkar, P. K., and Moscona, A. A. (1977) *Differentiation 7*, 75-82.

17. Patton, D., Kirk, D., and Moscona, A. A. (1969) *Exp. Cell Res. 54*, 181-186.

18. Garfield, S., and Moscona, A. A. (1973) *Biochem. Biophys. Res. Commun. 51*, 718-724.

19. Sarkar, P. K., and Moscona, A. A. (1973) *Proc. Natl. Acad. Sci. U.S.A. 70*, 1667-1671.

20. Soh, B-B., and Sarkar, P. K. (1978) *Develop. Biol. 64*, 316-328.

21. Morris, John E., and Moscona, A. A. (1970) *Science 167*, 1736-1738.

22. Moscona, A. A., and Piddington, R. (1967) *Science 158*, 496-497.

23. Shimada, Y., Piddington, R., and Moscona, A. A. (1967) *Exp. Cell Res. 48*, 240-243.

24. Moscona, M., and Moscona, A. A. (1979) *Differentiation 13*, 165-172.

25. Sarkar, P. K., Fischman, D. A., Goldwasser, E., and Moscona, A. A. (1972) *J. Biol. Chem. 247*, 7743-7749.

26. Ronzio, R. A., Rowe, W. B., Wilk, S., and Meister, A. (1969) *Biochemistry 8*, 2670-2679.

27. Deuel, T. F., Ginsburg, A., Yeh, J., Shelton, E., and Stadtman, E. R. (1970) *J. Biol. Chem. 245*, 5196-5205.

28. Sarkar, P. K., and Moscona, A. A. (1971) *Proc. Natl. Acad. Sci. U. S. A. 68*, 2308-2311.

29. Linser, P., and Moscona, A. A. (1980) *Proc. Natl. Acad. Sci. U. S. A. 76*, 6476-6480.

30. Riepe, R. E., and Norenberg, M. D. (1977) *Nature 268*, 654-655.

31. Riepe, R. E., and Norenberg, M. D. (1978) *Exp. Eye Res. 27*, 435-444.

32. Balazs, R., Patel, A. J., and Richter, D. (1972) in *Metabolic Compartmentation in the Brain* (Balazs, R., and Cremer, J. E., eds.), pp. 167, Wiley, New York.

33. Voaden, M. (1974) *Biochemical Society Transactions 2*, 1224-1227.

34. Olney, J. W., Ho, O. L., Rhee, V. (1971) *Exp. Brain Res. 14*, 61-76.

35. Pedersen, O., and Karlsen, R. L. (1979) *Exp. Eye Res. 28*, 569-575.

36. Kaplowitz, P. B., and Moscona, A. A. (1976) *Cell Differentiation 5*, 109-119.

37. Morris, John E., and Moscona, A. A. (1971) *Develop. Biol. 25*, 420-444.

38. Moscona, A. A. (1974) in *The Cell Surface in Development* (Moscona, A. A., ed.) pp. 67-99, Wiley, New York.

39. Moscona, A. A. (1961) *Exp. Cell Res. 22*, 455-475.

40. Hausman, R. E., and Moscona, A. A. (1979) *Exp. Cell Res. 119*, 191-204.

41. Mayerson, P. L., and Moscona, A. A. (1979) *Differentiation 13*, 173-184.

42. Strom, C. M., Moscona, M., and Dorfman, A. (1978) *Proc. Natl. Acad. Sci. U. S. A. 75*, 4451-4454.

17. Enhancement of Glutamine Excretion as Phenylacetylglutamine in the Treatment of Cases With Congenital Defects of Enzymes involved in Urea Biosynthesis

PHILIP P. COHEN

Department of Physiological Chemistry
University of Wisconsin, Madison, Wis. 53706

The presentations in this Symposium relating to glutamine metabolism in animal systems have been so comprehensive that there is very little to add. However, I would like to call attention to a biosynthetic pathway in humans (and other primates) involving glutamine which is normally not considered to be of major importance but which has been exploited in the treatment of certain clinical conditions of hyperammonemia. This pathway involves the conjugation of glutamine with phenylacetic acid to form phenylacetylglutamine which is excreted practically quantitatively in urine (1).

As has been pointed out in earlier presentations, glutamine is the most prevalent amino acid in mammalian blood plasma, and serves normally as the major form in which ammonia is transported in the body and as the major precursor of urinary ammonia.

Ammonia is relatively toxic to the central nervous system of mammals (and most terrestrial animals). The circulating level of metabolic ammonia in mammals is normally kept low by the efficient operation of the urea biosynthetic pathway in the liver. In clinical

ISBN: 0-12-506040-8

conditions in which liver dysfunction is seriously impaired or in cases of inborn deficiencies of the enzymes involved in urea biosynthesis, hyperammonemia with resultant effects on the central nervous system is a prominent feature.

Brusilow *et al.* (2) have recently reviewed the therapeutic procedures which have been used to treat hyperammonemia in newborns and children who have a congenital deficiency of one of the enzymes involved in urea biosynthesis. Therapy has been directed toward reduction of nitrogen intake by the following means: 1) low protein diets; 2) mixtures of essential amino acids in place of protein; and 3) low protein diets plus α-keto acid analogues of five essential amino acids. While these procedures have been of some benefit in patients with partial enzyme defects of urea biosynthesis, the mortality and morbidity incidence in these conditions continues to be high.

Brusilow *et al* (2) describe new therapeutic approaches in treatment of these conditions directed toward enhancement of the excretion of nitrogen-containing metabolites which in the normal individual are metabolized to form nitrogen precursors of urea. Of relevance to this Symposium is their effective use of phenylacetic acid in enhancing the excretion of glutamine in the form of phenylacetylglutamine.

Brusilow *et al.* (2) describe a case of a 16-year old girl with carbamyl phosphate synthetase I (EC 6.3.4.16) deficiency who was fed 6.4 g of phenylacetic acid for 7 days. Urinary nitrogen excretion rose 45% all of which could be accounted for by an increase in phenylacetylglutamine. No side effects were noted.

Brusilow *et al.* (3), have recently presented detailed studies of amino acid acylation as a mechanism of nitrogen excretion in inborn errors of urea biosynthesis. The effective use of benzoic acid (to form hippuric acid) and phenylacetic acid (to form phenylacetylglutamine) is described.

The enzymes and the reaction mechanism of phenylacetylglutamine synthesis in human liver and kidney have been reported (4, 5). Because separate enzymes are involved in the human liver in the synthesis of hippuric acid and of phenylacetylglutamine (5), an additive effect of feeding benzoic and phenylacetic acids on nitrogen excretion can be predicted. Further, since the enzyme activity for synthesis of phenylacetylglutamine is much higher in human kidney than in liver (4), phenylacetic acid might prove to be more efficacious than benzoic acid in conditions where liver function is impaired.

REFERENCES

1. James, M. O., Smith, R. L., Williams, R. T., and Reidenberg, M. (1972), *Proc. R. Soc. Lond. B. 182,* 25-35.
2. Brusilow, S. W., Valle, D. L., and Batshaw, M. L. (1979), *Lancet 1,* 452-454.
3. Brusilow, S. W., Tinker, J., and Batshaw, M. L. (1980), *Science 207,* 659-661.
4. Moldave, K., and Meister, A. (1957), *J. Biol. Chem. 229,* 463-476.
5. Webster, L. T., Jr., Siddiqui, U. A., Lucas, S. V., Strong, J. M., and Mieyal, J. J. (1976), *J. Biol. Chem. 251,* 3352-3358.

18. *Special Lecture:* Glutamine Metabolism in the Animal Body

HANS KREBS

Metabolic Research Laboratory,
University of Oxford
Radcliffe Infirmary, Woodstock Road,
Oxford, U. K.

SUMMARY

This paper surveys the metabolism of glutamine in the animal body.

EARLY STAGES IN THE HISTORY OF GLUTAMINE METABOLISM

I propose to start my survey of glutamine metabolism with historical aspects and set out how knowledge in the field has developed.

The presence of glutamine in proteins was suspected long before it was directly demonstrated 40 years ago, because the ammonia released on acid hydrolysis of proteins was assumed to be derived from glutamine and asparagine. The occurrence of free glutamine in certain plants has been known for about 100 years, but the study of the details of glutamine metabolism is of more recent date. At the time when, in 1935, I reported the ready synthesis of glutamine from glutamate and ammonia in guinea pig and rabbit kidney (1), there was hardly any literature on glutamine metabolism. I came across the synthesis of glutamine in animal tissues not because I had been looking for it, but because, while studying the deamination of amino acids in kidney slices, I found that glutamate in rabbit or guinea pig kidney failed to produce any ammonia and even prevented the endogenous shedding of ammonia, although the added glutamate was rapidly oxidized. Even added ammonia was readily removed, and on studying the fate of the added ammonia I found that it reappears when the medium in which the slices are shaken is heated for 5 min in the presence of 2 N H_2SO_4. Glutamine was the only known cell constituent which releases ammonia under these conditions and this, together with the fact that ammonia was taken up in major amounts only in the presence of glutamate justified the conclusion that the tissue synthesizes glutamine. Studying systematically this disappearance of ammonia in the presence of glutamate, I found that brain cortex and retina also readily synthesize glutamine. The significance of the synthesis of glutamine for the body as a whole remained obscure at that time. In fact I never followed up this question because I did not know where to start; in other words I had no working hypothesis about the function of glutamine. No role for glutamine was known at that time other than that of serving as a protein constituent and being a partner in the conjugation and detoxication of at least one extraneous substance, phenylacetic acid. The idea suggested itself that in nervous tissue the binding of ammonia may be a detoxicating process because ammonia was known to damage the brain.

GLUTAMINE AS PRECURSOR OF URINARY AMMONIA

The first clue concerning major special functions of glutamine was revealed by the work of van Slyke *et al.* (2) eight years after the discovery of glutamine synthesis in animal tissues. They discovered that glutamine was the main source of urinary ammonia in acidosis. This established glutamine as an important link in the regulation of the acid-base balance.

RELATIONS BETWEEN RENAL AMMONIAGENESIS AND GLUCOGENESIS

Much later, in the early 1960s, I came across a phenomenon which we can now relate to the function of glutamine as the precursor of urinary ammonia. We found that kidney cortex has a high capacity for the synthesis of glucose from the breakdown products of glutamine (3, 4). Apart from the liver, kidney cortex is the only organ which has a high gluconeogenic capacity. Again I failed to appreciate the physiological significance of this phenomenon. Then in 1966 Goodman *et al.* (5) showed that the capacity for gluconeogenesis of the kidney increases under conditions of acidosis when urine needs ammonia. This led to the proper appreciation of the significance of gluconeogenesis in the kidney: it represents a mechanism whereby the carbon skeleton of glutamine, after the release of the two nitrogen atoms as ammonia, is recovered in the form of glucose when the kidney is not in need of this skeleton as a fuel. Thus renal ammoniagenesis and glucogenesis are interlinked. What fits in with the concept — apart from the increased gluconeogenic capacity in acidosis — is the fact that the glucogenic capacity in rat kidney is particularly high for the degradation products of glutamate, but low for other substrates such as alanine. Thus renal gluconeogenesis is to be looked upon as a salvage process saving the carbon skeleton of glutamine that remains after the release of ammonia, at least in some situations. It may be that renal glucogenesis has other functions. What is not explained by the concept of a salvage process is the increase in gluconeogenesis from lactate after regular severe exercise. W. Guder (personal communication) has suggested that a high rate of gluconeogenesis from lactate facilitates reabsorption of lactic acid when, on exercise, the primary urine may contain up to 10 mM lactate.

Perhaps a sound view to take is the assumption that renal gluconeogenesis may be useful from more than one point of view, as a salvage reaction and also as an accelerator for the reabsorption of metabolites, such as lactic acid.

GLUTAMINE AS A NITROGEN DONOR IN THE SYNTHESIS OF PURINES, PYRIMIDINES AND AMINO SUGARS

In the 1940s and 1950s further metabolic functions of glutamine came to light, namely, that it is a precursor of nitrogen atoms in the synthesis of purines and pyrimidines, as well as of amino sugars, a subject fully dealt with in standard textbooks.

GLUTAMINE AS A CARRIER OF NITROGEN BETWEEN TISSUES

In the late 1960s, measurements of arteriovenous differences in amino acids in the human forearm — made possible by the development of the catheterisation technique — revealed that muscle tissue may shed much more alanine than can be derived by the hydrolysis of protein (6). Muscle protein contains 7-9% alanine, but under some conditions alanine contributes 30% to the α-amino nitrogen released by the muscle. It was not noted in the early experiments of Cahill's team that glutamine is also formed in major quantities by the muscles of the human forearm, because the amino acid analyser did not separate glutamine properly (7). Later (8) the use of enzymic methods for the assay of glutamate and glutamine indicated that in the post-absorptive state, alanine and glutamine together may account for 50% of the total amino nitrogen released by muscle. The implication is that alanine and glutamine act as vehicles for amino nitrogen released during the breakdown of amino acids in muscle tissue. The amino acids concerned are mainly the branched-chain amino acids which readily transaminate in muscle tissue. This subject is fully discussed by A. J. Garber (Chapter 14, see also Felig, 9).

The alanine shed by muscle is eventually taken up by the liver and converted to glucose or oxidised to completion. Much of the glutamine released is removed by the epithelium of the intestinal wall, a matter to which I will presently refer in more detail (see also H. G. Windmueller, chapter 13). Under conditions of acidosis, some glutamine is degraded in the kidneys, and some fraction serves to maintain the relatively high concentration of glutamine in the plasma, 0.5 mM, higher than that of any other single amino acid. This relatively high concentration is presumably related to the fact that glutamine, in addition to acting as a precursor of protein, has several special functions in the animal body, discussed below.

KINETIC DIFFERENCES BETWEEN GLUTAMINASES OF DIFFERENT TISSUES

Now I must discuss kinetic properties of glutaminase, or rather glutaminases, the first enzyme involved in the degradation of glutamine. It has recently become clear that the kinetic properties of the glutaminases in liver, kidney and intestinal mucosa possess characteristic differences which relate to the physiological function of glutamine in these different tissues.

When I first came across the enzymic hydrolysis of glutamine in 1935 (1) I noted that the glutaminases in extracts of liver and kidney differed in their kinetic properties. They had different pH optima and behaved differently in respect to inhibition by glutamate; the kidney enzyme was inhibited, the liver enzyme was not. I speculated that the kidney enzyme might be connected with the synthesis of glutamine but I could offer no satisfactory explanation for the existence of two enzymes with different kinetic properties, catalysing the same reaction.

Recent work in the Oxford laboratory by Baverel and Lund (10) has provided additional information on the kinetic properties of the liver and kidney enzymes, and on the basis of these new findings it is possible to correlate some properties of the two enzymes to their physiological roles.

After my early work on what now would be called glutaminase isoenzymes, Greenstein (11), Katunuma *et al.* (12), Curthoys and Lowry (13), Curthoys *et al.* (14) and Meister (15) made important contributions to the study of glutaminases which I cannot discuss here. Some mammalian glutaminases have allosteric properties, as was first noted by Charles (16). Their activity is influenced by modifiers and under some conditions the activity *versus* substrate concentration curve has a sigmoid shape. What I must emphasise is that the work which I am to report was all carried out on freshly prepared hepatocytes or kidney tubules, as opposed to tissue extracts, tissue homogenates, purified enzymes, or tissue fragments such as mitochondria. This meant that enzyme activities were measured under near-physiological conditions: as far as the modifiers are concerned the enzymes were present in their natural intracellular environment. The only major unphysiological aspect was the concentration of the added glutamine in the suspension fluid which was 5 mM, against about 0.5 mM in blood plasma (but intracellular hepatic glutamine is near 5 mM).

Most of the glutaminase activity under the experimental conditions is probably due to the glutaminase I of Greenstein, which Meister considers to be the only true glutaminase. This is the mitochondrial enzyme which Katunuma named phosphate-dependent glutaminase. The other enzymes which exhibit glutaminase activity in animal tissues catalyse primarily other reactions such as γ-glutamyl transfer (γ-glutamyl transpeptidase of Tate and Meister), or transamination of glutamine with α-oxoacids, especially α-oxoglutarate, to form α-oxoglutaramate which is hydrolysed to ammonia and α-oxoglutarate by an ω-amidase.

The new experiments show that the concentration of bicarbonate and CO_2 (the criterion of acidosis), affects hepatic and renal glutaminases in strikingly different ways. The experiments were carried out at constant pH, the concentrations of bicarbonate and CO_2 being varied in a parallel manner. As shown in Table I, increasing the bicarbonate buffer concentration substantially raises the rate of removal of glutamine by rat hepatocytes and this is paralleled by an increased formation of glucose and of urea. Some glutamate is also formed. The data are in quantitative agreement with the assumption that glutamate, glucose and urea are the major products of glutamine degradation. Thus much glutamine is broken down beyond the stage of glutamate. Acidification by doubling the CO_2 concentration at 25 mM bicarbonate (last horizontal line) decreases glutamine removal.

TABLE I

Effect of [HCO_3^-] on hepatic glutamine metabolism

Hepatocytes (125 mg wet wt.) incubated 30 min with 5 mM glutamine. Values are μmol, corrected for control. (For details see Ref. 10).

[HCO_3^-] mM	[CO_2] mM	pH	Glutamine removed	Glutamate formed	Glucose formed	Urea formed
11.5	0.5	7.4	2.1	1.1	0.6	1.1
25	1.1	7.4	3.1	1.5	0.9	2.0
50	2.2	7.4	3.8	1.8	1.0	2.9
25	1.1	7.1	1.6	0.9	0.5	0.9

The effects of the bicarbonate concentration are also found in the presence of 0.1 mM leucine (10), an effector both of glutaminase activity, and glutamate dehydrogenase activity. That leucine activates glutamate dehydrogenase was first shown by Yielding and Tomkins (17). That leucine also activates glutaminase is new.

As shown in Table II, renal glutamine removal, in contrast to hepatic glutamine removal, is not much affected by variations in the concentrations of the bicarbonate buffer; if anything there was a slight increase when [HCO_3^-] decreased.

The glutaminase of the epithelium of the small intestine – the enterocytes – is different again from those of either liver or kidney. In enterocytes glutamine is a major fuel of respiration (see H. G. Windmueller, chapter 13, 18). Neither leucine nor cyclic AMP affects the activity of intestinal glutaminase. Glutamine yields more than one ammonia in the small intestine and this indicates that the cells also possess glutamate dehydrogenase activity. A major nitrogenous endproduct of glutamine degradation is alanine, formed by transamination from glutamate with pyruvate.

Bicarbonate has a slightly stimulating effect in enterocytes but much less than in liver.

TABLE II

Effect of [HCO_3^-] on renal glutamine metabolism

About 20 mg (dry wt.) renal tubules were incubated 30 min with 5 mM [I-^{14}C]-glutamine pH 7.4 (For full details see Ref. 10).

[HCO_3^-]	[CO_2]	Glutamine removed	NH_3 formed	$^{14}CO_2$ formed
mM	mM	μmol	μmol	μmol
10.5	0.46	15.8	27.5	11.2
25	1.1	15.6	27.8	10.4
47	2.1	15.6	26.3	10.5

The significance of these differences in the kinetics of hepatic and renal glutaminase activity may be interpreted as follows. The kidney's need for glutamine as a precursor of urinary ammonia in acidosis, that is at low bicarbonate concentrations, is met by decreased glutamine degradation in the liver; this preserves the glutamine synthesized by extrarenal tissues, such as the musculature. Thus glutamine becomes available for degradation in the kidneys where the rate of degradation is independent of the bicarbonate concentration and also occurs at low bicarbonate concentrations. On the other hand when, under non-acidotic conditions, ammonia formation by the kidney is not required, the rate of degradation of glutamine in the liver (and possibly other tissues) rises and the surplus glutamine is either converted into carbohydrate by gluconeogenesis, or serves as a source of energy.

We now know four effectors of tissue glutaminase activity which operate in the liver but, as far as has been tested, not in the kidney. They are bicarbonate, leucine, ammonium ions (16, 19, 20, 21, 22), and cyclic AMP (or glucagon which acts through the release of cyclic AMP). Of these four effectors, only the bicarbonate concentration seems to have a direct link with acidosis.

The accelerating effect of ammonium ions on glutamine degradation has been interpreted by Häussinger and Sies (23) as a positive feedback. Tissue glutamine is a store of nitrogen for the synthesis of purines, pyrimidines and amino sugars. When the stores are "full" and when there is a surplus of ammonia in the plasma or portal blood, glutamine reaching the liver must be degraded, whatever the state of the acid-base balance. Häussinger and Sies (23) found no degradation of glutamine by the perfused liver unless about 0.7 mM ammonia was present; ammonia appears to serve as a signal for hepatic glutamine degradation at physiological glutamine concentrations.

Thus the action of several of the effectors of hepatic glutaminases can be correlated to the physiological function of the enzyme; the differences between the glutaminases of different organs make physiological sense.

GLUTAMINE AS A RESPIRATORY FUEL

In recent years a new role of glutamine has come to light. It is a major source of energy in several specific tissues. These are the epithelium of the small intestine (24, 25, 26), reticulocytes (27), lymphoma cells (28) and other malignant cells (29), HeLa cells (30) and fibroblasts (31). These types of cell have one common feature: they are all cells that grow relatively rapidly, intestinal epithelium because these cells are very quickly regenerated, and the other types of cell are directly growing cells. The oxidation of glutamine in these cells is remarkable because in general amino acids as such are not a major fuel of respiration in peripheral tissues. There are only three amino acids —alanine, aspartate and glutamate— that are readily broken down in tissues other than liver and kidney. These amino acids are fed into the main pathways of metabolism by transamination which give pyruvate, oxaloacetate and α-oxoglutarate. All the other amino acids, with the exception of the branched-chain amino acids, are primarily attacked by the liver (and to a less extent by kidney) where most of them are partially broken down to products which enter the main pathways of degradation and lead to complete combustion, or gluconeogenesis or fatty acid synthesis. The branched-chain amino acids are converted in skeletal muscle, heart, kidney and other peripheral tissues into the α-oxo derivatives (32), but this is not a major energy supplying process because the bulk of the α-oxo acids is discharged into the circulation and metabolized in the liver (33).

I know of only one non-growing tissue which degrades glutamine. This is the lens – probably its epithelial cells (34).

I have already mentioned that the concentration of glutamine in blood plasma is higher than that in any other amino acid and this is true also of the glutamine content of animal tissues generally. Maybe the ready oxidation of glutamine is related to the fact that glutamine plays a major role in the synthesis of purines and pyrimidines. These syntheses occur when cells divide and this would account for the ready disposal of the carbon skeleton of glutamine when the nitrogen atoms have been used for the synthesis of purines and pyrimidines. Glutamine, it should be borne in mind, is also the precursor of the aspartate needed in both purine and pyrimidine synthesis.

EFFECT OF GLUTAMINE ON GLYCOGEN SYNTHESIS IN RAT LIVER

I must briefly refer to another function of glutamine which has come to light recently through the work of Katz (35). Glutamine promotes the synthesis of glycogen from gluconeogenic precursors such as lactate, dihydroxyacetone, glycerol or fructose. Glutamine increases the activity of glycogen synthase without having any effect on glycogen phosphorylase (36, 37). The mechanism of the action of glutamine on the synthase is not yet understood. The concentrations of glutamine used in the experiments of Katz were rather high (10 mM) but this may be connected with the fact that at low concentrations glutamine is not taken up by hepatocytes.

MULTIPLE FUNCTIONS OF GLUTAMINE IN THE ANIMAL BODY

Finally a few comments on the multiplicity of the functions of glutamine. To summarise its roles: apart from being a protein and peptide constituent, it plays a role in the acid-base balance as a precursor of urinary ammonia, it is a precursor of the purine and pyrimidine rings of nucleic acids and nucleotides, it is a precursor of amino sugars, it plays a role in detoxication, it is a nitrogen carrier between tissues, it is a regulator of hepatic glycogen synthesis and it is a respiratory fuel in certain tissues.

Most amino acids have multiple functions (Table III) but glutamine appears to be the most versatile. This versatility illustrates a general principle of biochemical evolution and of evolution generally. Once a potential has evolved, be it a substance, a chemical reaction mechanism or a structure like a limb, it is adapted to perform a variety of functions.

TABLE III

Multiple functions of amino acids in the mammalian body

Amino Acid	Precursor of
Glycine	Creatine, bile acids, porphyrines, glutathione, hippurate
Serine	Ethanolanine and choline i. e. of phospholipids, acetylcholine (neurotransmitter)
Cysteine	Taurine (bile acids), glutathione
Arginine	Creatine
Aspartate	Nucleic acids, nucleotides, β-alanine (anserine, carnosine)
Alanine	Nitrogen carrier between tissues
Tyrosine	Epinephrine, nor-epinephrine, thyrocine, melanine
Tryptophan	Serotonin
Methionine	Methyl group of creatine and choline

REFERENCES

1. Krebs, H. A. (1935) *Biochem. J. 29,* 1951-1969.
2. Van Slyke, D. D., Phillips, R. A., Hamilton, P. B., Archibald, R. M., Futcher, P. H. and Hiller, A. (1943) *J. Biol. Chem. 150,* 481-482.
3. Krebs, H. A. (1963) *Adv. Enz. Regulation 1,* 385-400.
4. Krebs, H. A., Hems, R. and Gascoyne, T. (1963) *Acta Biol. Med. German 11,* 607-615.
5. Goodman, A. D., Fuisz, R. E. and Cahill, G. F. (1966) *J. Clin. Invest. 45,* 612-619.
6. Marliss, E., Aoki, T. T., Felig, P., Pozefsky, T. and Cahill, G. F. Jr. (1970) *Adv. Enz. Regulation 8,* 3-11.
7. Krebs, H. A. (1970) *Adv. Enz. Regulation 8,* 335-353.
8. Aoki, T. T., Müller, W. A. and Cahill, G. F. Jr. (1972) *Adv. Enz. Regulation 10,* 145-151.
9. Felig, P. (1975) *Ann. Rev. Biochem. 44,* 930-945.
10. Baverel, G. and Lund, P. (1979) *Biochem. J. 184,* 599-606.
11. Greenstein, J. P. (1949) *Adv. Enzymol. 8,* 117-169.
12. Katunuma, N., Huzino, A. and Tomino, I. (1967) *Adv. Enz. Regulation 5,* 55-59.
13. Curthoys, N. P. and Lowry, O. H. (1973) *J. Biol. Chem. 248,* 162-168.
14. Curthoys, N. P. and Kuhlenschmidt, T. (1975) *J. Biol. Chem. 250,* 2099-2105.

15. Tate, L. S. and Meister, A. (1975) *J. Biol. Chem. 250*, 4619-4627.
16. Charles, R. (1968) Ph. D. Thesis, University of Amsterdam.
17. Yielding, K. L. and Tomkins, G. M. (1961) *Proc. Natl. Acad. Sci. U.S.A. 47*, 983-990.
18. Watford, M., Lund, P. and Krebs, H. A. (1979) *Biochem. J. 178*, 589-596.
19. Haussinger, D. and Sies, H. (1975) *10th FEBS Meeting*, Abstract Nº 1497.
20. Haussinger, D., Weiss, L. and Sies, H. (1975) *Eur. J. Biochem. 52*, 421-431.
21. Joseph, S. K. and McGivan, J. D. (1978a). *Biochim. Biophys. Acta 543*, 16-28.
22. Joseph, S. K. and McGivan, J. D. (1978b) *Biochem. J. 176*, 837-844.
23. Haussinger, D. and Sies, H. (1979) *Eur. J. Biochem. 101*, 179-184.
24. Neptune, E. M. Jr. (1965) *Am. J. Physiol. 209*, 329-332.
25. Windmueller, H. G. and Spaeth, A. E. (1974) *J. Biol. Chem. 249*, 5070-5079.
26. Windmueller, H. G. and Spaeth, A. E. (1975) *Arch. Biochem. Biophys. 171*, 662-672.
27. Rapoport, S., Rost, J. and Schultze, M. (1971) *Eur. J. Biochem. 23*, 166-170.
28. Lavietes, B. B., Regan, D. H. and Demopoulos, H. B. (1974) *Proc. Natl. Acad. Sci. U.S.A. 71*, 3993-3997.
29. Kovacevic, S. and Morris, H. P. (1972) *Cancer Res. 32*, 326-333.
30. Reitzer, L. J., Wice, B. M. and Kennell, D. (1979) *J. Biol. Chem. 254*, 2669-2676.
31. Donnelly, M. and Scheffler, I. E. (1976) *J. Cell Physiol. 89*, 39-51.
32. Krebs, H. A. (1972) *Adv. Enz. Regulation 10*, 397-420.
33. Livesey, G. and Lund, P. (1980) submitted to *Biochem. J.*
34. Trayhurn, P. and van Heyningen, R. (1973) *Exp. Eye Res. 17*, 149-154.
35. Katz, J., Golden, S. and Wals, P. A. (1976) *Proc. Natl. Acad. Sci. U.S.A. 73*, 3433-3437.
36. Golden, S., Wals, P. A., Okajima, F. and Katz, J. (1979) *Biochem. J. 182*, 727-734.
37. Katz, J., Golden, S. and Wals, P. A. (1979) *Biochem. J. 180*, 389-402.

INDEX

G

U